CONTROL AND DYNAMIC SYSTEMS

Advances in Theory and Applications

Volume 77

CONTRIBUTORS TO THIS VOLUME

STEVEN BECK
DAVID D. BENNINK
LARRY DEUSER
WOON S. GAN
JOYDEEP GHOSH
F. D. GROUTAGE
C. KOTROPOULOS
FU LI
YANG LU
I. PITAS
MICHAEL SMITH
PETER A. STUBBERUD
KAGAN TUMER
A. N. VENETSANOPOULOS
PAUL R. WHITE
JIE YANG

CONTROL AND DYNAMIC SYSTEMS

ADVANCES IN THEORY AND APPLICATIONS

Edited by

CORNELIUS T. LEONDES

School of Engineering and Applied Science
University of California, Los Angeles
Los Angeles, California

VOLUME 77: MULTIDIMENSIONAL SYSTEMS SIGNAL PROCESSING ALGORITHMS AND APPLICATION TECHNIQUES

ACADEMIC PRESS

San Diego New York Boston
London Sydney Tokyo Toronto

This book is printed on acid-free paper. ♾

Academic Press, Inc.
A Division of Harcourt Brace & Company
525 B Street, Suite 1900, San Diego, California 92101-4495

United Kingdom Edition published by
Academic Press Limited
24-28 Oval Road, London NW1 7DX

International Standard Serial Number: 0090-5267

International Standard Book Number: 0-12-012777-6

Transferred to digital printing 2006
96 97 98 99 00 01 QW 9 8 7 6 5 4 3 2 1

CONTENTS

CONTRIBUTORS

Numbers in parentheses indicate the pages on which the authors' contributions begin.

Steven Beck (301), *Tracor Applied Sciences, Austin, Texas 78725*

David D. Bennink (49), *Applied Measurement Systems, Inc., Bremerton, Washington 98380*

Larry Deuser (301), *Tracor Applied Sciences, Austin, Texas 78725*

Woon S. Gan (339), *Acoustical Services Pte. Ltd., Singapore 048429 Republic of Singapore*

Joydeep Ghosh (301), *Department of Electrical and Computer Engineering, College of Engineering, The University of Texas at Austin, Austin, Texas 78712*

F. D. Groutage (49), *Naval Surface Warfare Center, Carderock Division, Puget Sound Detachment, Bremerton, Washington 98314*

C. Kotropoulos (1), *Department of Electrical Engineering, University of Thessaloniki, Thessaloniki 54006, Greece*

Fu Li (149), *Department of Electrical Engineering, Portland State University, Portland, Oregon 97207*

Yang Lu (149), *Department of Electrical Engineering, Portland State University, Portland, Oregon 97207*

I. Pitas (1), *Department of Electrical Engineering, University of Thessaloniki, Thessaloniki 54006, Greece*

Michael Smith (225), *Department of Electrical and Computer Engineering, University of Calgary, Calgary, Alberta, Canada T2N 1N4*

Peter A. Stubberud (117), *Department of Electrical and Computer Engineering, University of Nevada Las Vegas, Las Vegas, Nevada 89154*

Kagan Tumer (301), *Department of Electrical and Computer Engineering, College of Engineering, The University of Texas at Austin, Austin, Texas 78712*

A. N. Venetsanopoulos (1), *Department of Electrical Engineering, University of Toronto, Toronto, Ontario, Canada M5S 1A4*

Paul R. White (193), *Institute of Sound and Vibration Research, University of South Hampton, Hants, United Kingdom*

Jie Yang (225), *Department of Electrical and Computer Engineering, University of Calgary, Calgary, Alberta, Canada T2N 1N4*

PREFACE

From about the mid-1950s to the early 1960s, the field of digital filtering, which was based on processing data from various sources on a mainframe computer, played a key role in the processing of telemetry data. During this period the processing of airborne radar data was based on analog computer technology. In this application area, an airborne radar used in tactical aircraft could detect the radar return from another low-flying aircraft in the environment of competing radar return from the ground. This was accomplished by the processing and filtering of the radar signal by analog circuitry, taking advantage of the Doppler frequency shift due to the velocity of the observed aircraft. This analog implementation lacked the flexibility and capability inherent in programmable digital signal processing technology, which was just coming onto the technological scene.

Powerful technological advances in integrated digital electronics coalesced soon after the early 1960s to lay the foundations for modern digital signal processing. Continuing developments in techniques and supporting technology, particularly very-large-scale integrated digital electronics circuitry, have resulted in significant advances in many areas. These areas include consumer products, medical products, automotive systems, aerospace systems, geophysical systems, and defense-related systems. Therefore, this is a particularly appropriate time for *Control and Dynamic Systems* to address the theme of "Multidimensional Systems Signal Processing Algorithms and Application Techniques."

The first contribution to this volume is "Techniques in Knowledge-Based Signal/Image Processing and Their Application in Geophysical Image Interpretation," by I. Pitas, C. Kotropoulos, and A. N. Venetsanopoulos. One of the most important applications of multidimensional signal processing is geophysical seismic interpretation and, in particular, geophysical oil prospecting. This contribution is an in-depth treatment of techniques for integrated, interactive, and intelligent computer-aided geophysical interpretation methods. As such it is a most appropriate contribution with which to begin this volume.

The next contribution is "The Foundations of Nearfield Acoustic Holography in Terms of Direct and Inverse Diffraction," by David D. Bennink and F. D. Groutage. In general terms, holography is an imaging method for reconstructing information concerning a three-dimensional wave field from data recorded on a two-dimensional surface. This contribution is a comprehensive review of this broad area and the many techniques involved in optical and digital processing.

In "A Design Technique for 2-D Linear Phase Frequency Sampling Filters with Fourfold Symmetry," Peter A. Stubberud discusses frequency sampling filters, one of the most efficient and effective classes of filters for 2-D signal or image processing. This contribution is an in-depth treatment of the issues involved in their realization, including techniques that control interpolation errors and optimization techniques for system error minimization.

"Unified Bias Analysis of Subspace-Based DOA Estimation Algorithms," by Fu Li and Yang Lu, provides a comparative analysis of various direction-of-arrival (DOA) algorithms with notes on the more popular ones. Increasing demands in applications such as radar and sonar detection, geophysical exploration, telecommunications, biomedical science, and other areas of great importance have made sensor array signal processing a very active research field for several decades. One of the principal tasks in array processing is to estimate directions of incoming signals impinging simultaneously on an array of sensors. Many DOA algorithms have been developed, and numerous examples that illustrate these methods are presented.

The approach taken in "Detection Algorithms for Underwater Acoustic Transients," by Paul R. White, is highly pragmatic, and the algorithm for this major problem is broadly applicable to other areas as well. The resulting algorithms are implementable on real-time signal processing chips working at reasonable sampling rates. The illustrative examples that are presented allow one to gauge how well these algorithms perform in realistic scenarios.

The next contribution is "Constrained and Adaptive ARMA Modeling as an Alternative to the DFT—With Application to MRI," by Jie Yang and Michael Smith. In many commercial and research applications, the use of the discrete Fourier transform (DFT) allows the transfer of data gathered in one domain (typically spatial) into another (frequency). This alternative representation often allows easier characterization or manipulation of the signal. For example, the removal of unwanted noise components is achieved more efficiently by multiplying the frequency domain signal by the desired filter response. However, the DFT can have serious drawbacks in important areas of major applied significance. This contribution presents several significantly effective alternate algorithms and exemplifies their effectiveness in such areas of applied significance as MRI (magnetic resonance imaging) in noninvasive diagnosis data and geological MRI data sets.

The identification and classification problem in multidimensional systems with low signal-to-noise ratios (SNRs), which can be a characteristic of many systems including the processing of underwater acoustic signals, calls for more effective processing techniques. "Integration of Neural Classifiers for Passive Sonar Signals," by Joydeep Ghosh, Kogan Tumer, Steven Beck, and Larry Deuser, reviews five different approaches and notes that integration techniques can significantly enhance system performance in this major area.

The final contribution to this volume is "Techniques in the Application of Chaos Theory in Signal and Image Processing," by Woon S. Gan. Chaos is a characteristic of nonlinear phenomena, and a wide spectrum of these phenomena is noted in this contribution. In particular, the techniques and applications of chaos theory in nonlinear digital signal and imaging processing are treated in depth. As such, this is a most appropriate contribution with which to conclude this volume.

This volume on multidimensional systems signal processing algorithms and application techniques clearly reveals the significance and power of the techniques available and, with further development, the essential role they will play in a wide variety of applications. The authors are all to be highly commended for their splendid contributions, which will provide a significant and unique reference for students, research workers, computer scientists, practicing engineers, and others on the international scene for years to come.

Techniques in Knowledge-Based Signal/Image Processing and Their Application in Geophysical Image Interpretation

I. Pitas   C. Kotropoulos

Department of Electrical Engineering
University of Thessaloniki
Thessaloniki 54006, GREECE

A.N. Venetsanopoulos

Department of Electrical Engineering
University of Toronto
Toronto M5S 1A4, CANADA

I. INTRODUCTION TO GEOPHYSICAL INTERPRETATION

Geophysical seismic interpretation is part of geophysical oil prospecting. It evaluates and analyses seismic reflection data aiming at the detection of the position of hydrocarbon reservoirs. This work requires considerable experience and knowledge and must be done by skillful interpreters. Therefore, it can not be automated easily. This chapter provides a review of the current efforts to automate, at least partially, seismic interpretation. As it will be shown, this research area is very active and it is a melting pot of various different approaches and techniques: artificial intelligence, pattern recognition, image processing, graphics, fuzzy set theory and of course, geophysics and geology.

Oil is found underground and it usually occurs in rocks between the sand grains in a sand stone, in cracks in a shattered rock and in little cavities in limestone. Much of the earth is covered with sedimentary basins

CONTROL AND DYNAMIC SYSTEMS, VOL. 77

that were once seas. Gravity compressed the sediments turning them into rock. The high pressure and the heat transformed the organic matter to oil and gas, which together with salt water saturated the porous rocks. Oil floats on water, so, if the layer is tilted, it will gradually creep upwards. Sometimes, as a result of stresses and deformations, there are some local high spots, where oil and gas are concentrated. Such oil reservoirs are shown in Figure 1.

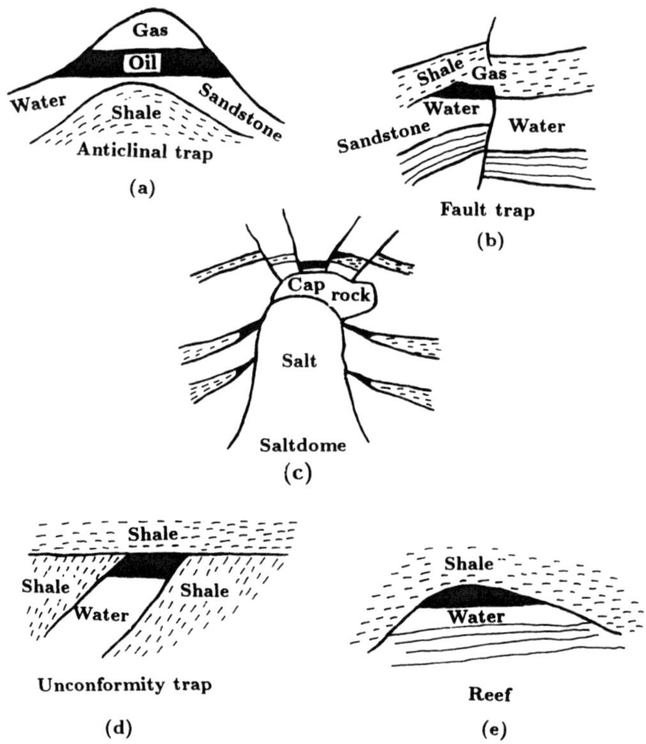

Fig. 1. Structural configurations for oil traps.

Reflection seismology [1,2] is a widely used method to construct an accurate profile of the subsurface geology. Seismic energy from an explosion or other artificial seismic source on the earth surface propagates downward through rock layers. If there are acoustic impedance variations between different rock layers of geologic strata, reflection of the seismic energy from the rock layer interfaces occurs and is detected at the surface receivers (geophones/hydrophones). A seismic trace is the output of a geophone. A seismic section is composed of many adjacent seismic traces. Seismic traces are processed extensively before being used for the interpretation of the earth subsurface. Typical processes are stacking, velocity analysis, deconvolution and migration [1]–[8]. The processed seismic sections provide a fairly accurate seismic image of the subsurface geology. The next step in oil prospecting is to interpret the seismic images. Seismic interpretation generally assumes that [6]:

- Coherent events seen on a seismic record or on processed seismic sections are reflections from acoustic impedance contrasts in the earth.
- Seismic detail (waveshape,amplitude etc.) is related to geological detail, that is to stratigraphy and the nature of the interstitial fluids.

A detailed analysis of seismic interpretation can be found in many books [1]–[7]. The description of [7] is well suited as an introduction for the non-geophysicist.

The first task of seismic interpretation is the so-called *structural interpretation*. An interpreter generally starts with the most obvious feature, usually the strongest reflection event or the event which possesses the most distinctive character and follows this event as long as it remains reliable. Such a lateral correlation of reflection events produces the so called *seismic horizons*. Each horizon has several attributes which depend on the nature of the rock layers, e.g. reflection amplitude and reflection signature (shape of the reflection wavelets) shown in Figure 2. It has also attributes that

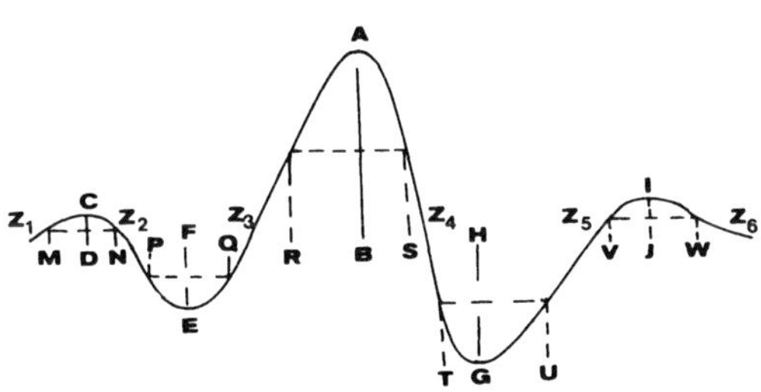

Fig. 2. Reflection skeletons.

depend on the geometrical position of rock layers in the earth, e.g. length, direction, curvature, abrupt changes in orientation. After following seismic horizons, the interpreter tries to identify the fairly large-scale features of the depositional structure of sedimentary rocks and the major deformation which has affected such rocks. These structures can be broadly classified as being either faulting or folding. Structures of interest are faults, anticlines, synclines, salt domes, unconformities etc. [2]–[7]. The detection of consistent ends of horizons indicates the presence of a fault. A geologic fault is also indicated by consistent abrupt changes of neighbor horizons. An anticline trap is indicated by convex horizons that are above a strong horizontal seismic horizon.

A second task of seismic interpretation is *seismic modeling* [6]. It includes the verification of the interpretation model by a computer simulation of the seismic experiment. The interpretation is successful if the simulation results match the seismic image.

The third step of interpretation is *seismic stratigraphy* [9,10]. A seismic facies unit is defined as a mappable group of reflections whose elements, such as reflection pattern, amplitude, continuity, frequency, interval velocity differ from the elements of adjacent units. The three principal types of reflection configuration are:

1. Reflection-free zones from areas where few reflecting surfaces exist. They are indicative of a uniform, single lithology or of intense postdepositional homogenization of multiple lithologies. It is characteristic of reefs.

2. Simple stratified patterns in which parallel or divergent reflections are present and have reasonable degree of continuity. Continuous reflections with uniform amplitude and frequency from trace to trace arise from rock layers that are uniform in their thickness and lithology over the region covered by the section. Parallel arrangements suggest uniform rates of deposition on a stable or uniformly subsiding surface. The divergent arrangements suggest areal variations in the rate of deposition, progressive tilting of the depositional surface. Complex stratified configuration include sigmoid and oblique arrangements, which occur in connection with progradational patterns on the shelf margin.

3. Chaotic patterns in which reflections are discontinuous and discordant. They suggest a disordered arrangement of reflection surfaces and are characteristic of diapiric cores.

Seismic interpretation is a difficult task, because the seismic data are usually fuzzy and noisy. Furthermore, it is heavily based on the available geological and geophysical knowledge of the region and on the expertise of the interpreter. It is difficult to be cast in a mathematical formulation (except perhaps seismic modeling) and, unlike other tasks of geophysical oil prospecting, has not been automated and it has not taken into advantage the digital data and signal processing techniques available to the scientific community in the past two decades. Therefore, it is a labor intensive task. However, there are some reasons for preferring computerized methods in seismic interpretation, namely speed, consistency and specification of recognition criteria [11]. There have been several approaches to automate geophysical interpretation. All of them use advanced data processing techniques which have been developed in the past twenty years and which have already been used in several other applications (e.g. biomedical signal and image processing, speech and image processing). The most common approaches are the following:

1. Seismic pattern recognition
2. Seismic image processing
3. Graphics
4. Geophysical and geologic expert systems

Each of these approaches includes several related techniques, which are usually results of independent researchers. Therefore, sometimes, there is no direct connection between the various proposed techniques. A review of these approaches will be presented in the subsequent sections.

II. SEISMIC PATTERN RECOGNITION

Pattern recognition [12,13] has been perhaps, the first approach to automate certain tasks of geophysical interpretation (e.g. horizon picking, remote correlation, recognition of the nature and boundaries of an oil or gas reservoir). The work of P. Bois [15]–[21] was pioneering in this area.

Horizon picking is the first task of geophysical interpretation which took advantage of pattern recognition techniques. The reason is that horizon picking is the first and fairly simple step in geophysical interpretation. A model of seismic reflections is usually needed for horizon picking. Seismic reflections are ideally quite similar to Ricker wavelets [14]. Therefore, they can be modeled by a set of parameters which take into account their spectrum and their character that may exist in their arches. Bois has proposed [15] the following set of parameters for a five-arch reflection shown in Figure 2:

a) the differences of the zero crossings $Z_2-Z_1, Z_3-Z_2, Z_4-Z_3, Z_5-Z_4, Z_6-Z_5$

b) the normalized amplitudes of the peaks $CD/AB, EF/AB, GH/AB, IJ/AB$

c) the distances MN, PQ, RS, TY, VW of the half-amplitude points of each arch.

Therefore, fourteen parameters are needed for the modeling of a seismic reflection. This choice of parameters is quite arbitrary, although it takes into account the spectral properties of a reflection.

Another approach is to model the seismic reflections by the coefficients b_i, a_k of its ARMA model [22]:

$$H(z) = \frac{B(z)}{A(z)} = G\frac{1+\sum_{i=1}^{p} b_i\, z^{-i}}{1+\sum_{k=1}^{q} a_k\, z^{-k}} \tag{1}$$

In this case $p+q$ parameters are needed for the description of a seismic reflection. Nine or fifteen coefficients have been used for the description of earthquake waves [22].

Syntactic methods can also be used for reflection modeling [23]. A syntactic pattern recognition approach that uses structural information of the wavelet to classify Ricker wavelets is proposed in [25]. Another scheme of syntactic pattern recognition employing Hough transform is proposed in [26].

Having defined a reflection model, horizon picking can be done in the following way: The peaks are determined on the first seismic trace and those are kept that are higher than a given threshold. The reflection parameters of these peaks are calculated and stored. The peaks of the second trace, which are close (within a window) to the peaks of the first trace, are kept and their reflection parameters are calculated and stored. This procedure is repeated until the whole seismic section is scanned. A horizon consists of a list of adjacent peaks. Finally, only the horizons consisting of coherent reflection parameters are kept. Seismic events are usually fuzzy and corrupted by noise. Therefore, fuzzy set theory [24] has been proposed for horizon picking [21]. In this approach, a horizon consists of a fuzzy set made up of N reflections, and the corresponding membership grades of their M reflection parameters. The horizons, which are finally picked, consist of reflections having the greatest *resemblance* evaluated by using the Hamming distances between the different reflections of the same fuzzy horizon.

The remote correlation is an extension of automatic picking, which is used when seismic horizons are interrupted by noisy (or blind) zones. In this case, the horizons on the two sides of the blind zone must be correlated. Automatic picking is used to determine the horizons on the two sides of the blind zone. The averages and the covariance matrix of the reflection parameters are calculated for every horizon. Thus, the problem of remote correlation is reduced to find the similarity of the horizons of the left and of the right side of the blind zone. One approach to this problem is to find the probabilities $p(I, J)$ for a horizon I on the left of the blind zone to be the continuation of horizon J on the right zone [17]. Mahalanobis distance [12] can be used as a measure of the difference between the averages of two populations of the reflection parameters of the horizons I, J. Another approach is to use discriminant factor analysis [17] to find the discriminant factorial axes, so that the distances between the horizons are maximal and the scattering of reflections inside the same horizon is minimal.

The recognition of the nature and the boundaries of a reservoir is of great economic importance. It can be done in the following way [19,20]: The trace sectors, which are bounded by horizons determined by roughly picking the contours of the reservoirs are modeled by an ARMA or an AR model. The parameters form a $(p+q)$-dimensional space, where each trace section corresponds to one point. The points belonging to traces inside the reservoir are well grouped in a cluster, whereas the points corresponding to traces on the reservoir boundary or outside the reservoir, tend to have must greater dispersion. Therefore, the boundaries of the reservoir can be easily determined. The points belonging to trace sectors of reservoirs of different nature tend to group in separate clusters. Therefore, if the nature of one reservoir is known (e.g. by drilling), the nature of another reservoir can be determined.

Fuzzy modeling is used for the verification of the results of seismic interpretation. If a geophysicist assumes that the geologic model $\mathcal{G}$ and the impulse $\mathcal{J}$ correspond to a seismic section $\mathcal{S}$, he forms a synthetic section $\mathcal{F}$:

$$\mathcal{F} = \mathcal{G} * \mathcal{J} \tag{2}$$

where $*$ is the convolution operator and tests the synthetic section against the actual section $\mathcal{S}$. Model (2) can be considered to be fuzzy. In this case, the membership function of the fuzzy set $\mathcal{F}$ is given by:

$$\mu_{\mathcal{F}}(t) = \min\{\mu_{\mathcal{G}}(t), \mu_{\mathcal{J}}(t)\} \tag{3}$$

where $\mu_{\mathcal{G}}(t), \mu_{\mathcal{J}}(t)$ are the membership functions of the fuzzy models $\mathcal{G}, \mathcal{J}$. The fuzzy synthetic section $\mathcal{F}$ is tested against the actual section $\mathcal{S}$. If the models are reasonable, there exists a good agreement between $\mathcal{S}, \mathcal{F}$. However, differences may exist between $\mathcal{S}$ and $\mathcal{F}$, which form the *anomaly fuzzy set* $\mathcal{A}$. An iterative technique has been proposed which modifies $\mathcal{G}$, so that the anomaly set $\mathcal{A}$ is minimized [21].

Finally, multidimensional data analysis methods such as clustering techniques and factor analysis are also proposed in [27]. One-step Markov chains are proposed in [28] to model different lithologies and discriminant analysis is used to give a fair idea about synthetic litho-stratigraphy.

III. SEISMIC IMAGE PROCESSING

Digital image processing techniques can be used for the processing of seismic images [29]-[33]. There are two tasks of geophysical interpretation where digital image processing techniques can be employed:

- horizon picking
- texture analysis of seismic images.

A. HORIZON PICKING

The simplest approach to horizon picking is to consider horizons as sequences of local extrema of reflection intensity in the seismic image. By this way, horizon picking can be made by contour following techniques [30] based on local decisions [34], or by the use of edge detector operators (e.g., the Laplacian operator [29]–[33] or edge detectors based on nonlinear filters [61]). More complicated techniques for horizon picking are performed by using neighborhood information based on Markovian image models [30,35] and dynamic programming. Another interesting approach of horizon picking based on heuristic techniques has been proposed by Keskes and Mermey [36]. An edge detector is applied to the seismic image. The local image edges (edge elements) are considered to be parts of seismic horizons. Each of them represents a node in a connection graph. The nodes are connected to each other by connection arrows. A connection cost is associated to each connection. A natural continuation of a local edge, must be an edge element lying at almost the same depth and having almost similar orientation. Two edge elements belonging to the same horizon, must have similar reflection amplitudes and signatures. The correlation coefficient of the traces which cross the two edge elements is a good indicator of their continuation. Thus, the connection cost depends on the difference of the orientation of the two edge elements, on the differences of the reflection amplitudes and on the correlation of the seismic traces which cross the two edges. The higher the cost is, the more difficult is the connection between the two corresponding edges as parts of the same horizon. If the first and the last node of an horizon are given by the interpreter, the system can decide which edges must be connected to form the horizon with the minimal cost or equivalently to find a path having minimal cost in the connection graph. There exist several solutions to this graph searching problem [30]. Such an algorithm can be found in [36]. The same algorithm can be used to find horizons in closed loops in 3-d seismic sections. The only difference from the 2-d case is that the start and end nodes lie at the same depth. A different approach to horizon detection which is based on a binary consistency checking scheme is described by Cheng [37]. Horizon picking will be discussed also in Section VIII.

B. TEXTURE ANALYSIS OF SEISMIC IMAGES

Texture information of the seismic images is directly related to the stratigraphic information. Chaotic or reflection-free or stratified patterns are simple texture patterns. Some techniques that have been proposed for seismic texture analysis will be presented briefly.

Template matching assumes that a seismic pattern can be represented by a set of matrices called templates. Each seismic region corresponds to a seismic pattern, which is described by a set of templates. These templates can be selected by an expert from an already interpreted seismic section. Another matrix (having equal dimensions with the template), contains the reflection coefficients around a pixel of the seismic image [38]. The projection vectors of this matrix on the templates and the projection angles can be used for the classification of a seismic image pixel to a region. The classification of a pixel to a region can be based either on a largest projection norm, or on the smallest projection angle. Template matching segmentation can be followed by relaxation labeling techniques to reduce the probability of misclassification of pixels of a seismic image [30].

Run length segmentation is based on the gray level run, which is defined as a collinear connected set of pixels, all having the same gray level [39]. The length of the run is equal to the number of its pixels. Before the application of the method, the seismic image can become binary, where 1s correspond to positive reflections and 0s to negative ones, for simplicity reasons. A reasonable assumption for seismic images is that every run is horizontal. The algorithm for run length segmentation is based on the calculation of the run lengths. It is composed of a "look-forward" and a "look-backward" loop. The final run length is the sum of the results of the two loops. This algorithm can be slightly modified to allow non-horizontal reflectors. If a segmentation of the seismic image is desired, the binary image is replaced by another one consisting of the run lengths. The RMS-average run length and the average vertical spacing between runs have been used for seismic image segmentation [38]. A modification of the above-described run length segmentation will be described in Section IX. Simaan et al. [40] have been found that the "texture energy measure" method developed by Laws [41] provides better discriminating power than methods based on template matching, run length, and cooccurrence matrix statistics [39].

A knowledge-based segmentation system for texture images has been proposed in [42,43,44]. This system is characterized by a control mechanism based on an *Iterative Linked Quadtree Splitting (ILQS)* scheme. The main advantages of ILQS scheme are:

1. The information collection and decision making processes for the segmentation of the seismic section are performed at different resolution

scales and in a cumulative fashion.

2. The classification process is balanced and less dependent on the order in which the image is processed.

3. Global information regarding the overall progress of segmentation is available so that knowledge which is more complicated than mere adjacency compatibilities can be utilized.

The performances of three knowledge-based texture segmentation systems are compared in [45]. The first system is based on a run length statistics algorithm extended by a decision process, which incorporates heuristic rules to influence the segmentation. The second and third systems are based on texture energy measure algorithms followed by two different knowledge-based classification processes. The knowledge-based process of the second system is controlled by a parallel region growing scheme and that of the third system is controlled by the above-mentioned ILQS scheme. It has been found that the second and third systems produce better segmentation than the first system.

Directional filtering is a technique for the decomposition of an image to regions having similar texture directionality [46]–[48]. Directional information about texture is contained in the power spectrum of a seismic image. Power concentrations on lines in the power spectrum of an image correspond to texture having perpendicular orientation to the spectral lines. Therefore, directional filters can be used for seismic texture segmentation. They are filters, whose passband covers a cone in the 2-d frequency domain [46]–[48]. A directional filter has a passband along its main direction and its output contains lines and texture features in the space domain that have perpendicular direction in the space domain. If a number of N directional filters covers the entire frequency domain, each filter has radial bandwidth equal to $2\pi/N$. The impulse response of a directional filter is multiplied by a Gaussian weighting function to avoid Gibbs oscillations along its main direction. A non-linear algorithm which improves the directional selectivity without producing the Gibbs phenomena is described in [48]. If a directional seismic image decomposition is desired, a set of directional filters is applied to the image, which covers the entire frequency domain. The filter outputs are a set of images having directional texture information plus an image which is the response to a lowpass filter. The sum of the whole set of images composes the entire input image. If both directional and radial frequency information is desired for seismic texture image segmentation, Gabor filters can be used [68]. They are filters which are both directional and bandpass [49]–[51]. The output of a Gabor filter is a directional image which contains also a specific radial frequency content. Both directional and Gabor filters can be applied either in the spatial or in the frequency domain. The results of the application of a set of six directional filters to a seismic image are shown in Figure 3.

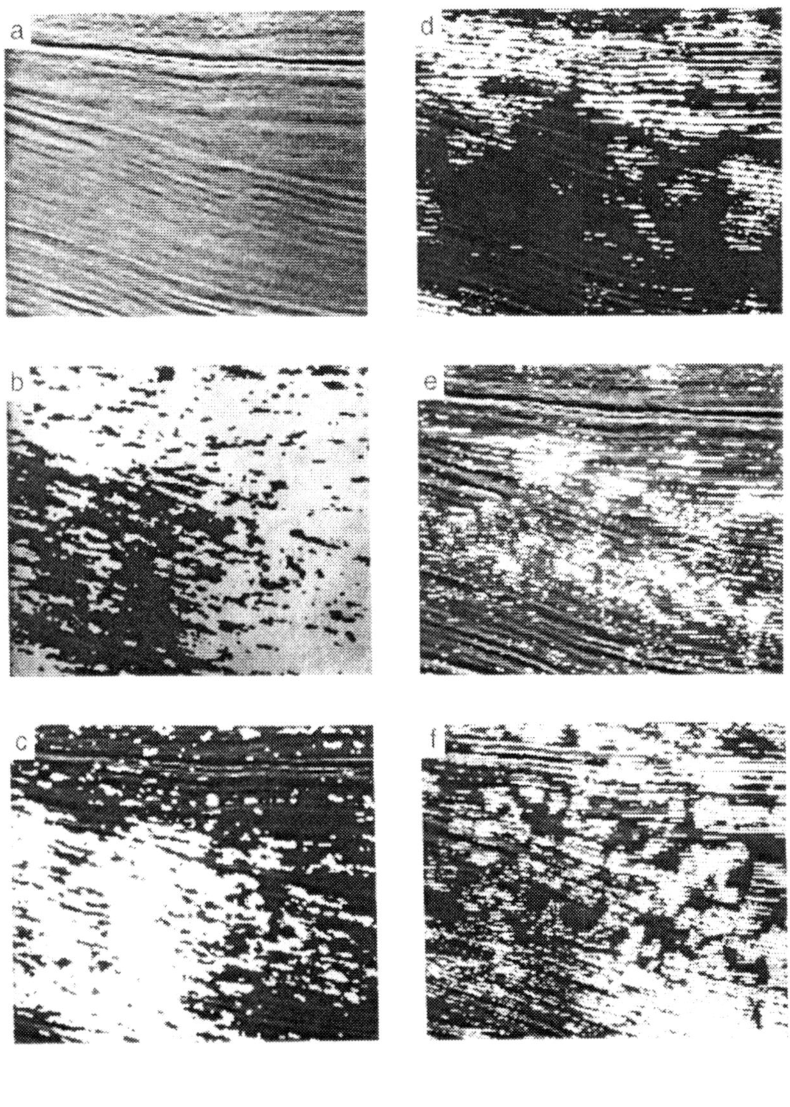

Fig. 3. Seismic image filtering by using directional and Gabor filters.

The original seismic image is shown in Figure 3a. The segmented output of the horizontal directional filter is shown in Figure 3b. The seismic image region having horizontal texture direction is shown in white. The seismic image regions having negative texture slope is shown in white in Figure 3c. The results of the application of a set of three Gabor filters with the same orientation (horizontal), but different radial frequencies are shown also in Figure 3. The seismic image regions having horizontal orientation and low, medium and high radial frequency content are shown in Figures 3d, 3e, 3f respectively.

IV. GRAPHICS

The use of graphics workstations has been one of the greatest advances towards automated interpretation. Nowadays the color pencils and the seismic section plots have almost been abandoned in seismic interpretation. The interpreter works interactively in front of his workstation [1]. Such workstations are usually connected to a mainframe or to a mini computer. They are usually supported by software which has extended filing and book-keeping utilities, very good human interface and impressive graphics and display capabilities. Volumes of 3-d seismic data can be displayed in various modes (e.g. chair display, concertina display, open cube display, variable density display, wiggle display, partial zoom of section display, 3-d loop display) by using pseudocolors. Also synthetic seismograms, log data and instantaneous frequencies can be displayed. In most cases interactive horizon picking routines and fault detection routines are supported. Therefore, graphics and image processing are powerful tools for geophysical interpretation.

V. GEOPHYSICAL AND GEOLOGIC EXPERT SYSTEMS

Geology and geophysics has been one of the first areas where expert systems have been applied. This comes from the fact that geologic and geophysical interpretation are heavily based on experience. Furthermore, the domain is at the appropriate stage of development. It has a vocabulary of basic concepts and useful rules of thumb. However, it possesses no general solution method. This is the main reason why the systems already developed do not have general characteristics and they are case studies.

The first geologic expert system is PROSPECTOR [52,53]. It is primarily concerned with hard-crystaline rocks. Therefore, we shall not analyze it further. The major attempts to build geophysical expert systems have been done in the domain of the interpretation of well log measurements. Logs are measurements, represented by curves, of characteristic properties (density, electrical resistivity, sound transmission, radio-activity) of the various rocks penetrated by the drill.

The DIPMETER ADVISOR [54,55] is an expert system developed by SCHLUMBERGER-DOLL(USA) and MIT for the interpretation of dipmeter measurements. The dipmeter tool measures the conductivity of the rock in a number of directions around the borehole. Variations in the conductivity can be correlated and combined with measurements of the inclination and orientation of the tool to estimate the magnitude and the azimuth of the *dip* or tilt of various layers. Sequences of dip estimates can be grouped together in patterns:

a) **Green pattern**: A zone of constant dip magnitude and azimuth.

b) **Red pattern**: A zone of increasing dip with constant azimuth over depth.

c) **Blue pattern**: A zone of decreasing dip magnitude with constant azimuth over depth.

From these patterns, a skilled interpreter is able to deduct the history of deposition, the composition and structures of the beds and (in connection with the seismic data) the locations for future wells. The DIPMETER ADVISOR emulates human interpretation of the dipmeter logs. The system is written in INTERLISP and interfaces to the user via a high resolution color display and a mouse. Its *knowledge base* has production rules of the form:

```
IF
there is a red pattern over a fault
and
the direction of the red pattern is perpendicular to the fault
and
the length of the red pattern is greater than 200 feet
THEN
the fault is a  growth fault.
```

In addition to the rules, the knowledge base also contains a few simple feature detectors.

The *data base* contains everything the system knows about the well. The *inference engine* controls the system, invokes its rules in data-directed fashion, matches them against patterns in the data base and adds new conclusions, whenever the rule matches successfully. The interpreter can modify any of the results of the system, in any phase. He can add his conclusions and he can revert to early phases of the analysis.

LITHO [56] is a system developed at SCHLUMBERGER (FRANCE) for the interpretation of log measurements of any type (sonic, resistivity etc.). The strength of its approach is its capacity to integrate different sources of information and to accommodate possible contradiction. It is written in INTERLISP and its knowledge base consists of production rules.

A different approach to the well-log interpretation is the use of automatic programming for the construction of log interpretation software. Such systems e.g. Φ_0 and Φ_{NIX} have been developed in SCHLUMBERGER-DOLL [58]. The primary conclusion of these efforts is that domain knowledge plays a critical role in the development of automatic programming systems.

Geophysical interpretation is closely related to geologic interpretation, i.e., the task of inferring from a description of a region the sequence of events which formed that region. The description of the region can be a diagram representing a cross-section of the region, which comes e.g. from the geophysical interpretation of a seismic cross-section, together with an identification of the rock types. Geologic interpretation is not static. It attempts to reconstruct the sequence of events which occurred, i.e., it converts the signal data from a spatial domain to the temporal domain of geologic processes. A system for automated interpretation has been constructed in MIT [57].

VI. TOWARDS AN INTEGRATED GEOPHYSICAL INTERPRETATION SYSTEM

As has been discussed in the previous sections, there have been several approaches to automated geophysical interpretation. All of them have their advantages and deficiencies. Pattern recognition and image processing can give excellent quantitative results in specific tasks. They are based on statistics and have rigorous mathematical background. The success of their application can be estimated in advance. However, they can not take easily into account symbolic information and experience, which is highly important for geophysical interpretation. Expert systems and artificial intelligence techniques can easily incorporate symbolic information and knowledge. A major drawback of these techniques is that arithmetic computations and mathematical analysis can not be merged smoothly with them. Thus, nobody can guarantee or predict the performance of an expert system, built with conventional techniques, in real-world cases. Graphics provide excellent human-machine interface, but nothing more than this. Based on this analysis, we think that a future geophysical interpretation system must have the following characteristics:

a) It should be *interactive* with excellent interpreter-machine interface. Geophysical interpretation is a difficult task and we are far away from building a completely automated expert system.

b) It should combine all the previously mentioned approaches (expert system techniques, image processing, graphics and pattern recognition).

c) It should possess multisensorial data processing capabilities (e.g. seismic, magnetic, gravitational, geologic, geographical and well log data).

d) It should have reasoning capabilities for incomplete, imprecise and fuzzy data and knowledge.

e) It should have learning capabilities and natural language understanding.

f) Its hardware support should be able to perform symbolic operations (for inferences) and arithmetic operations (for data processing and graphics) at very high speed. It should also have massive data storage and manipulation capabilities.

In the following, we shall describe a system called AGIS [62] having much less capabilities, which has been developed originally at the University of Toronto. A second version of this system is currently built at the University of Thessaloniki. It is limited to the seismic interpretation of 2-d seismic images. It is interactive and it incorporates artificial intelligence and image processing techniques. It has also fuzzy reasoning capabilities, as it will be shown in the next sections. Another knowledge-based system for stratigraphic interpretation of seismic data which follows the above-mentioned guidelines has been recently developed by Roberto et al. [59].

VII. STRUCTURE OF AGIS

AGIS consists of two separate parts [62]. The first part corresponds to the *low-level vision*. It is composed of global image processing routines, which perform image filtering, line detection, gap filling, texture analysis etc. This part of the system can be completely automated or become interactive with the interpreter. A complete description of the low-level vision part is included in the next section. The second part of the system corresponds to the *high-level vision*. It has a knowledge base which describes geologic formations. This knowledge base is used in the search of various elements of the seismic image (e.g. seismic horizons). The detected horizons are encoded in symbolic form and they are used as an input in the knowledge-based detection of more complicated geologic formations (e.g. faults, anticline traps, rock layers). The detected formation is stored in a symbolic form and it possesses a degree of certainty. The high level vision part of the system can be completely automated or can work interactively with the interpreter.

VIII. DESCRIPTION OF THE LOW-LEVEL VISION PART OF AGIS

The low-level vision part of the system includes the filtering, horizon extraction and texture analysis of the system. We shall analyze each of these tasks separately. Seismic cross-sections obtained from the seismic data are usually very noisy. However, most of the noise is removed during the seismic data processing phase [3,8], before the interpretation phase. AGIS employs several noise filtering techniques for further noise suppression. Linear 2-d lowpass filtering [29,32,33] can be mainly used to remove white background noise. Median filtering [29,32,33,66] can be used to enhance edges. Linear directional filtering [46]–[48] can be used to enhance lines along one dimension. Nonlinear statistical mean filters [60,66] and morphological filters [66] can be used to enhance and thin lines. These filters narrow the width of the reflection, thus facilitating the task of the line follower. This kind of filtering is sometimes a very important part of the preprocessing of the seismic cross-section.

Horizon following is another important task of the low-level vision part of AGIS. Automatic horizon following has been extensively treated in the literature [20,36,37,62,64]. The basic underlying idea is that horizon following is considered to be peak reflection picking for reflections which are stronger than a predetermined threshold. Some constraints are imposed that take into account the orientation, the distance between predecessors and successors and the reflection strength. A seismic horizon is described as a list which has a head of the form:

```
struct horizonhead {
   unsigned long      global_info;
   struct horizonpoint *nextpoint;
  };
```

where global_info about a horizon could be either average reflection strength, reflection variance, horizon length, global slope. Every pixel participating in a horizon is described as:

```
struct horizonpoint {
  int  dtime;  /* two way travel time */
  int  trace;  /* tracenumber */
  int  peak;   /* reflection intensity of the node */
  int  leflvalley;
           /* reflection intensity of the lower valley */
  int  rightvalley;
           /* reflection intensity of the upper valley */
  int  w1, w2, w3;  /* widths of upper, middle and lower
              lobes (see Figure 4) */
  unsigned long feature;
           /* feature assigned to every node */
  struct horizonpoint  *nextpoint;
           /* pointer to the next node */
  };
```

We assume that the source pulse traveling through the earth is the modified Ricker wavelet shown in Figure 4. When the characteristic wavelet cannot

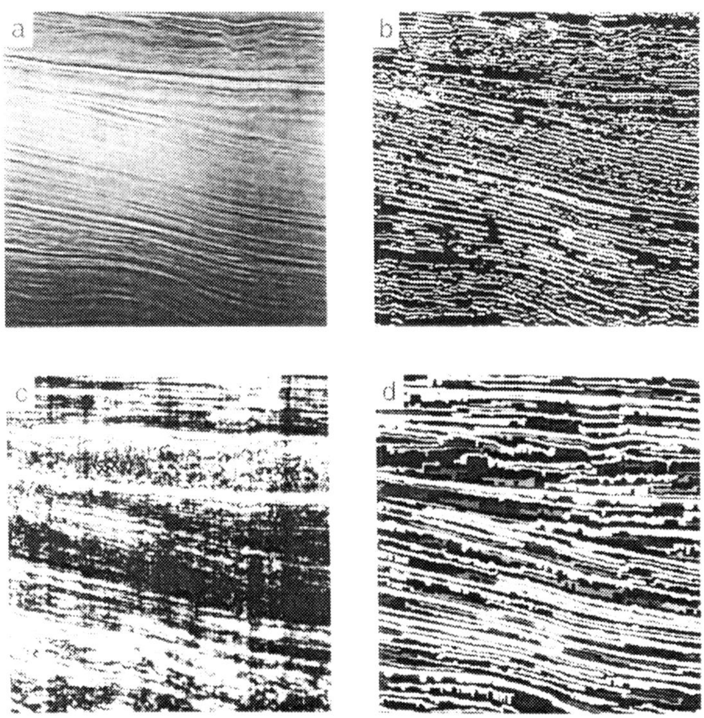

Fig. 4. Parameters of modified Ricker wavelet.

be identified (e.g., we may have a two-lobe pattern instead of a three-lobe) the undetermined quantities are assumed of *INFINITY* value. Information about local horizon features (e.g., local reflection intensity, local orientation etc.) at each horizon point are stored at each node of the horizon. The procedure of horizon picking is described below. First of all, we keep only those pixels whose gray level (e.g., reflection strength) is greater than a threshold, usually in the middle of the dynamic range. In other words, we define as event whatever is between two successive threshold crossings and has value greater than the threshold. The local extremum (peak) in the extent of the event is determined. A peak linking strategy will be applied to all peaks. Let us suppose that we follow a horizon and we are at a peak located at pixel (i, j). Let us also denote by $I(i, j)$ the image (reflection) intensity at this pixel. The first coordinate denotes trace number and the second one denotes two way travel time. The pixels $I(i+1, q)$ are examined, where $j-2 \leq q \leq j+2$, as can be seen in Figure 5. The following decisions

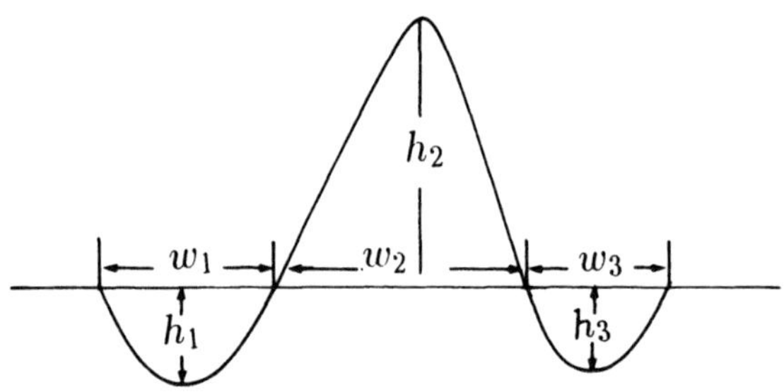

Fig. 5. Example of horizon following.

are made:

1. If there is only one peak at the next trace and in the defined interval for q, the horizon is expanded to the location of this peak.

2. If there are more candidate successors, the following steps are made.

    2.1 The local slope of the previous expansion (i.e., from trace $(i-1)$ to trace (i)) is calculated and the absolute differences of all possible current expansion slopes from the previous one are considered. We decide expansion to the more aligned candidate successor.

    2.2 If there is still ambiguity (i.e., more than one candidates) we apply the preceding step (2.1) considering global slopes (i.e., from the beginning of the horizon up to trace (i) and up to trace $(i+1)$) instead of local ones.

2.3 If we cannot find a solution still, we decide expansion to the peak having the maximal reflection strength.

3. If there is no peak at the next trace and in the specified interval for q, we give another chance to horizon to be expanded, repeating steps (1),(2) for trace $(i+2)$, two traces far away the current trace. Such a decision is justified theoretically, if we take into account the horizontal resolving power in the seismic section (i.e., the first Fresnel zone).

4. If the third step cannot lead to an expansion, the horizon picking is terminated. Short horizons are rejected.

After horizon picking, the local and global information about the horizon are calculated. Local information is stored at each horizon node, whereas global information is stored at the header of the horizon list. The computation of most horizon features is straightforward. Local horizon slope is calculated by finding a linear piecewise approximation of the horizon. The horizons, which have been followed on the seismic image of Figure 6a by the above-described technique are shown in Figure 6b.

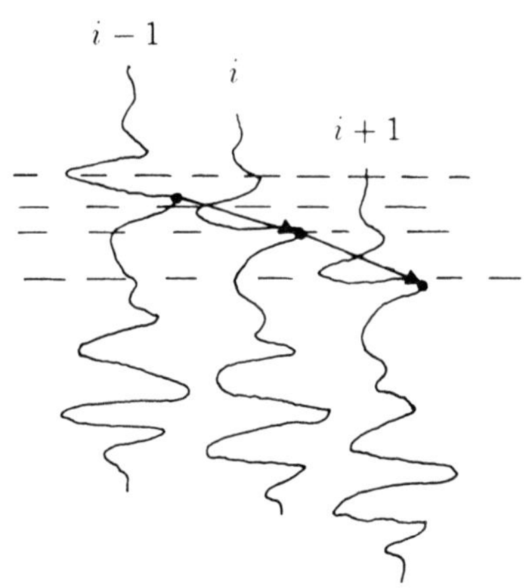

Fig. 6. (a) Original seismic image. (b) Detected horizons. (c) Hilbert transform magnitude. (d) Run length image.

Ideally, the horizons should be only one pixel wide. This is facilitated by prefiltering the seismic image [60,62]. The output of the horizon following is a binary image having ones at the horizon pixels. The horizon pixel coordinates are filtered to produce smooth horizons [62]. Another task of the low-level vision part is horizon gap filling. A mask is applied to each pixel of the image that contains the results of the horizon follower. If there are sufficient horizon points inside the mask in a specific direction and the actual pixel represents a gap, this gap is filled. Horizon coordinate filtering and gap filling can be repeated iteratively, until the results are acceptable. More sophisticated algorithms for gap filling that employ minimum entropy rule learning techniques are described in [63]. Horizon following, horizon coordinate filtering and gap filling are the first steps of the structural analysis of the seismic data. Their results are fed to the structural analysis, which is performed by the knowledge-based part of the system.

The last step of the low-level vision part of the system is texture analysis. A more detailed discussion on a texture-based approach to the segmentation of seismic images can be found in Section IX. Texture analysis tries to segment the seismic image in homogeneous regions. The texture properties of a seismic image can be used to partition image into regions which are characterized by a property consistent with the stratigraphic information and are contrasted to their adjacent regions. This means that the primitive features used in the segmentation must be such so that they have correspondence with the entities used by the interpreter. The features that are used for segmentation belong to three classes [64,65,67]: features that can be computed at every pixel, horizon-based features and features referred to pixels participating in runs. A feature which is already available with the seismic data, is the reflection strength which is essentially the intensity of the seismic image. Seismic image transformations can also be used in texture analysis [39]. Such a transform traditionally used in stratigraphic interpretation is the Hilbert transform. It produces additional texture features (instantaneous amplitude, phase, frequency). Geophysical interpretation is heavily based on the seismic horizons, their characteristics and their interrelationships, as has already been described. Therefore, a second category of features may be calculated on horizons. Such features are horizon length, mean reflection strength, signature, global slope and local slopes. A third category of features may be calculated on runs. The most straightforward feature is run length. Seismic texture features have the following geophysical significance [9]:

1. Local and global slope are related to stratification patterns.

2. Horizon and run lengths are related to the lateral continuity seismic events.

3. Reflection strength and Hilbert transform amplitude are related to acoustic impedance variations.

4. Instantaneous phase of Hilbert transform is related to seismic event continuity.

5. Instantaneous frequency of the Hilbert transform is related to frequency pattern variations according to changes in thickness and lithology.

Seismic image segmentation requires the use of a logical predicate (rule) which is based on the feature vector and is applied to the entire image to be segmented. Such rules are very difficult to be evaluated in seismic applications, because there exists no straightforward relation between the features and the seismic image regions that have some geophysical significance. It is very desirable to construct a system that can infer the rule from examples given by the interpreter. Learning techniques from examples can be used to provide interactive methods for texture image segmentation [63,65,67]. More specifically, image regions, which are representative of the different types of seismic textures, are chosen by the interpreter. These regions and the corresponding sets of feature vectors constitute the examples of the interpretation task. The appropriate seismic texture discrimination rule is created by using rule learning techniques based on the minimal entropy principle. The system has the ability to reject features possessing no discriminatory power. The rules derived by this system are able to discriminate, for example, regions having long and strong horizons from regions having short and weak horizons, or tilted regions from regions having approximately horizontal horizons. Thus, the derived rule predicates are of the form "feature value less or equal than a proper threshold". The derived rule consists of disjunctions of conjunctions of predicates. Thus, the segmentation rule splits the feature space in hypercubes. Each hypercube describes one seismic texture classes. This type of texture discrimination rule has been chosen, because it has close resemblance with the intuitive rules used by the human interpreter. Therefore, the interpreter can easily check the geophysical significance of the derived rule. Other types of rules (e.g., using discriminant surfaces [12]) can not be easily evaluated by the interpreter.

If the features can be calculated on every image pixel, the derived rule can be used directly for the segmentation of the entire seismic image [30,31]. However, if horizon/run features are used, only the image pixels corresponding to seismic horizons/runs can be segmented. Region growing techniques based on Voronoi tessellation [75] can be used for the segmentation of the entire seismic image, when horizon-based features are employed [64,67]. In other words, pixels can be assigned to seismic image regions by using the geometric proximity to the already segmented seismic horizons/runs. Therefore, the image domain is partitioned into regions such that all points in the same region have as their nearest neighbors reference points of a specific texture class. Such a segmentation corresponds to a generalized Voronoi tessellation of the image domain and is obtained by using mathematical morphology techniques [66,76].

IX. A TEXTURE-BASED APPROACH TO THE SEISMIC IMAGE SEGMENTATION

In this section we will describe briefly feature extraction, rule selection and the region growing techniques mentioned earlier.

A. CALCULATION OF SEISMIC TEXTURE FEATURES

Signal transformations are common in signal and texture analysis [39]. Hilbert transform analysis effects a natural separation of amplitude and phase information. Therefore, it has particular importance in seismic texture discrimination. It has already found several applications in seismic stratigraphy [71].

Hilbert transform is the basis of the mathematical procedure that creates a complex trace from a real one. Therefore, the corresponding analysis is also called complex trace analysis. Hilbert transform relations are relationships between the real and imaginary components of a complex sequence [69]. Let us define by $s(n)$ the complex trace:

$$s(n) = s_r(n) + \jmath s_i(n) \tag{4}$$

where $s_r(n)$ and $s_i(n)$ are real sequences. The real trace $s_r(n)$ is the already available seismic trace. The imaginary trace $s_i(n)$ is the Hilbert transform of the real seismic trace. The Hilbert transform is basically a special filter that shifts all positive frequencies of an input signal by $-90°$ and all negative frequencies by $+90°$. Therefore, the Fourier transforms $S_r(e^{\jmath\omega})$ and $S_i(e^{\jmath\omega})$ are directly related by:

$$S_i(e^{\jmath\omega}) = H(e^{\jmath\omega})S_r(e^{\jmath\omega}) \tag{5}$$

where:

$$H(e^{\jmath\omega}) = \begin{cases} -\jmath & 0 \leq \omega < \pi \\ \jmath & -\pi \leq \omega < 0 \end{cases} \tag{6}$$

The magnitude sequence $A(n)$ is given by:

$$A(n) = \sqrt{s_r^2(n) + s_i^2(n)} \tag{7}$$

and is called instantaneous amplitude. It is the envelope of the complex trace sequence $s(n)$. It may have its maximum at phase points other than the peaks or the valleys of the real trace, especially where an event consists of several reflections [70]. An interesting geophysical interpretation of the magnitude sequence is given in Hardage [71]: The real part of a complex trace can be used as an estimate of the kinetic energy of the earth system resulting from a seismic excitation, while the imaginary part can be used to measure the potential energy involved in the earth's elastic response to a seismic disturbance. The magnitude of the complex trace is a measure of the total energy involved in the seismic reflection response. The magnitude of the Hilbert transform of the seismic image of Figure 6a is shown in Figure 6c. The instantaneous phase is given by:

$$\phi(n) = \arctan \frac{s_i(n)}{s_r(n)} \tag{8}$$

and emphasizes the continuity of the events. It often makes weak coherent events clearer. In continuous time notation, the instantaneous frequency is defined as the rate of change of the time-dependent phase:

$$w(t) = \frac{d}{dt}\phi(t) \tag{9}$$

and by using (8):

$$w(t) = \frac{s_r(t)\frac{d}{dt}s_i(t) - s_i(t)\frac{d}{dt}s_r(t)}{s_r^2(t) + s_i^2(t)} \tag{10}$$

Since most reflection events are composed of individual reflections from a number of closely spaced reflectors, the frequency character of a composite reflection changes gradually as the sequence of layers changes in thickness and lithology. The instantaneous frequency can take negative values, according to its definition (10). Negative frequencies occur when an interference between two waveforms exists.

A seismic image can be easily binarized by using an intensity threshold: all pixels having intensity greater than this threshold take value 1, whereas the rest of them take value 0. The binary seismic images consists of binary seismic traces denoted by $B_j, j = 1, \ldots, N$ in this section. In geophysical applications it is important to follow nonhorizontal runs in seismic sections, because they have stratigraphic significance. The classical implementation of runs does not give such a possibility because it follows runs along the horizontal direction. A generalized run definition alleviates this difficulty [38]. It uses a binary operator $\star$ on two binary traces B_j and B_{j+1} whose operation is illustrated in Figure 7.

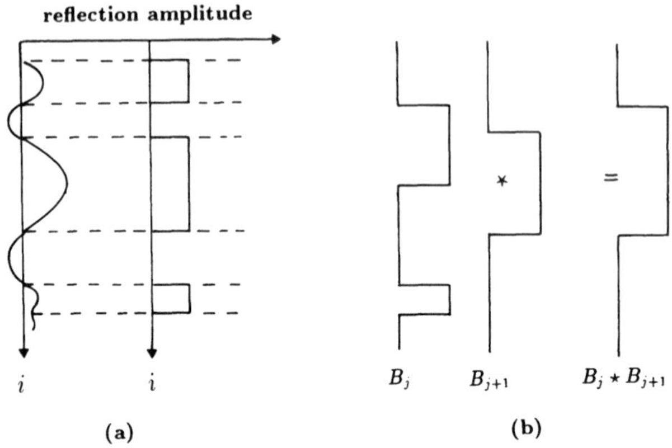

Fig. 7. (a) Binary seismic trace. (b) Example of the operation $B_j \star B_{j+1}$.

If parts of the binary signals overlap, the binary output has value 1 for all the duration of the overlapping, otherwise this operator performs as a binary *AND* operator. The difference between the classical and the generalized definition of runs is illustrated in Figure 8. Figure 8a shows

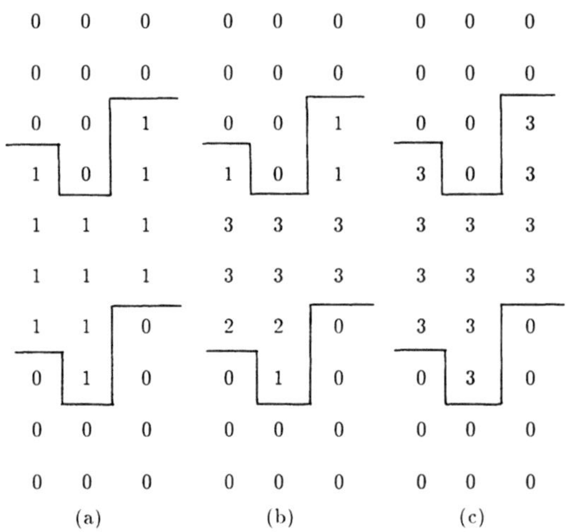

Fig. 8. (a) Three consecutive binary traces. (b) Run lengths by using the classical definition. (c) Run lengths by using the generalized definition.

three adjacent binary traces. Obviously, all 1s belong to the same event. The classical and the generalized definitions of run give the results shown in Figures 8b, 8c respectively. The generalized run identifies the region of 1s as a single entity having run length 3, whereas the classical definition fails to do so. The run following algorithm is based on the following simple idea:

1. A generalized run has to be followed, when overlapping between the part of 1s of the current trace and the part of 1s of the subsequent trace occurs. Run following continues until no overlap can be detected or the seismic section has been exhausted.

2. This process is repeated for all pixels having value 1 that have not been taken into account yet.

3. Pixels having value 0 are not taken into account.

A run is described as a list of head

```
struct runheader {
   int  length;
   struct rtail *rest;
   };
```

and nodes of the form

```
struct rtail {
   int trace;     /* tracenumber */
   int up;        /* leading edge of the pulse */
   int down;      /* lagging edge of the pulse */
   struct rtail  *nextpoints;    /* pointer to the next node */
  };
```

The run length is the same for all pixels in the extent of the pulse and for all overlapping pulses. The run length is directly related to the horizon length feature of a seismic image. The binarized seismic image can also be used to obtain other features, e.g., the local width of an image run and the average reflection strength of an image run. The above-described algorithm has been applied to the seismic image of Figure 6a. The run length image is shown in Figure 6d.

B. LEARNING TECHNIQUES IN THE DERIVATION OF TEXTURE DISCRIMINATION RULES

So far, we have described the texture of seismic images in terms of features that refer to every pixel (reflection strength, instantaneous amplitude and frequency) or in terms of features that characterize only the pixels that participate in horizons (length, reflection strength, local slope) or in terms of run lengths for the pixels participating in runs. At this stage, we have to form rules for texture discrimination based on these features. Examples of the form of predicates and rules were given in Section VIII. Now we proceed to the mathematical formulation of the problem.

Let us suppose that m features $x_1, x_2, \ldots, x_m$ are used in the description of seismic texture. The frequency of appearance of a specific feature m-tuple $(x_1, \ldots, x_m)$ creates the m-d histogram of this m-tuple. The domain of the m-d histogram constitutes the feature space. Let us suppose that we want to discriminate K different texture classes namely $C_1, C_2, \ldots, C_K$. A pixel characterized by a feature vector $\mathbf{x}$ is assigned to class C_k, if it satisfies a decision rule of the form:

$$\text{if} \quad L_k(\mathcal{P}_1, \mathcal{P}_2, \ldots, \mathcal{P}_m) \quad \text{then} \quad \mathbf{x} \in C_k \tag{11}$$

where L_k is a propositional logic formula and $\mathcal{P}_i, i = 1, \ldots, m$ are predicates of the form:

$$\mathcal{P}_i : x_i \leq T_{i,opt} \tag{12}$$

$x_i, T_{i,opt}, i = 1, \ldots, m$ are the features and their corresponding optimal thresholds. The choice of optimal thresholds in (12) and the optimal rule in (11) can be done automatically by a similar learning procedure described in [63]. Specifically, the optimal choice of thresholds (used in predicates) and rules from examples and counterexamples have been proposed by minimizing an entropy function. The rules have the form of disjunctions of conjunctions of predicates. The example and counterexample set are given by the interpreter. Therefore, such a system can learn rules from examples and it is based on an optimality criterion. The discrimination of two different classes ($K = 2$) will be treated first, in order to simplify the presentation. A modification of this scheme, capable of implementing a multiclass rule learning, will be described afterwards. Simultaneous optimization for the thresholds and the rule structure is very difficult. Therefore, the optimization is split in two suboptimal steps:

- minimum entropy threshold selection of $T_{i,opt}, i = 1, \ldots, m$
- minimum entropy rule selection of $L_k, k = 1, \ldots, K$.

First, minimum entropy threshold selection is described. The presentation of the minimum entropy rule selection and of the multiclass discrimination follows.

Let Ω_1, Ω_2 denote the sets of the examples and the counterexamples in a two-class discrimination problem. Both the sets are consisted of training feature vectors of the form $(x_1, x_2, \ldots, x_m)$.

The probability of correct classification of the example set is denoted by:

$$P^1_{i,T} = \text{Prob}\{x_i \leq T_i | \Omega_1\} \qquad i = 1, \ldots, m \tag{13}$$

Similarly, the probability of wrong classification of the counterexample set is given by:

$$P^2_{i,F} = 1 - \text{Prob}\{x_i \leq T_i | \Omega_2\} \qquad i = 1, \ldots, m \tag{14}$$

The thresholds T_1, T_2 are chosen in such a way, so that the polynomial counterpart of the entropy function is minimized [63]:

$$U(T_i) = -2\,(P^1_{i,T} + P^2_{i,F} - 1)^2 \;\rightarrow\; \min \tag{15}$$

We have m entropy functions (15), one for each threshold T_i, $i = 1, \ldots, m$. Therefore, the optimization can be done independently for each threshold.

The optimal rule will have the following canonical form [63,72]:

$$L_k = \bigvee_{I=0}^{2^m-1} a_I \; R^m_I \tag{16}$$

where the Boolean constants a_I are either 0 or 1 depending upon whether the corresponding product term R^m_I is to be excluded from or included in the rule. The product terms R^m_I are defined as:

$$R^m_I = \mathcal{P}^{i_m}_m \wedge \cdots \wedge \mathcal{P}^{i_2}_2 \wedge \mathcal{P}^{i_1}_1 \tag{17}$$

where

$$\mathcal{P}^{i_j}_j = \begin{cases} \mathcal{P}_j & \text{if } i_j = 1 \\ \overline{\mathcal{P}_j} & \text{if } i_j = 0 \end{cases} \tag{18}$$

The superscript m in (17) indicates the number of predicates and $i_m, \ldots, i_1$ are the binary digits of the decimal subscript I, i.e.:

$$(I)_{10} = (i_m\, i_{m-1}\, \ldots\, i_2\, i_1)_2 \tag{19}$$

We shall drop the superscript m in product terms for simplicity. The predicates in (18) are of the form (12). The probabilities of success of each product term R_I are defined as follows:

$$\begin{aligned} P^1_T(R_I) &= \text{Prob}\,\{\text{ success of } \; R_I \; \text{ over } \; \Omega_1\} \\ P^2_T(R_I) &= \text{Prob}\,\{\text{ success of } \; R_I \; \text{ over } \; \Omega_2\} \quad I = 0, \ldots, 2^m - 1 \end{aligned} \tag{20}$$

It has been proven that the optimal rule, in the minimum entropy sense, consists of the product terms for which the following relation holds [63]:

$$\text{if} \quad P^1_T(R_I) > P^2_T(R_I) \quad \text{then} \quad a_I = 1 \tag{21}$$

This relation is interpreted as follows: in the optimal rule contribute those product terms for which their success over examples is greater than than their success over counterexamples.

Thus far, an effective procedure for discriminating examples against counterexamples has been described. We shall iteratively use this scheme in a multiclass classification task. Let us denote by $C = \{C_1, C_2, \ldots, C_K\}$ the set of the K different classes and by $\Omega_l, l = 1, \ldots, K$ the corresponding set of examples for each class. The classification procedure used is hierarchical. Therefore, it corresponds to a binary decision tree. Each node has two branches:

- a stable one which corresponds to the example set that has been discriminated at this level
- an unstable branch which corresponds to the remaining unclassified example sets.

Obviously, a stable branch leads to a leaf and has no successors, whereas an unstable branch leads to a node that is split into a stable and an unstable branch at the next level. Each leaf corresponds to a class and to a recognition rule for this class. The rule which discriminates a class at a certain level is calculated as follows. One of the unclassified example sets, e.g., Ω_k is considered to be candidate for discrimination. The union Ω' of the rest of the unclassified example sets is considered to be the counterexample set. Thus, the problem is reduced to a two-class classification problem at each level. The derived rule L_k discriminates the class C_k against the rest of the unclassified classes. This procedure is repeated for every class C_k that has not been discriminated yet. The rule L_k that possesses the minimal probability of misclassification is chosen as the discrimination rule for this level. This procedure is repeated with the rest of the unclassified example sets iteratively, until all classes are discriminated.

We shall give an example to illustrate the application of the above-described rule learning procedure. Four representative seismic image regions having different seismic texture are chosen by the interpreter and are shown in Figure 9a. These image regions are denoted by R_i, $i = 1, \ldots, 4$. The corresponding feature classes are denoted by C_i, $i = 1, \ldots, 4$. The feature vectors corresponding to the horizon pixels in these regions consist the example sets of the learning procedure. The following features have been used for texture description [65,67]: mean reflection strength, horizon local slope and horizon length. Thus, the feature vector $\mathbf{x}$ is described as a triplet (x_1, x_2, x_3), where x_1 denotes mean reflection strength, x_2 denotes horizon partial slope and x_3 denotes horizon length. The results of the learning procedure are summarized in Table I. At the first step, the class C_3 is discriminated from the others. The corresponding rule having the minimal probability of misclassification is shown in Table I. The choice of the optimal threshold $T_1 = 137$ is found from the minimum of the entropy curve. At the second step, the class C_4 is discriminated from the classes C_1, C_2, as shown in Table I. At the last step the classes C_1, C_2 are discriminated from each other. Therefore, this strategy leads to a binary decision

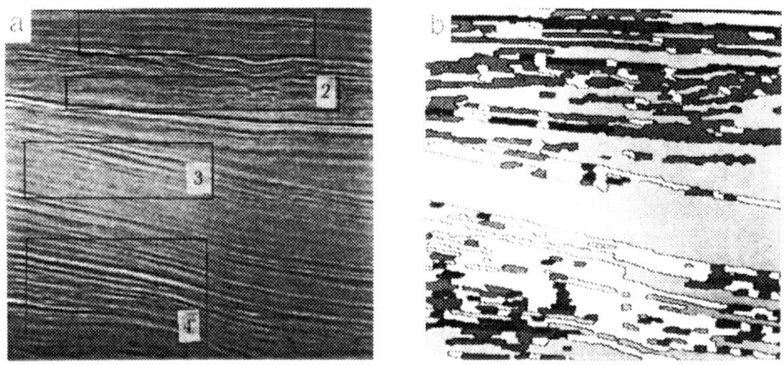

Fig. 9. (a) Examples of regions having different seismic texture. (b) Segmentation of the original seismic image in four regions.

tree of the form:

$$
\begin{array}{l}
\text{if} \quad (x_1 \leq 137) \qquad \mathbf{x} \in C_3; \\
\text{else} \;\; \{ \\
\qquad \text{if} \quad (x_2 \leq -12^\circ) \qquad \mathbf{x} \in C_4; \\
\qquad \text{else} \;\; \{ \\
\qquad\qquad \text{if} \;\; (x_1 \leq 142 \vee x_3 \leq 70) \qquad \mathbf{x} \in C_2; \\
\qquad\qquad \text{else} \qquad \mathbf{x} \in C_1; \\
\qquad \} \\
\}
\end{array}
\tag{22}
$$

where the mean reflection strength and the horizon length are scaled in the range [0,. . .,255] and the partial horizon slope is measured in degrees. This rule can be used to characterize only horizon pixels. Region growing techniques must be used for the segmentation of the entire image.

Table I. Learning procedure from training patterns on horizons

Step	class of examples	classes of counter-examples	Thresholds (T_1, T_2, T_3)	Rule	Probability of misclassification
1	C_1	C_2, C_3, C_4	$142, -6°, 128$	$x_2 > -6°$	0.202496
	C_2	C_1, C_3, C_4	$153, -12°, 70$	$x_2 > -12°$	0.388880
	C_3	C_1, C_2, C_4	$137, -12°, 104$	$x_1 \leq 137$	0.143421
	C_4	C_1, C_2, C_3	$137, -12°, 81$	$x_1 > 137$ $\wedge$ $x_2 \leq -12°$	0.167114
2	C_1	C_2, C_4	$150, -6°, 128$	$x_2 > -6°$	0.208602
	C_2	C_1, C_4	$157, -12°, 70$	$x_1 > 157$ $\vee$ $x_2 > -12°$	0.403871
	C_4	C_1, C_2	$149, -12°, 81$	$x_2 \leq -12°$	0.14
3	C_2	C_1	$142, -6°, 70$	$x_1 \leq 142$ $\vee$ $x_3 \leq 70$	0.30005

C. REGION GROWING

The problem of image segmentation is defined as follows [30,31]. Let $X \subset \mathcal{Z}^2$ be an image domain. A logical rule L is defined on the subsets X_k of X, such that $L(X_k)$ evaluates a property of the pixels in X which belong to X_k. Image segmentation is a partition of X into subsets or regions X_k, $k = 1, \ldots, K$ for some K such that:

$$
\begin{aligned}
X &= \bigcup_{k=1}^{K} X_k \\
X_k \cap X_l &= \emptyset \quad \text{for} \quad k \neq l \\
L(X_k) &= \text{TRUE} \quad \forall k \\
L(X_k \cup X_l) &= \text{FALSE} \quad \text{for} \quad k \neq l
\end{aligned}
\tag{23}
$$

If the features can be calculated at each image pixel, the learning procedure of the previous section infers the required rule L which defines the segmentation for the entire image. If one of the features used in the rule is the run length, only the pixels participating in runs can be segmented. If horizon features are used in the rule, the situation is even worse: only the pixels belonging to horizons can be segmented initially. In these two cases, region growing techniques must be applied, if segmentation of the entire image is required. We have to infer what information can be assigned to other pixels that do not participate in horizons, assuming that pixels close to horizons will behave similarly. Thus, we have to investigate for proximity rather than for similarity. A method for solving the problem under consideration is based on a modification of the Voronoi tessellation [75] and the mathematical morphology [76]. A second method based on a "radiation model" has been proposed in [64]. The horizon pixels or the pixels belonging to runs are used as "seeds" for the Voronoi tessellation of the seismic image. The Voronoi tessellation of a plane (in our case the image plane) results in a partition into regions, such that all points in the one region are nearest neighbors to a "seed". Thus, they are classified to the class of the "seed". Voronoi tessellation can be easily performed by a growing procedure [64]. The horizon pixels grow in successive steps until they cover the entire image. At each step, it is checked if regions belonging to the same cluster have common boundary. If this is the case, these regions are merged. The boundary (if any) between two different clusters is "frozen" at each step. The growing of the image regions is performed by conditional dilation[76]. Let X_k, $k = 1, \ldots, K$ be subsets of the image domain X, representing the image pixels which correspond to each texture region. Let also $X_k(i)$, $k = 1, \ldots, K$ be the sets representing the regions at step (i) of the growing procedure. At step (0), $X_k(0)$ contains the "seeds" corresponding to class C_k. The region growing which leads to a modified Voronoi tessellation of the image plane is based on mathematical morphology. Let B be a structuring element [66,76], whose size determines the size of the region growing at each step. Its shape governs the geometry of cluster growing. If uniform growing along all dimensions is required, the structuring element B must be a disk. However, in the Euclidean grid $\mathcal{Z}^2$ no exact representation of a disk can be found. Thus, the structuring element *CIRCLE* shown in Figure 10 has been used instead. It produces

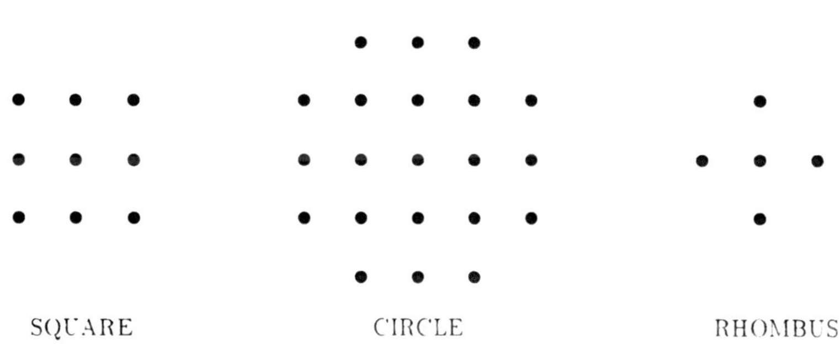

Fig. 10. Structuring elements.

an acceptable relatively uniform growing along all dimensions. The region growing at step (i) is given by the following recursive procedure:

$$X_k(i) = [X_k(i-1) \oplus B^s] \bigcap [\bigcup_{l=1,\, l\neq k}^{K} X_l(i-1)]^c \tag{24}$$

and

$$X_k(i-1) \oplus B^s = \bigcup_{b \in B} X_k(i-1)_{-b} \tag{25}$$

$$X_k(i-1)_{-b} = \{z \in \mathcal{Z}^2 \;:\; z = x + b \,,\; x \in X_k(i-1)\} \tag{26}$$

where $\oplus, \cap, \cup$, X^c denote set dilation, intersection union and complementary set [76,66]. Equation (24) permits the growing of the k-th cluster in the image regions which have not already been covered by other clusters. The main disadvantage of this approach is that it does not eliminate small patches corresponding to noise, which may be found inside much larger regions. Its advantage is that only a few recursions are required for the segmentation of the entire image.

Thus, the seismic image of Figure 6 can be segmented in the following way. Horizon pixels are clustered by using (22). Region growing (24)–(26) is applied afterwards. The final segmentation is shown in Figure 9b. The segmented regions, which correspond to the classes C_1, C_2, C_3, C_4 are shown by increasing brightness.

X. KNOWLEDGE REPRESENTATION

There were three key questions to be answered during the development of AGIS:

a) What kind of patterns should be searched for?

b) How the knowledge related to these patterns should be represented?

c) What control structure for pattern search should be used?

The first question arises from the fact that seismic data are usually very complicated, due to the complex nature of the earth structure. The patterns sought are sometimes fuzzy, imperfect, overlapping. Thus, some basic features (patterns) of interest must be chosen, which are simple enough and which can be elements of other more complicated geologic formations. The last two questions are encountered in the development of every expert system [52], and they are active research topics in artificial intelligence.

The following seismic patterns, which correspond to geologic formations, have been described in our system:

1. Horizons
2. Anticline traps
3. Faults
4. Salt domes and reefs
5. Layers

These are the basic geologic patterns that are searched in a seismic cross-section. The following remarks can be made for the seismic patterns:

a) More elementary patterns are parts of more complicated ones (in a semantic not in a geological sense). Thus, horizons are part of a fault, in the sense that a fault consists of horizons cracked in a certain direction.

b) Some patterns are special kinds of others (e.g. a strong horizontal horizon is a special kind of horizon).

c) Spatial relations are always required (e.g. neighborhood relations, parallelism relations).

The analysis of the seismic patterns shows that each of them possesses some attributes and it is related to the other ones by *generalization/specification*, *decomposition/aggregation* and *spatial relations*. The fact that each pattern has some attributes has led us to choose the frames [78] for the knowledge representation of each pattern. Furthermore, the need of relations between the patterns imposed the use of the semantic networks, whose nodes are frames representing seismic patterns [79]. We shall describe briefly these notions, although they are well-known in the literature. The idea to pack knowledge in a modular form is due to Minsky [78]. Each knowledge package is called *frame* [78] or *class* [79]. Each particular entity (called *token*) (e.g. a particular horizon) is member (called *instance*) of a class. The process of making a token instance of a class is called *instantiation*. A specific pattern is described in a class by its components and the relations it has to satisfy. The components are called *slots* of the class. Many times, certain constraints on the slots are to be satisfied, e.g. a horizon cannot have a vertical orientation. These constraints are described by a procedure which is attached to each class. This procedure instantiates a token by checking if the class slots are filled and if the class constraints are satisfied. Therefore, it is called instantiation procedure.

AGIS contains the following classes:

1. POINT

2. HORIZON-SEGMENT
3. HORIZON
4. STRAIGHT-HORIZON
5. CONVEX-HORIZON
6. CONCAVE-HORIZON
7. HORIZONTAL-HORIZON
8. ANTICLINE-TRAP
9. FAULT
10. SALT-DOME
11. REEF
12. ROCK-LAYER

The class HORIZON used for the representation of geologic horizons is described as follows:

```
class HORIZON
 {
 HORIZON-CERTAINTY;
 array of POINTS;        * POINT is a class *
 HORIZON-START;
 HORIZON-END;
 REFLECTION-STRENGTH;
 HORIZON-LENGTH;
 HORIZON-SLOPE;
 array of HORIZON-SEGMENTS;     *HORIZON-SEGMENT is a class *
 array of UPPER-NEIGHBOR horizons;
 array of LOWER-NEIGHBOR horizons;
 array of UPPER-PARALLEL horizons;
 array of LOWER-PARALLEL horizons; }
```

Its slots are the HORIZON-CERTAINTY (to be described in next section), an array of POINTS (which contains the horizon coordinates), the HORIZON-START and HORIZON-END (which are the horizon start and end coordinates), the REFLECTION-STRENGTH (which is the average reflection strength of the horizon), the HORIZON-LENGTH, the HORIZON-SLOPE, and an array of HORIZON-SEGMENTS. The members of this array are line segments produced by a linear piecewise approximation of the horizon. The arrays of UPPER-NEIGHBOR horizons, etc. describe spatial relations (to be defined later on in this section). The class POINT is of the form (i, j), where i is a trace number and j is reflection time.

The classes are connected to each other by relations and they form a semantic network. Our system includes generalization/specialization, decomposition/aggregation and spatial relations. The generalization/specialization relation is defined by the IS-A link between the classes [83,84]. The class STRAIGHT-HORIZON IS-A HORIZON. The classes have the slots of their IS-A parents plus some additional slots. The decomposition/aggregation relation is defined by a PART-OF link between classes [79]. The class LINE-SEGMENT is PART-OF the class HORIZON. The class CONVEX-HORIZON is PART-OF the class ANTICLINE-TRAP etc. An example of the PART-OF and the IS-A organization of the knowledge is shown in Figure 11.

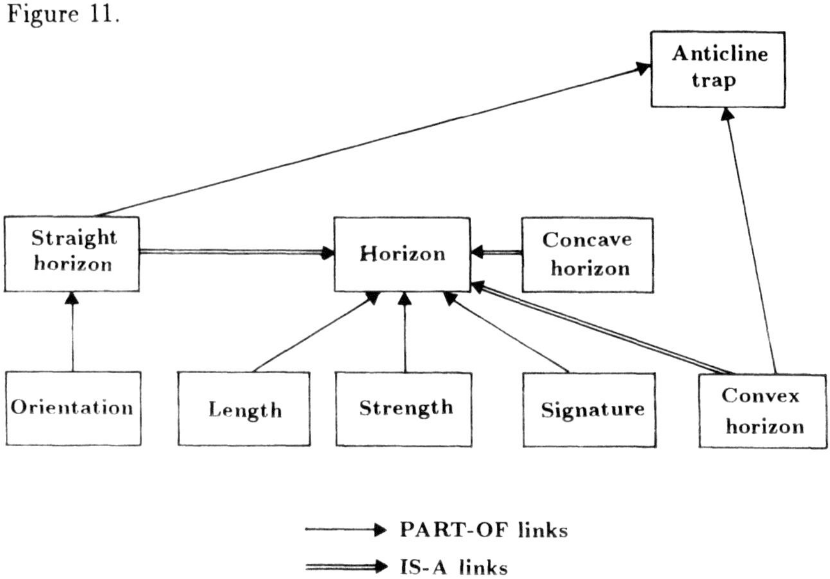

Fig. 11. PART-OF AND IS-A organization example.

Another important element of our system is the existence of spatial relations. Especially the following three relations are of vital importance:

a) UPPER-HORIZON-NEIGHBORHOOD, LOWER-HORIZON-NEIGHBORHOOD.

b) UPPER-HORIZON-PARALLELISM, LOWER-HORIZON- PARALLELISM.

c) UPPER-LAYER-NEIGHBORHOOD, LOWER-LAYER-NEIGHBORHOOD.

Two instances of the class HORIZON are connected by an UPPER- or LOWER-NEIGHBORHOOD relation if they are one above the other and no third horizon exists between them. Two instances of the class HORIZON are connected by an UPPER- or LOWER-HORIZON-PARALLELISM relation, if they are connected by an UPPER- or LOWER-HORIZON-NEIGHBORHOOD relation and their distance remains constant (i.e., within a certain threshold). Two instances of the class ROCK-LAYER are connected by an UPPER- or LOWER-LAYER-NEIGHBORHOOD relation, if they have a common upper or lower boundary. The system has procedures that detect the spatial relations between instances of the class HORIZON or ROCK-LAYER. Having defined our knowledge representation we shall now describe how it can be used in seismic pattern search.

XI. CONTROL STRUCTURE FOR SEISMIC PATTERN SEARCH

The knowledge representation in an expert system merely describes various patterns and facts and does not describe how it can be used for the search of these patterns in the data. The mechanism of pattern search is called *control structure*. *Hypothesize and test* [85] is the control mechanism used in our system. A hypothesis is formed, when we try to create a new instance of a class. The class instantiation procedure tries to verify the hypothesis. This means that it tries to fill all slots necessary and to test if appropriate slot constraints are valid. If the hypothesis ranking mechanism (to be described later) returns high hypothesis certainty, the hypothesis is instantiated, otherwise the instantiation is denied. For example, the instantiation procedure of the class ANTICLINE-TRAP searches for an instance of the class STRAIGHT-HORIZON and an instance of the class CONVEX-HORIZON that are connected by a HORIZON-NEIGHBORHOOD relation. If such a combination is found, its certainty is measured. If it is above a threshold (defined by the interpreter or by default), a new instance of the class ANTICLINE-TRAP is created. If the certainty is low (e.g. the slope of the instance of the class STRAIGHT-HORIZON is high), the insertion is denied.

Hypothesis ranking is a very important factor in the performance of an expert system. It measures the certainty of each hypothesis during the process of the hypothesis verification. Hypotheses with low certainty (below a threshold) may be rejected. The hypotheses with high certainty form the so-called *focus* of the system. The certainty of a hypothesis h is a number

$\mathcal{R}(h)$, $0 < \mathcal{R}(h) < 1$, which receives contributions from three sources:

1. Contribution from more general hypotheses along the IS-A hierarchy.
2. Contribution from its components along the PART-OF hierarchy.
3. Contribution from the successful matching of its internal constraints.

An example of the IS-A contribution is that a line cannot be more certainly "straight horizon" than "horizon". An example for the PART-OF contribution is that horizon certainty depends on its length and on its reflection strength (weak and short lines have less certainty to be real horizons). An example of the internal constraints contributions is that a vertical line has small certainty to be a horizon. The degree of the influence of these contributions is defined by the appropriate choice of the following factors called *compatibilities* [80,85,86]:

a) Self-Compatibility k_{self}

b) IS-A compatibility k_{is_a}

c) PART-Of compatibility k_{part_of}.

Compatibilities are chosen so that $-1 < k < 1$. Compatibilities are in the range [-1,0,1] corresponding to strongly incompatible hypotheses, independent, or strongly compatible respectively. The certainty of a hypothesis is described by the following equation:

$$\begin{aligned} \mathcal{R}(h) = \{ & w_{self}\, k_{self}\, \mathcal{R}_{self}(h) + \sum_{i \in N_{is\_a}(h)} w_{is\_a}(i)\, k_{is\_a}(i)\, \mathcal{R}_{is\_a}(h_i) \\ & + \sum_{j \in N_{part\_of}(h)} w_{part\_of}(j)\, k_{part\_of}(j)\, \mathcal{R}_{part\_of}(h_j) \} \\ & \times \min_{i \in N_{is\_a}(h)} \mathcal{R}(h_i) \end{aligned} \tag{27}$$

where $w_{self}, w_{is\_a}, w_{part\_of}$ are weights and $N_{is\_a}(h), N_{part\_of}(h)$ are the number of hypotheses which are IS-A and PART-OF respectively to hypothesis h. The multiplication by $\min_{i \in N_{is\_a}(h)} \mathcal{R}(h_i)$ in (27) is due to the fact that a hypothesis cannot have higher certainty than its IS-A ancestors (e.g. a line cannot be "more" straight horizon than "horizon"). The certainty $\mathcal{R}(h)$ given by (27) is a special form of the definition given in [86] and it has been successfully used in our system.

We shall give an example to illustrate the application of the concept of hypothesis certainty in the search of horizons. The certainty that an instance of the class HORIZON is a real seismic horizon depends on the horizon length, its reflection strength, its smoothness and its orientation.

Seismic horizons usually tend to have small slope, strong reflection amplitude and large length. Thus, horizon segments with large slope or small reflection amplitude have negative contribution to horizon certainty. Let us denote by $\mathcal{R}_{horizon}, \mathcal{R}_{seg}, \mathcal{R}_{length}, \mathcal{R}_{strength}, \mathcal{R}_{self}$ the certainties of an instance of the class HORIZON, of one of its segments, of its length, of its strength and of its smoothness respectively. $\mathcal{R}_{horizon}$ is given by the following relation:

$$\begin{aligned} \mathcal{R}_{horizon} &= w_{seg} \sum \mathcal{R}_{seg} k_{seg} + w_{length} \mathcal{R}_{length} + w_{strength} \mathcal{R}_{strength} + \\ &+ w_{self} \mathcal{R}_{self} \end{aligned} \quad (28)$$

where $w_{seg}, w_{length}, w_{strength}, w_{self}$ are weights. The compatibility k_{seg} is given by the relation:

$$k_{seg} = 1 - \frac{SEGMENT\_SLOPE}{\theta_1} \quad (29)$$

where θ_1 is a threshold. The certainty $\mathcal{R}_{seg}$ is given by the relation:

$$\begin{aligned} \mathcal{R}_{seg} &= w'_{length} \frac{SEGMENT\_LENGTH}{max\_length} \\ &+ w'_{strength} \frac{REFLECTION\_STRENGTH}{max\_reflection\_strength} \end{aligned} \quad (30)$$

where max_length and $max\_reflection\_strength$ are the maximum horizon lengths and the maximum reflection strengths found in the seismic image. The certainties $\mathcal{R}_{length}$, $\mathcal{R}_{strength}$ are given by the relations:

$$\mathcal{R}_{length} = \frac{HORIZON\_LENGTH}{max\_horizon\_length} \quad (31)$$

$$\mathcal{R}_{strength} = \frac{REFLECTION\_STRENGTH}{max\_reflection\_strength} \quad (32)$$

If $\Delta\theta$ is the maximal slope difference between two adjacent horizon segments of the same horizon and θ_2 is the maximum allowable slope difference, the certainty $\mathcal{R}_{self}$ is given by:

$$\mathcal{R}_{self} = 1 - \frac{\Delta\theta}{\theta_2} \quad (33)$$

XII. DESCRIPTION OF THE HIGH-LEVEL VISION PART OF THE SYSTEM

The aim of the high level vision part of AGIS is to perform a knowledge-based structural analysis of the seismic data and also to incorporate certain aspects of stratigraphic analysis (e.g. check of rock layer homogeneity).

The first task is horizon picking which is performed by the instantiation routine of the class HORIZON having as input the bilevel image produced by horizon following routine of the low vision part. The output of this task are instances of the class HORIZON. The instantiation procedure fills the list of horizon POINTS and the slots HORIZON-START and HORIZON-END by using the corresponding pixel coordinates. The average reflection strength along the horizon is calculated and the slot REFLECTION-STRENGTH and the slot HORIZON-LENGTH are filled. The horizon certainty is calculated according to (28)–(33), the slot HORIZON-SLOPE and the array of HORIZON-SEGMENTS are filled after the linear piecewise approximation of the horizon is performed.

The second task is to perform the detection of horizon features. It includes the instantiation routines of the classes STRAIGHT-HORIZON, CONVEX-HORIZON and CONCAVE-HORIZON with inputs the instances of the class HORIZON. The class STRAIGHT-HORIZON is created, if the certainty of the hypothesis is above a threshold. Otherwise, the horizon curvature is calculated. If it is positive or negative, the routine hypothesizes an instance of the class CONVEX-HORIZON or CONCAVE-HORIZON, respectively.

The third task is to define the spatial relations between the instances of the class HORIZON. These are: UPPER- or LOWER-HORIZON-NEIGHBORHOOD and UPPER- or LOWER-HORIZON-PARALLELISM. The inputs to the routine are the instances of the class HORIZON. The procedure HORIZON-NEIGHBORHOOD finds each HORIZON instance H_2 that is above or below a certain instance H_1 of the class HORIZON in the seismic image. It also measures the certainty of the UPPER- or LOWER-HORIZON-NEIGHBORHOOD relations. The certainty receives negative contribution from the average geometrical distance $d(H_1, H_2)$ between the two horizon instances and positive contribution from the length $l(H_1, H_2)$ of the neighbor segments of the two horizon instances. The HORIZON instances that are connected by UPPER- or LOWER-HORIZON-NEIGHBORHOOD relations are inputs to the HORIZON-PARALLELISM procedure. The HORIZON-PARALLELISM certainty receives negative contribution from the variance of the geometrical distance $d(H_1, H_2)$. Furthermore, it is always less than the HORIZON-NEIGHBORHOOD certainty. This corresponds to the fact that HORIZON-PARALLELISM relation IS-A HORIZON-NEIGHBORHOOD relation.

The fourth task is to detect seismic features. For this reason, the instantiation procedures of the classes ANTICLINE-TRAP, SALT-DOME, REEF and FAULT are invoked and new instances of these classes are produced. For example, instances of the class STRAIGHT-HORIZON that have zero HORIZON-SLOPE, strong REFLECTION-AMPLITUDE and are connected with UPPER-HORIZON-NEIGHBORHOOD relation to an instance of the class CONVEX-HORIZON are these that are checked to be instances of the class ANTICLINE-TRAP. If H_1, H_2 are instances of the STRAIGHT-HORIZON and of the CONVEX-HORIZON respectively and $\mathcal{R}(H_1), \mathcal{R}(H_2)$ their certainties, the certainty $\mathcal{R}_{anti}$ of the instance of the ANTICLINE-TRAP is given by:

$$\begin{aligned} \mathcal{R}_{anti} &= w_1 k_1 \mathcal{R}(H_1) + w_2 k_2 \mathcal{R}(H_2) + w_3 k_3 \mathcal{R}_{slope}(H_1) + w_4 k_4 \\ &\times \mathcal{R}_{amp}(H_1) + w_5 k_5 \mathcal{R}_{neigh}(H_1, H_2) \\ &\qquad k_1, k_2, k_4, k_5 > 0, \; k_3 < 0 \end{aligned} \tag{34}$$

In other words, (34) states that ANTICLINE-TRAP certainty receives positive contribution of all its slots (except of the HORIZON-SLOPE(H_1)) and of the HORIZON-NEIGHBORHOOD relation between H_1, H_2. The HORIZON-NEIGHBORHOOD contribution is a special case of the self contribution described in (27), because it is associated with the satisfaction of constraints between the slots of the class ANTICLINE-TRAP.

The instantiation procedure of the class FAULT searches for instances of the class HORIZON that are connected to each other by HORIZON-NEIGHBORHOOD relations. The same class is created, if there are instances of the class HORIZON that have abrupt changes in their HORIZON-SEGMENT slopes along a certain direction and that are connected to each other by HORIZON-NEIGHBORHOOD relations. An example of detected faults is shown in Figure 12b.

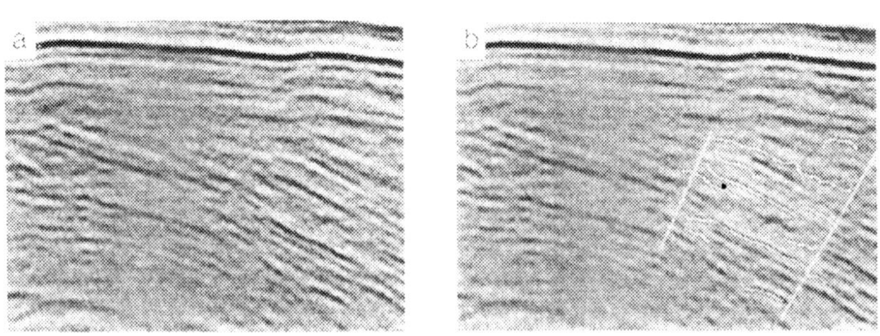

Fig. 12. (a) Seismic image. (b) Detected faults.

The fifth task of the high level vision part of AGIS is the creation of instances of the class ROCK-LAYER. The instantiation procedure of this class searches for couples of instances of the class HORIZON that have large HORIZON-LENGTH and that are connected to each other by HORIZON-NEIGHBORHOOD relation with high certainty. If the region between the two HORIZON instances is reflection-free, the two HORIZON instances are considered to be the boundaries of a rock layer. The ROCK-LAYER certainty receives positive contribution from the two HORIZON instances and from their HORIZON-NEIGHBORHOOD relation and negative contribution from the inhomogeneities in the texture of the region between the two HORIZON instances.

AGIS can work in an unsupervised (automatic) mode and in a supervised (interactive) mode. In the second case, the results (e.g. horizons found) are stored and displayed (on printer or on a high resolution monitor). An interpreter can interact with the system in two ways. The first is changing some parameters such as thresholds, window-lengths etc and apply the routines repeatedly until the results can be considered to be good, according to his experience. The second way of interaction is to modify the results of the search routine, in case he is not completely satisfied . For example, he can add a new horizon, or delete a horizon with low certainty.

XIII. CONCLUDING REMARKS

In this chapter we give a review of various techniques that are used in knowledge-based seismic interpretation. The use of pattern recognition techniques, digital image processing/analysis, graphics and expert systems have been presented. The presentation could not cover all existing techniques, due to diversity of the subject under discussion. Examples of the application of knowledge-based digital image analysis for seismic image interpretation that has been done at the University of Toronto, Canada and at the University of Thessaloniki, Greece are presented.

In conclusion, we can say that all recent advances show that integrated interactive and intelligent computer-aided geophysical interpretation is emerging. Such systems having good human-machine interface, excellent graphics and some intelligent interpretation mechanism exist already in the international market and greatly facilitate geophysical interpretation. The construction of full-scale intelligent geophysical interpretation is difficult but feasible. It will require extended hardware resources and the cooperation of scientists different background (e.g. geophysicists, geologists, knowledge engineers, graphics and image processing specialists).

XIV. REFERENCES

1. H.R. Nelson Jr., *New Technologies in Exploration Geophysics*, Gulf Publishing Co. (1983).
2. R.M. McQuillin, M. Bacon, and W. Barclay, *An Introduction to Seismic Interpretation*, Graham & Trotman Ltd. (1979).
3. M.B. Dobrin, *Introduction to Geophysical Prospecting*, McGraw Hill, New York (1976).
4. J.A. Coffeen, *Seismic Exploration Fundamentals*, PCC Books (1978).
5. K.H. Waters, *Reflection Seismology*, J. Wiley, New York (1981).
6. R.E. Sheriff, and L.P. Geldart, *Exploration Seismology*, Vol. 2, Cambridge University Press, Cambridge, England (1983).
7. N.A. Anstey, *Simple Seismics*, International Human Resources Development Corp. (1982).
8. E.A. Robinson, and S. Treitel, *Geophysical Signal Analysis*, Prentice Hall, Englewood Cliffs, N.J. (1980).
9. J.B. Sangree, and J.M. Widmier, "Interpretation of Depositional Facies from Seismic Data", *Geophysics* **44**, pp. 131–160 (1979).
10. M.M. Roksandic, "Seismic Facies Analysis Concepts", *Geophysical Prospecting* **26**, pp. 383–398 (1978).
11. I.B. Hoyle, "Computer Techniques for Zoning and Correlation of Well Logs", *Geophysical Prospecting* **34**, pp. 648–664 (1986).
12. J.T. Tou, and R.C. Gonzalez, *Pattern Recognition Principles*, Addison-Wesley, Reading MA (1974).
13. F. Aminzadeh, (ed.), *Pattern Recognition and Image Processing*, Geophysical Press, Amsterdam, Holland (1987).
14. N. Ricker, "The Form and Laws of Propagation of Seismic Wavelets", *Geophysics* **18**, pp. 10–40 (1953).
15. P. Bois, "Determination de l' Impulsion Sismique", *Geophysical prospecting* **16**, pp. 4–20 (1968).
16. P. Bois, and M. La Porte, "Pointe Automatique", *Geophysical Prospecting* **19**, pp. 592–611 (1971).

17. P. Bois, "Reconnaissance des Horizons Sismiques par Analyse Factorielle Discriminante", *Geophysical Prospecting* **24**, pp. 696–718 (1976).

18. P. Bois, "Autoregressive Pattern Recognition Applied to the Delimitation of Oil and Gas Reservoirs", *Geophysical prospecting* **28**, pp. 572–591 (1980).

19. P. Bois, "Determination of the Nature of Reservoirs by Use of Pattern Recognition Algorithms with Prior Learning", *Geophysical Prospecting*, **29**, pp. 687–701 (1981).

20. P. Bois, "Some Applications of Pattern Recognition to Oil and Gas Exploration", *IEEE Transactions on Geoscience and Remote Sensing* **21**, pp. 416–426 (1983).

21. P. Bois, "Fuzzy Seismic Interpretation", *IEEE Transactions on Geoscience and Remote Sensing* **22**, pp. 692–697 (1984).

22. C.H.Chen,"Application of Pattern Recognition to Seismic Wave Interpretation", in *Applications of Pattern Recognition*, K.-S. Fu (ed.), CRC Press (1982).

23. K.-S. Fu, *Syntactic Methods in Pattern Recognition*, Academic Press (1974).

24. J.C. Bezdek, *Pattern Recognition with Fuzzy Objective Function Algorithms*, Plenun (1981).

25. K.-Y. Huang, and K.-S. Fu, "Syntactic Pattern Recognition for the Classification of Ricker Wavelets", *Geophysics* **50**, pp. 1548–1555 (1985).

26. K.-Y. Huang, K.-S. Fu, S.-W. Cheng, and Z.-S. Lin, "Syntactic Pattern Recognition and Hough Transformation for Reconstruction of Seismic Patterns," *Geophysics* **52**, pp. 1612–1620 (1987).

27. J. Dumay, and F. Fournier, "Multivariate Statistical Analyses Applied to Seismic Facies Recognition", *Geophysics* **53**, pp. 1151–1159 (1988).

28. A. Sinvhal, and K. Khattri, "Application of Seismic Reflection Data to Discriminate Subsurface Lithostratigraphy," *Geophysics* **48**, pp. 1498–1513 (1983).

29. W.K. Pratt, *Digital Image Processing*, J. Wiley, New York (1978).

30. D.H. Ballard, and C.M. Brown, *Computer Vision*, Prentice Hall, Englewood Cliffs, N.J. (1982).

31. M.D. Levine, *Vision in Man and Machine*, McGraw-Hill, New York (1985).

32. R.C. Gonzalez, and P. Wintz, *Digital Image Processing*, Addison-Wesley, Reading MA (1987).

33. A.K. Jain, *Fundamentals of Digital Image Processing*, Prentice Hall, Englewood Cliffs, N.J. (1989).

34. N. Keskes, A. Boulanmar, and O. Faugeras, "Application of Image Analysis Techniques to Seismic Data", in *Proc. of IEEE Int. Conf. on Acoustics, Speech and Signal Processing*, Paris (1982).

35. D.B. Cooper, H. Elliot, F.S. Cohen, and P.G. Symosek, "Stochastic Boundary Estimation and Object Recognition", *Computer Vision, Graphics and Image Processing*, pp. 326–355 (1980).

36. N. Keskes, and P. Mermey, "Seismic Horizon Extraction by Heuristic Methods," in *Proc. of IEEE Conf. on Pattern Recognition*, pp. 95–100 (1984).

37. Y.-C. Cheng, and S.-Y. Lu, "Binary Consistency Checking Scheme and its Applications to Seismic Horizon Detection", *IEEE Trans. on Pattern Analysis and Machine Intelligence* **11**, pp. 439–447 (1989).

38. P.L. Love, and M. Simaan, "Segmentation of a Seismic Section Using Image Processing and Artificial Intelligence Techniques", *Pattern Recognition* **18**, pp. 409–419 (1985).

39. R. Haralick, "Statistical and Structural Approaches to Texture", *Proceedings of the IEEE* **67**, pp. 786–804 (1979).

40. M. Simaan, Z. Zhang, and P.L. Love, "Artificial Intelligence and Expert Systems for Seismic Data", in *Pattern Recognition and Image Processing*, F. Aminzadeh (ed.), pp. 389–425, Geophysical Press, Amsterdam, Holland (1987).

41. K.I. Laws, "Texture Image Segmentation", Ph.D. dissertation, USC (1980).

42. Z. Zhang, and M. Simaan, "A Rule-Based Interpretation System for Segmentation of Seismic Images", *Pattern Recognition* **20**, pp. 45–53 (1987).

43. Z. Zhang, and M. Simaan, "A Knowledge-Based System Controlled by an Iterative Quadtree Splitting Scheme for Segmentation of Seismic Sections", *IEEE Trans. on Geoscience and remote sensing* **26**, pp. 518–524 (1988).

44. Z. Zhang, and M. Simaan, "Knowledge-Based Texture Image Segmentation Using Iterative Linked Quadtree Splitting", in *Proc. of IEEE Int. Conf.on Acoustics, Speech and Signal Processing*, pp. 2321–2324, Albuquerque (1990).

45. M. Simaan, and Z. Zhang, "Automatic Segmentation of Seismic Data via Knowledge-Based Image Processing Techniques, in *Proc. of IEEE Int. Conf. on Acoustics, Speech and Signal Processing*, pp. 1949–1952, Albuquerque (1990).

46. A. Ikonomopoulos, and M. Kunt, "Directional Filtering, Zero Crossing Edge Detection and Image Coding", in *Signal Processing II: Theories and Applications*, H.W. Schuessler (ed.), Elsevier (1983).

47. A. Ikonomopoulos, and M. Unser,"A Directional Approach to Texture Discrimination", in *Proc. of IEEE Int. Conf. on Acoustics, Speech and Signal Processing*, pp. 87–89, (1984).

48. A. Ikonomopoulos, and M. Kunt, "High Compression of Image Coding via Directional Filtering", *Signal Processing* **8**, pp. 179–203 (1985).

49. N. Gopal, T. Emmoth, and A.C. Bovik, "Channel Interactions in Visible Pattern Analysis", in *Proc. of IEEE Int. Conf. on Acoustics, Speech and Signal Processing*, pp. 1643–1646, Glasgow, U.K. (1989).

50. A.C. Bovik, M. Clark, and W.S. Geisler, "Multichannel Texture Analysis Using Localized Spatial Filters", *IEEE Trans. on Pattern Analysis and Machine Intelligence* **12**, pp. 55–73 (1990).

51. A.C. Bovik, "Analysis of Multichannel Narrow-Band Filters for Image Texture Segmentation", *IEEE Trans. on Signal Processing* **39**, pp. 2025–2043 (1991).

52. A. Barr, P.R. Cohen, and E.A. Feigenbaum, *The Handbook of Artificial Intelligence*, Vols. I, II, III, Heuristech Press (1981).

53. R.O. Duda, "Development of the Prospector Consultation System for Mineral Exploration", (Final report), Artificial Intelligence Center, SRI Int'l (1978).

54. R. Davis, H. Austin, I. Carlborm, B. Frawley, P. Pruchmik, R. Sneiderman, and J.A. Gilreath, "The DIPMETER ADVISOR: Interpretation of geologic signals", in *Proc. Int. Joint Conf. Artificial Intelligence*, pp. 846–849 (1981).

55. R. Smith, and J.D. Baker, "The DIPMETER ADVISOR System: A Case Study in Commercial Expert System Development", in *Proc. Int. Joint Conf. Artificial Intelligence*, pp. 125–129 (1983).

56. A. Bonnet, and C. Dahan, "Oil-Well Data Interpretation Using Expert System and Pattern Recognition Technique", in *Proc. Int. Joint Conf. Artificial Intelligence*, pp. 185–189 (1983).

57. R.G. Simmons, "Representing the Reasoning About Change in Geologic Interpretation", M.Sc. Thesis, MIT (1983).

58. D. Barstow, "A Perspective on Automatic Programming", *AI magazine*, pp. 5–26, Spring (1984).

59. V. Roberto, L. Gargiulo, and A. Peron, "A Knowledge-Based System for Geophysical Interpretation", in *Proc. of the IEEE Int. Conf. on Acoustics, Speech, and Signal Processing*, pp. 1945–1948, Albuquerque (1990).

60. I. Pitas, and A.N. Venetsanopoulos, "Nonlinear Mean Filters in Image Processing", *IEEE Trans. on Acoustics. Speech and Signal Processing* **34**, pp. 573–584 (1986).

61. I. Pitas, and A.N. Venetsanopoulos, "Edge Detectors Based on Nonlinear Filters", *IEEE Trans. Pattern Analysis and Machine Intelligence* **8**, pp. 538–550 (1986).

62. I. Pitas, and A.N. Venetsanopoulos, "Towards a Knowledge-Based System for Automated Geophysical Interpretation of Seismic Data (AGIS)", *Signal Processing* **13**, pp. 229–253 (1987).

63. I. Pitas, E. Milios, and A.N. Venetsanopoulos, "A Minimum Entropy Approach to Rule Learning from Examples", *IEEE Transactions on Systems, Man and Cybernetics* **SMC-22**, pp. 621–635 (1992).

64. I. Pitas, and C. Kotropoulos, "Texture Analysis and Segmentation of Seismic Images", in *Proc. of IEEE Int. Conf. on Acoustics, Speech and Signal Processing*, pp. 1437–1440, Glasgow, U.K. (1989).

65. C. Kotropoulos, and I. Pitas, "Texture Description Rules for Geophysical Image Segmentation", in *Signal Processing V: Theories and Applications*, L. Tores *et al.* (Eds.), pp. 597–600, Elsevier (1990).

66. I. Pitas, and A.N. Venetsanopoulos, *Nonlinear Digital Filters: Principles and Applications*, Kluwer Academic,Dordrecht, Holland (1990).

67. I. Pitas, and C. Kotropoulos, "A Texture-Based Approach to the Segmentation of Seismic Images", *Pattern Recognition* **25**, pp. 929–945 (1992).

68. I. Pitas, and A.N. Venetsanopoulos, "Knowledge-Based Image Analysis for Geophysical Interpretation", *Journal of Intelligent and Robotic Systems*, in press (1992).

69. A.V. Oppenheim, and R.W. Schafer, *Digital Signal Processing*, Prentice Hall, Englewood Cliffs, N.J. (1975).

70. M.T. Taner, F. Koehler, and R.E. Sheriff, "Complex Seismic Trace Analysis", *Geophysics* **44**, pp. 1041–1063 (1979).

71. B.A. Hardage, (ed.), *Handbook of Geophysical Exploration: Seismic Stratigraphy*, Geophysical press (1987).

72. F. Mowle, *A Systematic Approach to Digital Logic Design*, Addison-Wesley, Reading, MA (1976).

73. G.B. Coleman, and H.C. Andrews, "Image Segmentation by Clustering", *Proceedings of the IEEE* **67**, pp. 773–785 (1979).

74. S. Zucker, "Region growing: Childhood and Adolescence", *Computer Graphics and Image Processing* **5**, pp. 382–399 (1976).

75. F. Preparata, and M.I. Shamos, *Computational Geometry: An Introduction*, Springer-Verlag, Berlin (1985).

76. J. Serra, *Image Analysis and Mathematical Morphology*, Academic Press (1982).

77. Special Issue on Knowledge Representation, *Computer* **16** (1983).

78. M. Minsky, "A Framework for Representing Knowledge", in *Psychology of Computer Vision*, Windston (ed.), McGraw Hill (1975).

79. H. Levesque, and J. Mylopoulos, "A Procedural Semantics for Semantic Networks", in *Representation and Understanding: Studies in Cognitive Science*, D. Bodrow and A. Collins (eds.), Academic Press, New York (1979).

80. J.K. Tsotsos, "Temporal Event Recognition: An Application to Left Ventricular Performance Evaluation", in *Proc. Int. Joint Conf. Artificial Intelligence* (1981).

81. J. Mylopoulos, T. Shibahara, and J.K. Tsotsos, "Building Knowledge-Based Systems: The PSN Experience", *Computer* **16**, pp. 83–89, (1983).

82. T. Shibahara, *et al.*, "CAA: A Knowledge-Based System with Casual Knowledge to Diagnose Rhythm Disorders in the Heart", in *Proc. Canadian Soc. Computational Studies Intelligence Conf.* (1982).

83. G. McGalla, and N. Cercone, "Approaches to Knowledge Representation", *Computer* **16**, pp. 12–18 (1983).

84. R.J. Brachman, "What IS-A Is and Isn't: An Analysis of Taxonomic Links in Semantic Networks", *Computer* **16**, pp. 30–36 (1983).

85. J.K. Tsotsos, "Knowledge Organization: Its Role in Representation, Decision-Making and Explanation Schemes for Expert Systems", Technical Report LCM-TR83-3, University of Toronto (1983).

86. J.K. Tsotsos , "Representational Axes and Temporal Cooperative Processes", in *Vision, Brain and Cooperative Computation*, M. Arbib and A. Hanson (eds.), MIT Press (1987).

The Foundations of Nearfield Acoustic Holography in Terms of Direct and Inverse Diffraction

David D. Bennink
Applied Measurement Systems, Inc.
Bremerton, WA 98380

F. D. Groutage
Naval Surface Warfare Center, Carderock Division
Puget Sound Detachment
Bremerton, WA 98314-5215

I. INTRODUCTION

The basic principles of holography were first described and demonstrated by Gabor [1,2]. In general terms, holography is an imaging method for reconstructing information concerning a three dimensional wave field from data recorded on a two dimensional surface. It follows that holography is a two-step process: a measurement or recording stage followed by a reconstruction or imaging stage. The field to be imaged is called the object wave. In the conventional approach introduced by Gabor, the measured data represents the spatial interference pattern between this object wave and a suitable reference wave. An interference pattern is recorded because it allows phase information to be included that might otherwise be lost, since detectors capable of directly measuring phase are not available for many wave fields. The actual record of the interference pattern is termed a hologram. Once a hologram is produced, it can be used to create an image through the process of direct diffraction (the solution of a direct boundary value problem for the wave equation). This is accomplished by using the hologram to modulate a suitable reconstruction wave, thus producing the necessary equivalent sources or boundary values on the hologram surface. The reconstruction wave must be appropriately related to the initial reference wave, and the image field is created on the far side of the hologram.

The nature of the boundary values on the hologram surface determines what the image field will represent in the reconstruction. Given that the hologram surface intercepted the original object wave as it propagated away from its source, two types of image field are of particular interest [3]. The first represents a continuation of the original object wave as it would have propagated without intervention. Such an image field would ideally be an exact reproduction of the object field from the location of the hologram surface onward, and could therefore be viewed and processed beyond the hologram exactly as the original object field. The second would represent the original object field from the location of the hologram surface backward to the source (excluding the volume that contained the original source since the object field would not be defined there in general). Such an image would still appear on the far side of the hologram.

The first type of reconstruction, that of determining the object field in the direction away from the source, will be called forward propagation. The second type of reconstruction, that of determining the object field in the direction toward the source, will be called backward propagation. The images for these two types of reconstruction are often referred to by other names, such as correct, true or virtual for forward propagation, and twin, conjugate or real for backward propagation. Both forms of image are produced in conventional holography. In fact, they are generated simultaneously in the reconstruction process, and may disturb one another unless separated [4,5]. Both images are produced, as well as a modulated version of the reconstruction wave and other higher order terms, because the hologram is the recording of an interference pattern. For example, if the reconstruction wave was modulated directly by the complex amplitude of the object wave at the hologram surface then only the forward propagated field would be generated [6].

The images produced by conventional holography are only approximations to the correct forward and backward propagated fields. In particular, backward propagation is approximated through the process of direct diffraction. Since it is strictly necessary to correct for the effects experienced by the object wave in forward propagating from the source to the hologram surface, backward propagation is properly a problem in inverse diffraction [7]. The necessary correction can only be approximated using direct diffraction (it would strictly require the generation of an image field with sinks at unknown locations). These limitations on the conventional approach present only minor difficulties for optical holography, where the wavelengths are small compared to typical physical dimensions and the wave front propagation is essentially geometric in nature. They can cause considerably more difficulty in acoustical holography, where source size to wavelength ratios can be on the order of unity or less.

In addition, since the general idea of acoustical holography is to enable the visualization of sound fields, only the measurement step will be acoustic in nature. The reconstruction step must generate an image field that can be viewed directly. If this is done with optical processing, then increased aberrations and distortions will occur in the optical image due to the generally large difference in the wavelengths used for the two steps [6,8]. Such optical reconstructions do have the advantage of speed (real-time capability) and high capacity (large image to pixel size ratios), and many techniques have been developed for producing acoustical holograms that can be used for optical imaging [8]. All such techniques can conveniently be termed as conventional acoustic holography.

Alternate advantages, such as increased flexibility, can be gained by replacing the optical imaging methods in the conventional approach with suitable digital processing [6,9]. The data must obviously be in discrete form for such processing. If this form corresponds to direct samples of the wave field, then digital methods can be realized that go beyond the imaging techniques of conventional holography. This is possible in acoustics because both amplitude and phase information can be directly measured for a time harmonic field. Thus, it is not necessary to measure a spatial interference pattern in order to record phase information, rather the complex amplitude of the object wave can be directly recorded over the measurement surface. Such an approach eliminates the simultaneous generation of multiple images associated with hologram based reconstructions.

The flexibility gained with digital processing includes the measurement surface shape and location, which is generally restricted in conventional theory to be planar and many wavelengths from the source. For example, the theory of generalized holography describes mathematically direct diffraction imaging for arbitrary measurement surfaces, and includes a formula for backward propagation that can be implemented digitally [10]. This is again an approximation, as in conventional holography, since an exact correction for diffraction effects is not applied. However, with the digital processing of direct wave field samples it is possible to correct for diffraction in backward propagation in a more exact manner. The general approach that has developed for this in acoustics is the subject of the present work. The method will be referred to in general as nearfield acoustic holography (NAH).

Nearfield acoustic holography involves the transformation of the field between the measurement surface and a second, reconstruction surface. The transformation is based on the field satisfying the homogeneous wave equation in the volume between the two surfaces. Direct diffraction is therefore used for forward propagation, when the reconstruction surface is exterior to the measurement surface, while inverse diffraction is used for backward propagation, when the reconstruction surface is interior to the measurement surface. The first developments in this direction were naturally for field transformations between parallel planar surfaces [11,12]. For such surfaces, the processing is based on the Fourier transform, and can therefore be implemented efficiently using the fast Fourier transform (FFT) [13,14].

It has been recognized that backward propagation in the manner of NAH can provide resolution beyond the usual wavelength limitation applicable to conventional holography [15]. In principle, such enhanced resolution is possible for inverse diffraction regardless of the measurement surface location, since even farfield data is theoretically sufficient for an exact reconstruction of the acoustic field everywhere exterior to the source. In practice, enhanced resolution is obtainable with NAH because the technique can include in the processing at least some of the evanescent wave components. These evanescent wave components carry information related to the higher spatial frequencies in the field and decay in amplitude as they propagate away from the source. Therefore, in order to retain as much of this information as possible, the measurement surface in NAH is usually located close to the source, often within the extreme nearfield.

It has also been recognized that the method of NAH for planar surfaces can be appropriately extended to other suitable surface shapes, and that it can provide information concerning such derived quantities as vector intensity and power [16]. The most direct of these extensions is for cylindrical surfaces, for which the method can again be efficiently implemented using the FFT [17]. Both planar and cylindrical surfaces are, however, inherently infinite in extent. A spherical measurement surface yields the simplest application of NAH for a finite surface. Since all of these versions of NAH can be formulated as straightforward applications of the method of separation of variables, other surfaces for which the Helmholtz equation is separable in an appropriate coordinate system can be treated similarly. The technique must be extended to more general surfaces if the full potential of NAH is to be realized. This can be accomplished by using integral equation methods and the singular value decomposition (SVD) to develop and apply the forward propagator, an operator which represents the solution to direct diffraction [18,19]. Such an approach yields a general formulation of forward and backward propagation and the necessary extension of NAH to arbitrary surfaces.

As described above and in the sections to follow, forward and backward propagation with NAH corresponds to a particular approach to the problems of direct and inverse diffraction. The approach begins with the forward propagator, the basis for which is developed in Section II where an integral representation is derived for the acoustic field. This representation is in terms of a Green's function and yields the forward propagator when the Green's function is chosen appropriately. Other properties of acoustic fields that are important for the development of NAH are also discussed in Section II. These are used in Section III to conclude that the forward propagator is a compact operator when the measurement and reconstruction surfaces are bounded. This result allows the introduction of the SVD, from which the general formula for forward and backward propagation are developed. The behaviour of the singular values then brings in quite naturally the concept of evanescent waves and the need for regularization in backward propagation.

Actual implementations of the general formula are presented in Section IV. This includes the basic forms of NAH for planar, cylindrical and spherical surfaces. For more general surfaces it is usually possible to obtain only a matrix approximation to the forward propagator and the techniques used for this are also discussed. The SVD is applied directly to the approximating matrix in such a situation. Since NAH is investigated in the following solely as a data processing technique, aspects concerning the actual measurement of an acoustic field are not considered. Futhermore, no attempt is made at providing either an exhaustive set of references or a comprehensive discussion of all the included material. Rather it is hoped that this work will provide a reasonably complete, unified development of NAH in its most general form. The analytical formulation, in particular one capable of handling arbitrary surfaces, is therefore covered in depth, whereas only the most basic aspects concerning implementation are considered.

II. PROPERTIES OF ACOUSTIC FIELDS

The main objective of this section is to formulate the process of direct diffraction in terms of an integral representation for the acoustic field. Such an integral representation is a mathematical statement of Huygens' principle. Some fundamental properties of acoustic fields such as uniqueness and existence will also be discussed, since they play a role in the development of NAH and help to understand its limitations.

A. The Helmholtz Equation

The acoustic field within a homogeneous inviscid fluid is represented by the excess pressure p. For harmonic fields the excess pressure satisfies the constant frequency form of the homogeneous wave equation, and for this work the harmonic time dependence $\exp(-i\omega t)$ will be assumed. This explicit time dependence will be suppressed for convenience and the manifest dependence of p on frequency ω will be left implicit. The governing equation is therefore written as

$$\left(\nabla^2 + k^2\right)p(\mathbf{r}) = 0 \tag{1}$$

where the propagation constant or wave number is $k = \omega / c$ and c is the speed of sound. Equation (1) is the Helmholtz equation and holds for $\mathbf{r} \in V$ where V is the fluid volume. The particle velocity $\mathbf{v}$ is related to the pressure by

$$\mathbf{v}(\mathbf{r}) = \frac{1}{i\omega\rho_o}\nabla p(\mathbf{r}) \tag{2}$$

where ρ_0 is the fluid density. The total volume for which an acoustic field is required or defined to be a solution of the Helmholtz equation will be termed the field's domain, or more precisely its defined domain. For any field there is also a natural domain, which is the total volume for which it actually exists as a solution of the Helmholtz equation. The two need not be identical, since an acoustic field often extends as a solution of the Helmholtz equation beyond its defined domain [20]. Here the term domain will almost always be synonymous with the fluid volume. In the following it will also be convenient to use the term region to refer to the volume interior to a single, finite closed surface.

The total surface which bounds the fluid volume V will be denoted by S, and may contain any number of disjoint components. For a connected fluid, the bounding surface S will always contain as a component an enclosing surface S_e. This enclosing surface may be unbounded or infinite in size, although all the remaining components of S will be finite. The enclosing surface is unbounded whenever the fluid volume extends to infinity in some direction. For such cases the special suface S_∞ is introduced. The surface S_∞ actually represents the spherical surface S_R of radius R in the limit $R \to \infty$. Thus any integral over S_∞ should be interpreted as

$$\int_{S_\infty} dS(\mathbf{r})\, f(\mathbf{r}) = \lim_{R\to\infty} \int_{S_R} dS(\mathbf{r})\, f(\mathbf{r}). \tag{3}$$

The surface S_∞ is located at infinity in all directions. Since the enclosing surface can have parts that cut through finite space (a simple example is an infinite cone), S_e may not correspond to all of S_∞. In the following section two specific cases will be discussed where this holds, namely nearfield acoustic holography for infinite planar and cylindrical surfaces. However, our interest will mainly be in situations for which S_e is either bounded or identically S_∞. The equations which will subsequently be developed will be based on this assumption. When necessary to handle both of these cases simultaneously, the finite components of S will collectively be denoted by $\overline{S}$.

The acoustic fields to be considered here are assumed to be classical or strong solutions of the Helmholtz equation. In order to be a strong solution p must be a twice continuously differentiable function of the spatial coordinates and must satisfy the Helmholtz equation at every point within its domain. Further assumptions are made concerning the behavior of p as $\mathbf{r}$ tends to the bounding surface S. For example, it is necessary to specify the behavior of the field at infinity when V is unbounded. This is usually done by requiring the field to satisfy the Sommerfeld radiation condition over S_∞,

$$\lim_{r\to\infty}\left[r\left(\frac{\partial}{\partial r} p(\mathbf{r}) - ikp(\mathbf{r}) \right)\right] = 0 \tag{4}$$

uniformly for all $\mathbf{n}_r$ with $\mathbf{r} = r\,\mathbf{n}_r$. The additional constraint

$$\lim_{r\to\infty}\left[r\left| p(\mathbf{r})\right| \right] < \infty \tag{5}$$

is often included as part of the radiation condition [21 Sec. 1.31], although it is not required since Eq. (4) alone is sufficient to completely characterize the behavior of the field at infinity [22 Sec. 3.2]. Since it is often convenient to consider as a group solutions of the Helmholtz equation with similar properties, a field will be called singular if its domain is unbounded and it satisfies the Sommerfeld radiation condition. Singular fields represent radiating solutions to the Helmholtz equation. A field will be called regular if it satisfies the Helmholtz equation throughout a volume interior to a single closed surface. This surface may be located at infinity, in which case the field is called entire and the radiation condition is not enforced. In fact, an entire field cannot satisfy the radiation condition without vanishing identically, and thus a singular field cannot be entire. However, a singular field can be classified as regular depending on the volume under consideration. This is because the term regular can be used to refer to either the local or the global behavior of a field, so that the volume to which the term is being applied must be specified. In particular, any acoustic field is regular within any region of its domain. Later this will be shown to be equivalent to the field being analytic within any such region.

The behavior of the field as $\mathbf{r}$ tends to any of the finite components of the bounding surface must also be specified. Since there are many solutions to the Helmholtz equation, it is necessary to select the particular solution of interest. For a boundary value problem, the boundary conditions determine this solution. Specific boundary conditions which allow a unique determination will be considered later, for now it is only assumed that the limit

$$p_{\mathrm{n}}(\mathbf{r}) = \lim_{\varepsilon \to +0} p(\mathbf{r} - \varepsilon\,\mathbf{n}(\mathbf{r})) \tag{6}$$

exists for $\mathbf{r} \in \overline{S}$ and that $p_{\mathrm{n}} \in \mathrm{C}(\overline{S})$. The quantity p_{n} represents the boundary value of the pressure over the surface $\overline{S}$, and $\mathbf{n}$ denotes the outward normal to V on S. The notation $\mathrm{C}^n(D)$ is used for the class of all functions that are n times continuously differentiable for $\mathbf{r} \in D$. This refers to functions of the appropriate surface coordinates when D is a surface, and the superscript is dropped when $n = 0$. Thus a strong solution of the Helmholtz equation is such that $p \in \mathrm{C}^2(V)$. It is also assumed that the limit

$$\upsilon_{\mathrm{n}}(\mathbf{r}) = -\lim_{\varepsilon \to +0} \mathbf{n}(\mathbf{r}) \cdot \mathbf{v}(\mathbf{r} - \varepsilon\,\mathbf{n}(\mathbf{r})) \tag{7}$$

exists for $\mathbf{r} \in \bar{S}$ and that $\upsilon_{\rm n} \in {\rm C}(\bar{S})$. Due to the minus sign the quantity $\upsilon_{\rm n}$ represents the boundary value of the inward normal velocity over the surface $\bar{S}$, since $\mathbf{n}$ is the outward normal to the fluid volume. The boundary values $(p_{\rm n}, \upsilon_{\rm n})$ will obviously appear in any discussion concerning boundary value problems. For convenience, the class of all strong solutions of the Helmholtz equation for a given fluid volume V with continuous boundary values as defined in Eqs. (6) and (7) will be denoted by ${\rm H}(V)$.

In general, solutions of the Helmholtz equation will have singularities on the boundary if the bounding surface has edges and corners [21 Sec. 9.2, 23]. The finite components of S will therefore be restricted in their regularity properties to ensure that $p \in {\rm H}(V)$. In particular, it will be assumed that each component of $\bar{S}$ is a smooth surface. A surface is considered to be smooth if the mapping which takes a local surface patch in three dimensions to an open region in two dimensions is twice continuously differentiable. That is, for any given point $\mathbf{r}' \in \bar{S}$ a local parameterization of the surface exists such that $\mathbf{r} = \mathbf{X}(u_1, u_2)$ for all $\mathbf{r} \in \bar{S}$ and sufficiently near $\mathbf{r}'$, and where the components of the vector $\mathbf{X}$ are ${\rm C}^2$ functions of the parameters u_1 and u_2. Although it is possible to deal with nonsmooth surfaces, the restriction to smooth surfaces is ultimately not a limitation for the work considered here.

B. Green's Theorem and The Radiation Condition

Green's theorem provides an important tool in the study of acoustic fields. In the following section it will be used to obtain an integral representation for solutions of the Helmholtz equation. For unbounded domains this will require an implication of the radiation condition which will be derived in this section. The necessary result will follow from the first form of Green's theorem,

$$\int_V dV[\Phi\nabla^2\Psi + \nabla\Phi\cdot\nabla\Psi] = \int_S dS\,\mathbf{n}\cdot[\Phi\nabla\Psi], \tag{8}$$

which is obtained by applying Gauss' divergence theorem to the identity

$$\nabla\cdot[\Phi\nabla\Psi] = \Phi\nabla^2\Psi + \nabla\Phi\cdot\nabla\Psi. \tag{9}$$

The restriction to smooth bounding surfaces is sufficient to guarantee the validity of Eq. (8), as well as the second form of Green's theorem

$$\int_V dV[\Phi\nabla^2\Psi - \Psi\nabla^2\Phi] = \int_S dS\,\mathbf{n}\cdot[\Phi\nabla\Psi - \Psi\nabla\Phi], \tag{10}$$

for Φ and $\Psi \in \mathrm{H}(V)$ [22 Sec. 3.2]. Equation (10) is an obvious extension of Eq. (8), and both can be applied to unbounded volumes when S_∞ is included as a component of S and interpreted according to Eq. (3). For singular fields this integration over S_∞ can be evaluated for Eq. (10) from the radiation condition. In particular, it will now be shown that

$$\int_{S_\infty} dS\,\mathbf{n}\cdot[\Phi\nabla\Psi - \Psi\nabla\Phi] = 0 \tag{11}$$

for any singular fields Φ and Ψ.

Equation (11) would follow directly from the radiation condition for the singular fields Φ and Ψ if they were already known to be square integrable on S_∞. To see why this is true, Eq. (11) is first rewritten as

$$\int_{S_\infty} dS\,\mathbf{n}\cdot[\Phi\nabla\Psi - \Psi\nabla\Phi] = \int_{S_\infty} dS[\mathbf{n}\cdot\nabla\Psi - ik\Psi]\Phi - \int_{S_\infty} dS[\mathbf{n}\cdot\nabla\Phi - ik\Phi]\Psi. \tag{12}$$

Schwarz's inequality can now be applied to each integral on the right to show that they vanish as a result of the radiation condition. For example, provided that

$$\int_{S_\infty} dS|\Phi|^2 < \infty \tag{13}$$

then

$$\left|\int_{S_\infty} dS[\mathbf{n}\cdot\nabla\Psi - ik\Psi]\Phi\right|^2 \leq \int_{S_\infty} dS|\mathbf{n}\cdot\nabla\Psi - ik\Psi|^2 \int_{S_\infty} dS|\Phi|^2 = 0 \tag{14}$$

by the radiation condition on Ψ since the normal over S_∞ is in the radial direction. Thus Eq. (11) follows if it can be shown that singular fields are square integrable over S_∞. Equation (13) would be a direct result of Eq. (5) if this additional constraint had been included as part of the radiation condition. However, the result in Eq. (5) will be obtained here as a consequence of Eqs. (10) and (11), so that including (5) would be redundant. Therefore it is necessary to prove Eq. (13) in the manner that follows [22 Sec. 3.2, 24]. The radiation condition causes the first factor on the right in Eq. (14) to vanish, which may be rewritten when expressed in terms of Φ as

$$\int_{S_\infty} dS|\mathbf{n}\cdot\nabla\Phi|^2 + k^2\int_{S_\infty} dS|\Phi|^2 + 2k\int_{S_\infty} dS\,\mathbf{n}\cdot\mathrm{Im}\{\Phi\nabla\Phi^*\} = 0 \tag{15}$$

where Φ^* is the complex conjugate of Φ and k is assumed real. Setting $\Psi = \Phi^*$ in the first form of Green's theorem, and taking the imaginary part of the result, yields

$$\int_S dS\,\mathbf{n}\cdot\mathrm{Im}\{\Phi\nabla\Phi^*\} = 0. \tag{16}$$

Equation (16) will prove useful when the uniqueness properties of boundary value problems are examined later. Here it is combined with Eq. (15) to give

$$\int_{S_\infty} dS|\mathbf{n}\cdot\nabla\Phi|^2 + k^2\int_{S_\infty} dS|\Phi|^2 = 2k\int_{S-S_\infty} dS\,\mathbf{n}\cdot\mathrm{Im}\{\Phi\nabla\Phi^*\}. \tag{17}$$

The right hand side of Eq. (17) is bounded while both terms on the left are nonnegative. Thus it follows that each term on the left is also bounded, showing that both Φ and its normal derivative are square integrable on S_∞. Equation (11) will be used in the process of obtaining an integral representation for singular fields from the second form of Green's theorem.

C. Green's Functions and The Representation Integral

In order to use Green's theorem to obtain an integral representation for solutions of the Helmholtz equation, it is necessary to introduce the concept of a Green's function [25 Chapt. 7]. A Green's function is the solution for a point source and therefore satisfies the following inhomogeneous version of the Helmholtz equation,

$$\left(\nabla^2 + k^2\right)G(\mathbf{r}|\mathbf{r}') = -\delta(\mathbf{r}-\mathbf{r}'). \tag{18}$$

Equation (18) must hold for $\mathbf{r}\in V$ and for all finite $\mathbf{r}'$. The notation $G(\mathbf{r}|\mathbf{r}')$ is used to emphasize that the first argument is the field point in the Helmholtz equation while the second argument is the location of the point source, and thus a parameter in Eq. (18). Such a distinction between the two arguments is often unnecessary since G can be shown to satisfy the reciprocity relation

$$G(\mathbf{r}|\mathbf{r}') = G(\mathbf{r}'|\mathbf{r}) \tag{19}$$

for $\mathbf{r}, \mathbf{r}' \in V$. Equation (19) is based on the general homogeneous boundary condition

$$A(\mathbf{r})G(\mathbf{r}|\mathbf{r}') + B(\mathbf{r})\mathbf{n}(\mathbf{r}) \cdot \nabla G(\mathbf{r}|\mathbf{r}') = 0 \tag{20}$$

imposed on G over the bounded components of S. If V is unbounded, then the Sommerfeld radiation condition,

$$\lim_{r \to \infty} \left[r \left(\frac{\partial}{\partial r} G(\mathbf{r}|\mathbf{r}') - ikG(\mathbf{r}|\mathbf{r}') \right) \right] = 0 \tag{21}$$

uniformly over S_∞, is also imposed for all finite $\mathbf{r}'$. At this point it is assumed that A and B in Eq. (20) are known continuous functions which do not vanish simultaneously. Later these functions will be seen to come from the boundary condition imposed on p.

The source term in Eq. (18) is a three-dimensional Dirac delta function. The delta function is actually a distribution or generalized function defined by the property

$$\int_V dV(\mathbf{r}) \Phi(\mathbf{r}) \delta(\mathbf{r} - \mathbf{r}') = \begin{cases} \Phi(\mathbf{r}') & , \ \mathbf{r}' \in V \\ 0 & , \ \mathbf{r}' \notin V \end{cases} \tag{22}$$

It therefore follows that G should be considered in general as a distribution [26]. However, it also is possible to deal with G as an ordinary function. From Eq. (18), the domain of G cannot include the point $\mathbf{r} = \mathbf{r}'$ when $\mathbf{r}' \in V$. In order to remove the point $\mathbf{r}'$ from direct consideration, V is replaced by the punctured volume V' where $V' = V - V_\varepsilon(\mathbf{r}')$ and $V_\varepsilon(\mathbf{r}')$ is the sphere centered at $\mathbf{r}'$ having arbitrarily small radius ε. Accordingly, Eq. (18) is replaced by

$$\left(\nabla^2 + k^2\right) G(\mathbf{r}|\mathbf{r}') = 0 \tag{23}$$

for $\mathbf{r} \in V'$. However, it is still necessary to account for the point source at $\mathbf{r}'$, and this is done by requiring that

$$\int_{S_\varepsilon(\mathbf{r}')} dS\,\mathbf{n}\cdot\nabla G(\mathbf{r}|\mathbf{r}') = -1 \tag{24}$$

for $\mathbf{r}' \in V$, where S_ε is the spherical surface bounding V_ε. Equations (23) and (24) are collectively equivalent to Eq. (18). That G is a normal function within the punctured volume V' will follow from the standard decomposition in terms of a particular solution G_0 and a homogeneous solution F,

$$G(\mathbf{r}|\mathbf{r}') = G_0(\mathbf{r}|\mathbf{r}') + F(\mathbf{r}|\mathbf{r}'). \tag{25}$$

For the Helmholtz equation, the particular solution for a point source is given by

$$G_0(\mathbf{r}|\mathbf{r}') = \frac{e^{ik|\mathbf{r}-\mathbf{r}'|}}{4\pi|\mathbf{r}-\mathbf{r}'|} \tag{26}$$

and is called the fundamental solution. Since it is not difficult to show that G_0 satisfies the radiation condition (21) and Eqs. (23) and (24) for V equal to all of three-dimensional space, it is also called the free-space Green's function. The remaining term F must satisfy the homogeneous equation in (23) for all $\mathbf{r} \in V$, but with the inhomogeneous boundary condition

$$A(\mathbf{r})F(\mathbf{r}|\mathbf{r}') + B(\mathbf{r})\mathbf{n}(\mathbf{r})\cdot\nabla F(\mathbf{r}|\mathbf{r}') = -A(\mathbf{r})G_0(\mathbf{r}|\mathbf{r}') - B(\mathbf{r})\mathbf{n}(\mathbf{r})\cdot\nabla G_0(\mathbf{r}|\mathbf{r}') \tag{27}$$

over the bounded components of S, and the radiation condition over S_∞ if V is unbounded. Thus F can be considered the boundary component of G, while G_0 is the source component. Equation (26) shows that $G_0 \in \mathrm{C}^\infty$ for $\mathbf{r} \neq \mathbf{r}'$ and therefore G is a normal function within V' provided that a solution F exists to the Helmholtz equation under the boundary condition in Eq. (27). Note that the right hand side of Eq. (27) is at least continuous, and this is assumed to be sufficient to ensure that $F \in \mathrm{H}(V)$. The existence and uniqueness of such a solution, and therefore of G itself, will be considered in Section II, E.

Not surprisingly, if an appropriate G does exist then it can be used to determine the general solution of the corresponding boundary value problem for the Helmholtz equation. The solution is provided by Green's representation integral, a result obtained by applying Green's theorem in Eq. (10) to the punctured domain V' with $\Phi = G(\mathbf{r}|\mathbf{r}')$ and $\Psi = p(\mathbf{r})$. Since both G and p

satisfy the homogeneous Helmholtz equation in V' the volume integral vanishes,

$$\int_{V'} dV[\Phi\nabla^2\Psi - \Psi\nabla^2\Phi] = 0 \tag{28}$$

for $\mathbf{r}' \notin S$. The surface integral separates into integrations over the finite components of S, an integration over S_ε about the deleted point $\mathbf{r}'$ provided $\mathbf{r}' \in V$, and an integration over S_∞ if V is unbounded. Since G is a singular field, the integral over S_∞ for V unbounded vanishes as a result of Eq. (11) if p is a singular field. The integration over S_ε for $\mathbf{r}' \in V$ can be separated into two integrals. The first, involving G and the normal derivative of p, vanishes according to

$$\int_{S_\varepsilon(\mathbf{r}')} dS\, G(\mathbf{r}|\mathbf{r}')\mathbf{n}(\mathbf{r})\cdot\nabla p(\mathbf{r}) \le \max_{\mathbf{r}\in S_\varepsilon}\left|\mathbf{n}(\mathbf{r})\cdot\nabla p(\mathbf{r})\right| \int_{S_\varepsilon(\mathbf{r}')} dS\left|G(\mathbf{r}|\mathbf{r}')\right| \to 0 \tag{29}$$

as $\varepsilon \to 0$. In Eq. (29), the first term on the right remains bounded since $p \in \mathrm{H}(V)$, while the second tends to zero from

$$\int_{S_\varepsilon(\mathbf{r}')} dS\left|G(\mathbf{r}|\mathbf{r}')\right| \le \int_{S_\varepsilon(\mathbf{r}')} dS\left|G_0(\mathbf{r}|\mathbf{r}')\right| + \int_{S_\varepsilon(\mathbf{r}')} dS\left|F(\mathbf{r}|\mathbf{r}')\right| = O(\varepsilon), \tag{30}$$

which follows from Eqs. (25) and (26) and $F \in \mathrm{H}(V)$. The second integral over S_ε, involving p and the normal derivative of G, can be rewritten as

$$\int_{S_\varepsilon(\mathbf{r}')} dS\, \mathbf{n}(\mathbf{r})\cdot\nabla G(\mathbf{r}|\mathbf{r}')p(\mathbf{r}) = -p(\mathbf{r}') + \int_{S_\varepsilon(\mathbf{r}')} dS\, \mathbf{n}(\mathbf{r})\cdot\nabla G(\mathbf{r}|\mathbf{r}')\Delta p, \tag{31}$$

where Eq. (24) has been used and $\Delta p = p(\mathbf{r}) - p(\mathbf{r}')$. The remaining integral on the right in Eq. (31) vanishes according to

$$\int_{S_\varepsilon(\mathbf{r}')} dS\, \mathbf{n}(\mathbf{r})\cdot\nabla G(\mathbf{r}|\mathbf{r}')\Delta p \le \max_{\mathbf{r}\in S_\varepsilon}\left|\Delta p\right| \int_{S_\varepsilon(\mathbf{r}')} dS\left|\mathbf{n}(\mathbf{r})\cdot\nabla G(\mathbf{r}|\mathbf{r}')\right| \to 0 \tag{32}$$

as $\varepsilon \to 0$. In Eq. (32), the first term on the right tends to zero since $p \in \mathrm{H}(V)$, while the second can be shown to remain bounded following Eq. (30). Combining Eqs. (28)-(32) yields Green's representation integral

$$\int_{\bar{S}} dS(\mathbf{r})\mathbf{n}(\mathbf{r})\cdot[G(\mathbf{r}|\mathbf{r}')\nabla p(\mathbf{r}) - p(\mathbf{r})\nabla G(\mathbf{r}|\mathbf{r}')] = \begin{cases} p(\mathbf{r}') & , \mathbf{r}' \in V \\ 0 & , \mathbf{r}' \notin V \cup S \end{cases} \tag{33}$$

where no particular boundary condition has as yet been imposed on G for $\mathbf{r} \in \bar{S}$. When the domain is unbounded, Eq. (33) holds only for singular fields, since the radiation condition was assumed in its derivation. Equation (33) also shows that there are no nontrivial singular fields which are entire, since the only component of S for such a field would be S_∞ and thus the surface integral in (33) would be identically zero. To solve a specific boundary value problem, the functions A and B in Eq. (20) come from the inhomogeneous boundary condition imposed on p

$$A(\mathbf{r})p(\mathbf{r}) + B(\mathbf{r})\mathbf{n}(\mathbf{r})\cdot\nabla p(\mathbf{r}) = f(\mathbf{r}). \tag{34}$$

Equations (20) and (34) can be combined with Eq. (33) to give the desired integral representation of the solution

$$p(\mathbf{r}) = \int_{\bar{S}} dS(\mathbf{r}')K(\mathbf{r},\mathbf{r}')f(\mathbf{r}') \tag{35}$$

for $\mathbf{r} \in V$, where the kernel K is given by

$$K(\mathbf{r},\mathbf{r}') = \begin{cases} -\mathbf{n}(\mathbf{r}')\cdot\nabla' G(\mathbf{r}'|\mathbf{r}) / A(\mathbf{r}') \ , & A \neq 0 \\ G(\mathbf{r}'|\mathbf{r}) / B(\mathbf{r}') \ , & B \neq 0 \end{cases} \tag{36}$$

The development of NAH in Section III will essentially be based upon Eq. (35), along with some subsequent results concerning the continuity properties of acoustic fields.

D. The Kirchhoff-Helmholtz Integral

The boundary conditions satisfied by G are not used in the derivation of Green's representation integral, with the exception that the radiation condition is assumed for V unbounded. Any Green's function with a domain that contains V can therefore be used in Eq. (33). In particular, G can be taken as the free-space Green's function G_0 to obtain the Kirchhoff-Helmholtz integral

$$p(\mathbf{r}) = \int_{\bar{S}} dS(\mathbf{r}')\mathbf{n}(\mathbf{r}')\cdot[G_0(\mathbf{r}'|\mathbf{r})\nabla' p(\mathbf{r}') - p(\mathbf{r}')\nabla' G_0(\mathbf{r}'|\mathbf{r})] \tag{37}$$

for $\mathbf{r} \in V$. Since G_0 is infinitely differentiable for $\mathbf{r} \neq \mathbf{r}'$, it follows from Eq. (37) that $p \in C^\infty(V)$ even though it was only assumed initially that $p \in C^2(V)$.

This surprising regularity property is not restricted to the Helmholtz equation, but is shared by many other elliptic partial differential equations [27]. In addition, since G_0 is analytic for $\mathbf{r} \neq \mathbf{r}'$, Eq. (37) actually defines an analytic function of the spatial coordinates for $\mathbf{r} \in V$ [22 Theorem 3.5]. Such an analytic function has an absolutely convergent local power series about any point in V. If the origin is located at this point for convenience then the local power series has the form

$$p(\mathbf{r}) = \sum_{n=0}^{\infty} \left\{ \sum_{l=0}^{n} \sum_{m=0}^{l} C_{lmn} x^{n-l} y^{l-m} z^{m} \right\} \tag{38}$$

and is uniformly convergent for $r \leq R$ for any $R < R_{min}$, where R_{min} is the distance from the origin to the closest point in the boundary of V. Equation (38) may also be derived by directly substituting appropriate power series expansions for both $G_0(\mathbf{r}'|\mathbf{r})$ and $\nabla' G_0(\mathbf{r}'|\mathbf{r})$ into Eq. (37) and then integrating term by term. The procedure is valid due to the continuity assumptions on p and the surface of integration, together with the uniform convergence of the expansions for $G_0(\mathbf{r}'|\mathbf{r})$ and $\nabla' G_0(\mathbf{r}'|\mathbf{r})$ for $r < r'$.

The convergence region of any local power series may in general cover a large part of the domain of p, but cannot be expected to cover the entire domain. A collection of such power series can, however, completely represent the acoustic field. The process by which local information is extended to determine the field everywhere is called analytic continuation [25 Chapt. 4]. For Eq. (38), term by term differentiation can be justified and the resulting power series will also converge uniformly for $r < R_{min}$ [28]. This process can be repeated to obtain any derivative, and evaluating the resulting series at the origin leads to the equation

$$(n-l)!(l-m)!m!C_{lmn} = \left. \frac{\partial^n p(\mathbf{r})}{\partial x^{n-l} \partial y^{l-m} \partial z^m} \right|_{x,y,z=0} \tag{39}$$

for the coefficients C_{lmn}. If two local power series exist and converge to the same function, then they must have the same derivatives and therefore the same coefficients. Hence, Eq. (39) shows that a local power series about a given point is unique. Equation (39) also shows that such a local power series can be used to generate additional power series about nearby points, since the derivatives required in (39) can be determined by differentiating the initial series. Although the expansion point for the new series must lie in the convergence region of the first series, the convergence region of the new series will in general extend beyond that of the first. For example, assume the coefficients are known for a local power series about $\mathbf{r}_0$ which converges for $|\mathbf{r}-\mathbf{r}_0| < R_0$. Call this convergence region V_0 and select a new point $\mathbf{r}_1 \in V_0$, as shown in Figure 1. The coefficients for the power series about $\mathbf{r}_1$ can be evaluated from the initial series based on Eq. (39). This new series will converge for $|\mathbf{r}-\mathbf{r}_1| < R_1$, or $\mathbf{r} \in V_1$. The region $V_0 \cup V_1$ is in general larger than either V_0 or V_1 separately. The process can be repeated for a new point $\mathbf{r}_2 \in V_0 \cup V_1$ and continued until p is represented for all $\mathbf{r}$ in its domain by a power series about some nearby point $\mathbf{r}_n$. An acoustic field is therefore determined in principle once enough information is known to obtain a convergent power series about some point in its domain. In particular, an acoustic field which vanishes in some region of its domain, no matter how small, must vanish identically.

The Kirchhoff-Helmholtz integral is obviously useful for developing many of the important properties of harmonic acoustic fields. As already indicated the results are derived based on the properties of the free-space Green's function. In particular, many expansions of the acoustic field can be found similar to the power series in Eq. (38) by expanding the Green's function in various functional series. For example, the Green's function can be written and expanded as

$$G_0(\mathbf{r}|\mathbf{r}') = \frac{e^{ikr}}{4\pi r} \frac{e^{-ikr'w(t)}}{1-tw(t)} = \frac{e^{ikr}}{4\pi r} \sum_{n=0}^{\infty} D_n(\Theta) t^n \tag{40}$$

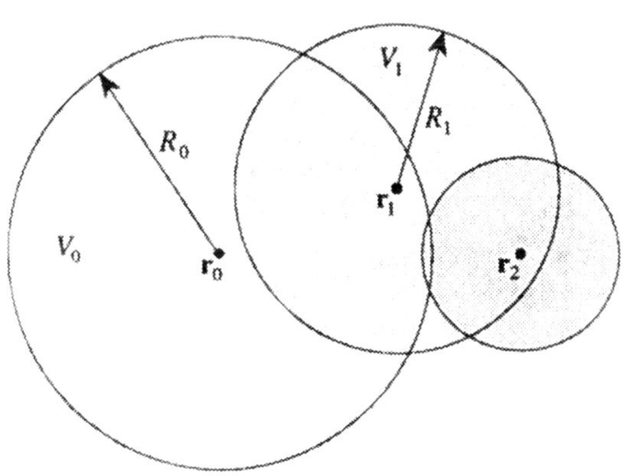

Figure 1. Analytic continuation by power series.

where $t = r'/r$, $\mathbf{r}\cdot\mathbf{r}' = rr'\cos\Theta$,

$$w(t) = t^{-1}\left(1-\sqrt{1-2t\cos\Theta+t^2}\right) \xrightarrow[t\to 0]{} \cos\Theta, \tag{41}$$

and the power series in t converges absolutely and uniformly for $t<1$. Equation (40) can be combined with Eq. (37) for singular fields to yield the expansion

$$p(\mathbf{r}) = \sum_{n=0}^{\infty} F_n(\theta,\phi)\frac{e^{ikr}}{r^{n+1}} \tag{42}$$

which is absolutely and uniformly convergent for $r > R_{max}$, where R_{max} is the distance from the origin to the farthest point in $\bar{S}$ [22 Sec. 3.2]. Equation (5) can be shown to be a result of Eq. (42), verifying that Eq. (4) alone is sufficient as the radiation condition. The first coefficient F_0 is termed the farfield amplitude and is found from Eqs. (37) and (40) to be given by

$$F_0(\theta,\phi) = \frac{1}{4\pi}\int_{\bar{S}} dS(\mathbf{r}')\mathbf{n}(\mathbf{r}')\cdot[\nabla' p(\mathbf{r}') + ik\mathbf{n}_r(\theta,\phi)p(\mathbf{r}')]e^{-ik\mathbf{n}_r(\theta,\phi)\cdot\mathbf{r}'} \tag{43}$$

with $\mathbf{r} = r\,\mathbf{n}_r(\theta,\phi)$. The higher coefficients F_n for $n>0$ can be determined recursively from F_0 using

$$2iknF_n(\theta,\phi) = n(n-1)F_{n-1}(\theta,\phi) + \mathbf{L}\{F_{n-1}(\theta,\phi)\}, \tag{44}$$

where the differential operator $\mathbf{L}$ is given by

$$\mathbf{L} = \frac{1}{\sin\theta}\frac{\partial}{\partial\theta}\left(\sin\theta\frac{\partial}{\partial\theta}\right) + \frac{1}{\sin^2\theta}\frac{\partial^2}{\partial\phi^2}. \tag{45}$$

Equations (42) and (44) show that the farfield amplitude alone is sufficient to determine p for $r > R_{max}$, and thus for all $\mathbf{r}$ by analytic continuation. In fact, Eq. (43) defines an analytic function of θ and ϕ, so that analytic continuation can also be used in principle to completely determine F_0 for all angles once it is known within any limited solid angle Ω. However, it is also clear from the form of Eq. (44) that the determination of the higher coefficients F_n will be very sensitive to the presence of errors in F_0. This sensitivity to errors is a basic aspect of ill-posed problems.

E. Boundary Value Problems

Various forms of information concerning p have been seen to be sufficient to uniquely determine the field everywhere within its domain. For example, the given data can represent the farfield amplitude within a solid angle Ω or the field itself throughout a small volume V_0 contained within the domain. These examples are possible because p is required to satisfy the homogeneous Helmholtz equation and is therefore analytic. For a boundary value problem the information necessary to pick out a particular solution of the Helmholtz equation is given as an inhomogeneous boundary condition, generally in the form of Eq. (35). This boundary condition is of course specified over the boundary of the volume for which the field is to be determined. For problems on unbounded domains the radiation condition (4) is usually applied over S_∞. The term exterior is often applied to such problems. Figure 2 shows the geometry to be considered from this point onwards. The volume V is unbounded and exterior to the single, finite closed surface S_0. The bounding surface of V is therefore $S = S_0 + S_\infty$ and the finite component of S is $\bar{S} = S_0$. Thus an exterior problem would involve finding p for $\mathbf{r} \in V$ given boundary conditions over S_0. The surface S_0 encloses the bounded volume V_0. The outward normal to V_0 over S_0 is denoted by $\mathbf{n}_0$, so the outward normal to V over $\bar{S}$ is $\mathbf{n} = -\mathbf{n}_0$. The term interior is generally applied to problems on bounded domains. Given boundary conditions over S_0, an interior problem would therefore involve finding p for $\mathbf{r} \in V_0$. Of particular interest for NAH are the Dirichlet and Neumann boundary

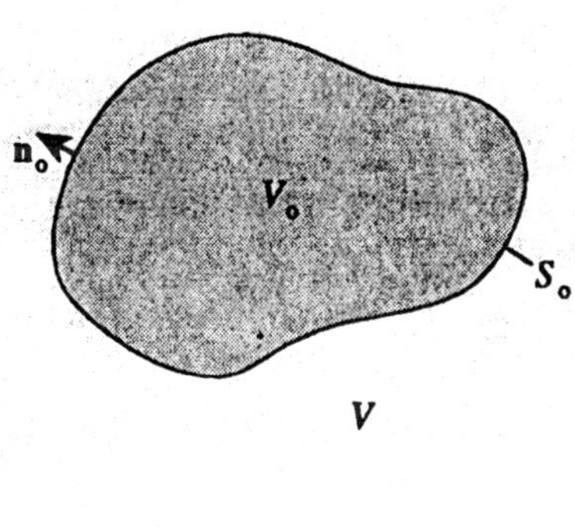

Figure 2. Geometry for exterior boundary value problem.

value problems. The Dirichlet boundary condition is given by Eq. (34) with $A=1$ and $B=0$, while the Neumann boundary condition has $A=0$ and $B=1$. The Dirichlet problem therefore involves specifying the boundary value p_n over the surface S_0, while the Neumann problem involves specifying v_n.

In the previous section, an integral representation was developed for the general solution of the Helmholtz equation under the boundary condition in Eq. (34). We have thus succeeded in representing direct diffraction in terms of an integral operator, a result that will be used in Section III. However, this representation is itself based on being able to solve the Helmholtz equation for the boundary component F of the Green's function. It is therefore necessary to investigate whether such a solution exists. In general, an analytical investigation of a boundary value problem is aimed at determining whether or not the problem is well-posed. A problem is called well-posed if a solution exits, is unique, and depends continuously on the given data. Uniqueness means that the given data is sufficient to pick out one and only one solution. For the Helmholtz equation we have already seen that several different types of data provide uniqueness. Once a particular type of data is chosen, whether or not a solution actually exists will depend on the properties of the data. In particular, it can often be shown that a solution is guaranteed to exist provided the data is known to have certain general properties, such as continuity. The third condition, that of continuous dependence on the data or stability, is very important from a practical standpoint. It means that small changes in the given data produce only comparable changes in the solution. Thus, when there are errors present in the data, or equivalently when there are errors in matching the data, the resulting errors in the solution remain bounded. Alternately, just as uniqueness requires that the solution be identically zero if the given data is zero, stability requires that the solution be near zero if the data is near zero. A problem is called ill-posed if it does not have the properties of uniqueness, existence, and stability [29 Sec. 4.1]. Whether or not a problem is well-posed has important consequences concerning the implementation of any solution method. In particular, ill-posed problems are ill-conditioned when discretized for numerical solution.

Boundary value problems for the Helmholtz equation are typically well-posed. For the exterior Dirichlet and Neumann problems, uniqueness follows directly from Eq. (16) with $\Phi = p$. From the geometry in Figure 2, the integration surface S separates into S_o and S_∞. The integration over S_∞ can be evaluated as

$$\int_{S_\infty} dS\,\mathbf{n}\cdot[p\nabla p^*] = -ik\int_{4\pi} d\Omega\left|F_0(\theta,\phi)\right|^2 \tag{46}$$

using the farfield expansion in Eq. (42), while the integration over S_o can be written as

$$\int_{S_o} dS\,\mathbf{n}\cdot[p\nabla p^*] = i\omega\rho_o\int_{S_o} dS\,p_n(\mathbf{r})\,v_n^*(\mathbf{r}) \tag{47}$$

in terms of the boundary values on S_o using Eq. (7). These results can be combined with Eq. (16) to yield

$$\int_{4\pi} d\Omega\left|F_0(\theta,\phi)\right|^2 = \rho_o c\int_{S_o} dS\,\mathrm{Re}\{p_n(\mathbf{r})\,v_n^*(\mathbf{r})\}. \tag{48}$$

Equation (48) shows that either $\mathrm{Re}\{p_n v_n^*\} > 0$ somewhere on S_o or $\mathrm{Re}\{p_n v_n^*\} \equiv 0$ everywhere on S_o and F_0 vanishes. Now the result

$$\mathrm{Re}\{p_n v_n^*\} = \begin{cases} -\omega\rho_o|v_n|^2\,\mathrm{Im}\left\{\dfrac{B}{A}\right\} + \mathrm{Re}\left\{\dfrac{v_n^* f}{A}\right\}, & A \neq 0 \\[2ex] \dfrac{|p_n|^2}{\omega\rho_o}\mathrm{Im}\left\{\dfrac{A}{B}\right\} - \dfrac{1}{\omega\rho_o}\mathrm{Im}\left\{\dfrac{p_n^* f}{B}\right\}, & B \neq 0 \end{cases} \tag{49}$$

follows from the general boundary condition in Eq. (34). Thus, with the given data $f = 0$, Eq. (49) shows that if at each point of S_o either $A = 0$, $B = 0$ or $\mathrm{Im}\{B/A\} > 0$ then F_0 must vanish. Under these conditions uniqueness follows since F_0 has already been shown to uniquely determine p. Solutions to the exterior Dirichlet problem ($B \equiv 0$) and the exterior Neumann problem ($A \equiv 0$) are therefore unique. Solutions to the exterior impedance problem, defined by the boundary condition

$$p_n(\mathbf{r}) + Z_n(\mathbf{r})\,v_n(\mathbf{r}) = f(\mathbf{r}) \tag{50}$$

where $|Z_n| < \infty$, are also unique provided $\mathrm{Re}\{Z_n\} \geq 0$ everywhere on S_o. In contrast, the interior problems do not in general have unique solutions. For these problems there exist eigenfrequencies k_n at which the Helmholtz equation admits solutions satisfying the homogeneous boundary condition.

Existence is generally a more difficult property to prove than uniqueness. Since solutions of the Helmholtz equation can be represented in integral form, one might assume that methods based on integral equations would prove effective. Indeed, for smooth surfaces and continuous boundary data Colton and Kress use integral equation methods to prove the existence of solutions for the exterior Dirichlet and Neumann problems [22 Theorem 3.21 and 3.25], and also to show that the solutions depend continuously on the data [29 Theorem 3.9 and 3.10]. Furthermore, if the given data for the Neumann problem $f = v_{\mathrm{n}} \in \mathrm{C}(S_{\mathrm{o}})$ then the resulting boundary pressure $p_{\mathrm{n}} \in \mathrm{C}(S_{\mathrm{o}})$ and the unique solution p is in $\mathrm{H}(V)$. Since the right hand side of Eq. (27) is certainly continuous for S_{o} smooth, it follows that a solution $F \in \mathrm{H}(V)$ exists. From Eq. (35) the solution p can therefore be written as

$$p(\mathbf{r}) = \int_{S_{\mathrm{o}}} dS(\mathbf{r}')N(\mathbf{r},\mathbf{r}')v_{\mathrm{n}}(\mathbf{r}') \tag{51}$$

where the Neumann kernel N is given by

$$N(\mathbf{r},\mathbf{r}') = -i\omega\rho_{\mathrm{o}}G_N(\mathbf{r}'|\mathbf{r}) \tag{52}$$

in terms of the Green's function G_N which satisfies the Neumann boundary condition. For the Dirichlet problem, assuming that the data $f = p_{\mathrm{n}} \in \mathrm{C}(S_{\mathrm{o}})$ does not ensure that the normal velocity $v_{\mathrm{n}} \in \mathrm{C}(S_{\mathrm{o}})$. The unique solution p is therefore not necessarily in $\mathrm{H}(V)$ as defined. However, because G_{o} is infinitely differentiable and S_{o} is assumed smooth, the right hand side of the inhomogeneous boundary condition for F in Eq. (27) is in $\mathrm{C}^2(S_{\mathrm{o}})$ for the Dirichlet problem. This is sufficient to ensure that $F \in \mathrm{H}(V)$ as stated, and therefore the integral representation

$$p(\mathbf{r}) = \int_{S_{\mathrm{o}}} dS(\mathbf{r}')D(\mathbf{r},\mathbf{r}')p_{\mathrm{n}}(\mathbf{r}') \tag{53}$$

from Eq. (35) is well defined, where the Dirichlet kernel D is given by

$$D(\mathbf{r},\mathbf{r}') = \mathbf{n}_{\mathrm{o}}(\mathbf{r}')\cdot\nabla' G_D(\mathbf{r}'|\mathbf{r}) \tag{54}$$

in terms of the Green's function G_D which satisfies the Dirichlet boundary condition. Equation (53) is valid for any p_{n} which is sufficiently smooth to ensure that $p \in \mathrm{H}(V)$, and it extends to the case when p_{n} is only continuous.

Since the exterior Dirichlet and Neumann problems are well-posed, it follows that p_n and v_n cannot both be arbitrarily specified over the boundary. This is the motivation for the general boundary condition in Eq. (34). In essence only one piece of information can be given at each point of the boundary; anything more results in an over specification. However, this assumes that information will be given for every point on the boundary. If data are to be given for only part of the boundary, then specifying either p_n or v_n alone is insufficient. Specifying both p_n and v_n over a surface is called a Cauchy problem, and the two quantities are collectively referred to as the Cauchy data. The Cauchy problem for the entire boundary is ill-posed since arbitrary Cauchy data are unlikely to be related as required to correctly define an acoustic field. It follows that the surface over which the Cauchy data are specified should be open. If this Cauchy surface is in fact completely contained within the domain of the field, then one can attempt to use Eq. (39) to develop a local power series solution. The required derivative values can be determined by differentiating as necessary both the Cauchy data and the Helmholtz equation, since the Cauchy surface is assumed to be in the domain of p. Uniqueness of the result follows directly from the solution process: if both p_n and v_n vanish, the derivatives must also vanish, and the series coefficients are all zero. Existence requires that the resulting power series actually converge. For analytic data, the Cauchy-Kowalewski theorem ensures the existence of such a locally convergent power series [30 Chapt. 1]. Analytic continuation can then be used to define p everywhere. Holmgren's theorem extends uniqueness to any smooth open surface, even one that is a segment of the boundary [30 Chapt. 6] . However, the Cauchy problem for an open surface is still ill-posed, as a simple example will demonstrate. Consider the pressure field

$$p(\mathbf{r}) = e^{-\kappa(2b - x - y)} e^{iz(k^2 + 2\kappa^2)^{1/2}} \tag{55}$$

for which

$$\left(p_n, v_n\right) = \left(1, \frac{(k^2 + 2\kappa^2)^{1/2}}{\omega\rho_0}\right) e^{-\kappa(2b - x - y)} \tag{56}$$

is the Cauchy data for the disk $z = 0$, $x^2 + y^2 < a^2$ taken as the Cauchy surface. Since this Cauchy data is analytic, the local power series derived from it will converge. In fact, the power series will converge everywhere to Eq. (55). Now, it follows from Eq. (55) that $|p(b,b,z)|=1$ for all κ, and yet for $b>a$ Eq. (56) shows that $(p_n, v_n)\rightarrow 0$ on the Cauchy surface as $\kappa\rightarrow\infty$. The problem is therefore unstable, since vanishingly small Cauchy data can correspond to a finite solution, and the determination of an acoustic field from data on an open surface is ill-posed [25 Chapt. 6].

III. PRINCIPLES OF NEARFIELD ACOUSTIC HOLOGRAPHY

In the previous section the process of direct diffraction was formulated in terms of an integral representation for the acoustic field. The principles of NAH will be developed in this section by treating this representation as an integral operator which maps field data from one surface to another. This integral operator can be termed the forward propagator, since it can be used to directly perform forward propagation. From the properties of the Green's function, it will follow that the forward propagator is a compact operator. The singular value decomposition of this operator then leads to the equations of NAH for forward and backward propagation. However, a regularization of the equation for backward propagation becomes necessary due to the behavior of the singular values. First, what is meant by forward and backward propagation in this context will be presented.

A. Forward and Backward Propagation

For forward and backward propagation, a measurement surface S_m is assumed to separate the fluid volume V into two disjoint volumes, as illustrated in Figure 3. The interior volume is labeled V_i and the exterior volume V_e. Note that the interest is now focused on singular acoustic fields for the unbounded volume V of Figure 2. Such an acoustic field is radiated by some collection of sources located within the source volume $V_s = V_0 \cup S_0$. For scattering these are secondary sources induced by the presence of an incident field. The measurement surface is taken to enclose this source volume in general, but does not have to be finite in extent. For example, an infinite cylindrical surface or two planar parallel surfaces which straddle V_s can be used. The exterior volume consists of those components of V partitioned by S_m that do not contain V_s. In the case of two planar surfaces, the exterior volume V_e therefore has two components, which may be treated separately. The interior

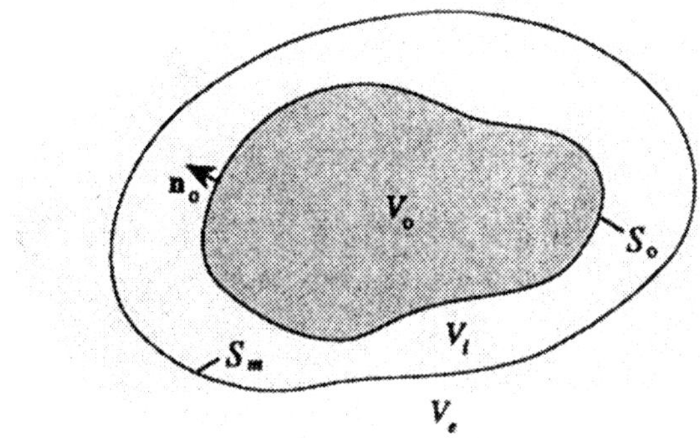

Figure 3. Geometry for forward and backward propagation.

volume is the remaining partition of V that contains V_s. Given data measured over the surface S_m, forward propagation refers to the determination of p for $\mathbf{r} \in V_e$, while backward propagation refers to the determination of p for $\mathbf{r} \in V_i$.

The measured data will consist of some part of the Cauchy data $(p_\mathrm{n}, \upsilon_\mathrm{n})$ over each section of S_m, and will be denoted by $(p_\mathrm{m}, \upsilon_\mathrm{m})$ for convenience. Since the measurement surface is within the fluid volume, the Cauchy data will be smooth over S_m for a smooth surface. Considering that the data will ultimately be measured at a finite number of discrete locations, a smooth interpolation between these locations can be taken to represent the actual measurement surface. Thus, there is no loss of generality in assuming S_m to be smooth at the start. In fact, S_m could just as well be taken to be infinitely smooth, where infinitely smooth refers to the existence of continuous derivatives of all order. Spherical, cylindrical and planar surfaces are for example infinitely smooth. The smoothness of the Cauchy data for a smooth surface follows from the pressure field p being analytic throughout V. Of course, since the measured data is discrete, it may also be assumed whenever necessary that p_m and υ_m represent sufficiently smooth interpolations of this discrete data.

Forward propagation may be formulated in general as a direct boundary value problem. For example, if p_m is measured, then the forward propagated field can be determined by solving the Helmholtz equation for a singular field $\hat{p}$ in V_e satisfying the boundary condition

$$\hat{p}_\mathrm{n}(\mathbf{r}) = p_\mathrm{m}(\mathbf{r}) \tag{57}$$

for $\mathbf{r} \in S_m$. The symbol $\hat{p}$ will be used to represent forward and backward propagated fields in order to distinguish them from the actual field p. Such a Dirichlet problem for $\hat{p}$ is well-posed, so that

$$|\hat{p}(\mathbf{r}) - p(\mathbf{r})| \leq C \max_{\mathbf{r}' \in S_m} |p_{\mathrm{m}}(\mathbf{r}') - p_{\mathrm{n}}(\mathbf{r}')| \tag{58}$$

for $\mathbf{r} \in V_e$, where C is some constant [29 Theorem 3.1 and Sec. 3.2]. Equation (58) shows that for discrete measurements $\hat{p}$ will approach p uniformly as the measurement density is increased, assuming a reasonable interpolation scheme and the absence of measurement errors. This is not true of the backward propagated field. Backward propagation cannot be formulated as a direct boundary value problem, but is instead an inverse problem. The boundary of the interior volume V_i for backward propagation is given by $S_i = S_m + S_o$ and boundary data is known only for the component S_m by measurement. In contrast, the exterior volume V_e for forward propagation has the boundary $S_e = S_m + S_\infty$ and the radiation condition supplies the necessary boundary condition over S_∞.

B. General Formulation of NAH

Nearfield acoustic holography accomplishes forward and backward propagation by transforming the field between surfaces based on direct and inverse diffraction. The general principles involved are most easily developed by treating direct diffraction as an integral operator, termed the forward propagator, which maps field data from one surface to another. For example, a Dirichlet operator **D** can be defined based on Eq. (53) and used to represent the forward propagated field as

$$\hat{p}(\mathbf{r}) = \mathbf{D}p_{\mathrm{m}}(\mathbf{r}) = \int_{S_m} dS(\mathbf{r}')\, D(\mathbf{r},\mathbf{r}')\, p_{\mathrm{m}}(\mathbf{r}') \tag{59}$$

for $\mathbf{r} \in S_r$ where the reconstruction surface S_r is any surface exterior to S_m, as shown in Figure 4a. Likewise, a Neumann operator **N** can be defined based on Eq. (51) and used to represent $\hat{p}$ as

$$\hat{p}(\mathbf{r}) = \mathbf{N}v_{\mathrm{m}}(\mathbf{r}) = \int_{S_m} dS(\mathbf{r}')\, N(\mathbf{r},\mathbf{r}')\, v_{\mathrm{m}}(\mathbf{r}'). \tag{60}$$

By definition, the kernels D and N solve the direct boundary value problem for either a Dirichlet or Neumann boundary condition on S_m respectively. Thus the operators **D** and **N** solve the direct diffraction problem for the forward propagated field with either $\hat{p}_{\mathrm{n}} = p_{\mathrm{m}}$ or $\hat{v}_{\mathrm{n}} = v_{\mathrm{m}}$. Backward propagation can also be formulated in terms of the operators **D** and **N** as the solution of either

$$p_{\mathrm{m}}(\mathbf{r}) = \mathbf{D}\hat{p}_{\mathrm{n}}(\mathbf{r}) = \int_{S_r} dS(\mathbf{r}')D(\mathbf{r},\mathbf{r}')\hat{p}_{\mathrm{n}}(\mathbf{r}') \tag{61}$$

or

$$p_{\mathrm{m}}(\mathbf{r}) = \mathbf{N}\hat{v}_{\mathrm{n}}(\mathbf{r}) = \int_{S_r} dS(\mathbf{r}')N(\mathbf{r},\mathbf{r}')\hat{v}_{\mathrm{n}}(\mathbf{r}') \tag{62}$$

for $\mathbf{r} \in S_m$, where S_r is now any surface interior to S_m but exterior to the source volume, as shown in Figure 4b. In Eqs. (61) and (62), the kernels D and N now solve the direct boundary value problem for either a Dirichlet or Neumann boundary condition on S_r. Furthermore, $\hat{p}_{\mathrm{n}}$ and $\hat{v}_{\mathrm{n}}$ represent the unknown boundary data for a backward propagated field defined exterior to S_r that equates to p_{m} for $\mathbf{r} \in S_m$. Equations (61) and (62) therefore represent the problem of inverse diffraction.

In Eqs. (59)-(62) **D** and **N** are to be interpreted as operators that map field data from a given surface to a second, exterior surface. Equation (35) provides the basis for a more general integral operator **K** which encompasses both **D** and

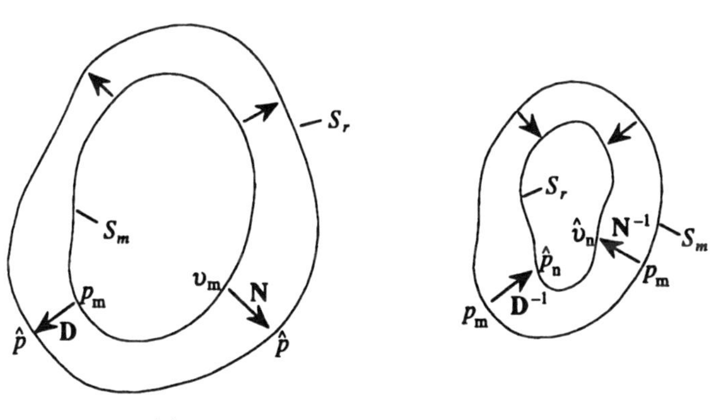

Figure 4. (a) Forward propagation by direct diffraction, (b) backward propagation by inverse diffraction.

N. To avoid having to deal with direct and inverse diffraction separately, the general linear operator equation is written as

$$v(\mathbf{x}) = \mathbf{K}u(\mathbf{x}) = \int_{S_u} dA(\mathbf{x}')K(\mathbf{x},\mathbf{x}')u(\mathbf{x}'). \tag{63}$$

The change in notation from **r** to **x** has been made to emphasize that this is to be considered as a mapping between surfaces. To be specific, **K** maps the field data u on the surface S_u into the field data v on the surface S_v. Thus **x** is to be understood as a general surface coordinate and dA as a general area element, not necessarily with the dimensions of area. For example, in the case of planar, cylindrical or spherical surfaces **x** represents the coordinates (x,y), (ϕ,z), or (θ,ϕ), for which dA is $d\mathbf{x} = dxdy$, $d\phi dz$, or $d\Omega = \sin\theta\, d\theta d\phi$ respectively. The function u that **K** operates on represents the boundary data on S_u, and therefore is either the measured values of p_{m} or v_{m} for forward propagation or the unknown values of $\hat{p}_{\mathrm{n}}$ or $\hat{v}_{\mathrm{n}}$ for backward propagation. In either case, $u \in \mathrm{U}$ where U is the set of all allowed boundary data for the operator **K**. Up to this point only boundary data that is at least continuous has been considered. For now we therefore choose the function space $\mathrm{U} = \mathrm{C}(S_u)$, which is a Banach space when associated with the norm

$$\|u\|_{\mathrm{U}} = \max_{\mathbf{x}\in S_u} |u(\mathbf{x})|. \tag{64}$$

For forward propagation, v represents the pressure $\hat{p}$ evaluated on the surface S_v. In this case, v is certainly continuous since an acoustic field is a C^∞ function of space and the surface S_v is assumed smooth. For backward propagation, v represents the measured data p_{m}, where it is assumed for convenience that backward propagation will be based only on the measured pressure and not on v_{m}. In this case, the function space chosen for v must be suitable for describing the measured data. To handle both situations it is reasonable to take $\mathrm{V} = \mathrm{C}(S_v)$ and associate with it the norm of Eq. (64) with S_u replaced by S_v.

The operator **K** can be termed the forward propagator, since by definition it solves the direct boundary value problem. Forward propagation therefore amounts to an application of the forward propagator, and is essentially solved when the kernel K, and thus the operator **K**, is determined. The kernel K is determined once the Green's function G is known, which in turn is found by solving the Helmholtz equation for F in Eq. (25) with the inhomogeneous boundary condition of Eq. (27). Thus F can be interpreted as an acoustic field. With G_0 given by Eq. (26), and both $\mathbf{r}$ and $\mathbf{r}'$ within the domain of F, it follows that G is an analytic function of the first argument $\mathbf{r}$, excluding the point $\mathbf{r} = \mathbf{r}'$. Furthermore, both G and its normal derivative are continuous functions of $\mathbf{r}$ over the boundary. The reciprocity relation in Eq. (19) extends these results to G as a function of the second argument $\mathbf{r}'$. In particular, when $\mathbf{r}$ is restricted to the boundary, both G and its normal derivative are analytic functions of $\mathbf{r}'$ off the boundary. Since S_υ is assumed to be smooth, the kernel K is therefore a continuous function of both its arguments. This is true as long as S_u and S_υ are separated everywhere by some distance. It follows from the continuity of the kernel that the operator $\mathbf{K}$ from U to V is bounded,

$$\|\upsilon\|_{\mathrm{V}} \le C_K \|u\|_{\mathrm{U}} \tag{65}$$

where

$$C_K = \max_{\mathbf{x} \in S_\upsilon} \int_{S_u} dA(\mathbf{x}') |K(\mathbf{x}, \mathbf{x}')|. \tag{66}$$

The well-posed nature of direct diffraction is evident from Eqs. (63) and (65). Existence and uniqueness follow from the fact that any continuous function has a unique definite integral. For discrete measurements, existence is therefore ensured by a proper interpretation of the measurement data. Stability follows from the fact that $\mathbf{K}$ is bounded, which may also be written as

$$|\upsilon(\mathbf{x}) - \upsilon'(\mathbf{x})| \le \|\upsilon - \upsilon'\|_{\mathrm{V}} \le C_K \|u - u'\|_{\mathrm{U}} = C_K \max_{\mathbf{x}' \in S_u} |u(\mathbf{x}') - u'(\mathbf{x}')|. \tag{67}$$

Equation (67) shows that υ depends continuously on u [31 Theorem 2.5], and is equivalent to Eq. (58) with $\mathbf{r}$ restricted to S_υ. More generally, Eq. (58) follows from Eq. (67) with C given by the maximum C_K value for all possible surfaces S_υ, provided this set of C_K values is bounded. However, it must be remembered that Eqs. (63) and (65) are based on the existence of a unique kernel K with the continuity properties described, which in turn goes back to the existence of a unique solution to the boundary value problem for F. Thus the validity of Eq. (63) is itself contingent upon the well-posed nature of direct diffraction. The principal advantage of Eq. (63) is its use in understanding the problem of inverse diffraction.

From Eq. (63) the inverse diffraction problem is solved if the inverse of **K** can be determined. Whether or not inverting the forward propagator is a well-posed problem can be determined from the results of Section II. The inverse of **K** will exist provided the first two conditions for being well-posed are satisfied. That is, for every $v \in \mathrm{V}$ there exists a unique $u \in \mathrm{U}$ such that $\mathbf{K}u = v$. Uniqueness is guaranteed since $v = 0$ is equivalent to $p = 0$ for $\mathbf{r} \in S_v$ and this is sufficient to ensure that p vanish everywhere. Thus $\mathbf{K}u = 0$ if and only if $u = 0$. Existence is not guaranteed, however, since the set of all continuous surface pressures V is larger than the range of **K** (the range R being the set of all $\mathbf{K}u$ for $u \in \mathrm{U}$). An element $v \in \mathrm{R}$ represents the evaluation on the surface S_v of an acoustic field that is due to sources interior to S_u. Since acoustic fields are C^∞ functions within their domains, the differentiability of v is therefore the same as that of S_v, and since S_v was assumed smooth, $\mathrm{R} \subseteq \mathrm{C}^2(S_v)$. Thus $\mathrm{R} \subset \mathrm{V}$, since $\mathrm{V} = \mathrm{C}(S_v)$, and the inverse of **K** does not exist over V. The inverse diffraction problem is therefore ill-posed. The property of existence can be restored by restricting V to be R, since the inverse of **K** does exist over R. However, this would require not only a method for determining if the given data is in R but also a means of forcing it into R if it is not. In any case, it is known that an integral in the form of Eq. (63) with a continuous kernel generates a compact operator [31 Theorem 2.22], and that a compact operator cannot have a bounded, continuous inverse [31 Theorem 15.4]. Thus the inverse diffraction problem is unstable, and therefore still ill-posed, even if V is restricted to R.

In order to continue with the development of NAH, it is necessary to extend both U and V. In particular, it will now be assumed that $\mathrm{U} = \mathrm{L}_2(S_u)$, where $\mathrm{L}_2(S_u)$ is the Hilbert space of functions square-integrable on S_u with the inner product

$$(f,g)_{\mathrm{U}} = \int_{S_u} dA(\mathbf{x}) f(\mathbf{x}) g^*(\mathbf{x}). \tag{68}$$

The space U now includes functions which are discontinuous, and even allows some forms of singularity. The norm induced by (68) is

$$\|u\|_{\mathrm{U}} = \sqrt{(u,u)_{\mathrm{U}}}, \tag{69}$$

and in this setting the norm equivalence of two functions f and g,

$$\|f - g\|_{\mathrm{U}} = 0, \tag{70}$$

only requires that they be equal on a point by point basis almost everywhere. Since K is a continuous function of both its arguments, Schwarz's inequality can be used to show that Eq. (65) holds with

$$C_K^2 = \int_{S_\upsilon} dA(\mathbf{x}) \int_{S_u} dA(\mathbf{x}') |K(\mathbf{x}, \mathbf{x}')|^2 \tag{71}$$

and $\mathrm{V} = \mathrm{L}_2(S_\upsilon)$. That is, $\mathbf{K}u$ is square-integrable on S_υ for $u \in \mathrm{U}$, and thus V can be taken as $\mathrm{L}_2(S_\upsilon)$ with an inner product in the form of Eq. (68), with S_u replaced by S_υ. The well-posed nature of the direct diffraction problem extends to **K** as a mapping from $\mathrm{L}_2(S_u)$ to $\mathrm{L}_2(S_\upsilon)$. In particular, **K** is bounded and compact. The ill-posed nature of inverse diffraction also extends, but the inner product in (68) provides the additional structure necessary to introduce the singular value decomposition of **K**.

For any compact operator like **K**, there exists a sequence of positive numbers σ_n and functions $\{u_n\} \in \mathrm{U}$ and $\{\upsilon_n\} \in \mathrm{V}$ such that

$$\mathbf{K}u(\mathbf{x}) = \sum_n \sigma_n (u, u_n) \upsilon_n(\mathbf{x}) \tag{72}$$

for any $u \in \mathrm{U}$ [32 Theorem 4.14]. The σ_n are called the singular values of **K** and satisfy

$$\sigma_1 \geq \sigma_2 \geq \sigma_3 \geq \cdots > 0. \tag{73}$$

The function sequences $\{u_n\}$ and $\{\upsilon_n\}$, which may be termed the left and right singular vectors, are orthonormal,

$$(u_m, u_n) = (\upsilon_m, \upsilon_n) = \delta_{mn}, \tag{74}$$

and from Eq. (72) satisfy

$$\mathbf{K}u_n(\mathbf{x}) = \sigma_n \upsilon_n(\mathbf{x}). \tag{75}$$

The family $\{\sigma_n, u_n, \upsilon_n\}$ is collectively called a singular system of **K**. The number of singular values and singular vectors may be either finite or infinite, depending on **K**. However, any $u \in \mathrm{U}$ has the expansion

$$u(\mathbf{x}) = \sum_n (u, u_n) u_n(\mathbf{x}) + u_0(\mathbf{x}) \tag{76}$$

where u_o is such that $\mathbf{K}u_o = 0$. Since $\mathbf{K}u_o = 0$ if and only if $u_o = 0$, it follows that the sequence $\{u_n\}$ is complete in U, and since U has infinite dimension, **K** has an infinite number of singular values. In the following section, the existence of an infinite number of singular values will be linked to the ill-posed nature of inverse diffraction.

The general equations of NAH for forward and backward propagation are developed from the singular value decomposition of **K**. From Eq. (76), any $u \in \mathrm{U}$ has an expansion in the form

$$u(\mathbf{x}) = \sum_n a_n u_n(\mathbf{x}) \tag{77}$$

where the a_n are given by $a_n = (u, u_n)$. Equation (77) is the Fourier series of u with respect to the system $\{u_n\}$, and the a_n are called the Fourier coefficients. Since the orthonormal sequence $\{u_n\}$ is complete in U, the norm of u is given in terms of the Fourier coefficients as

$$\|u\|_{\mathrm{U}}^2 = \sum_n |(u, u_n)|^2 = \sum_n |a_n|^2 \tag{78}$$

by Parseval's theorem [33 Theorem A8]. Equation (72) now shows that for v in the range of **K**,

$$v(\mathbf{x}) = \sum_n b_n v_n(\mathbf{x}), \tag{79}$$

where

$$b_n = (v, v_n) = \sigma_n (u, u_n) = \sigma_n a_n \tag{80}$$

and

$$\|v\|_{\mathrm{V}}^2 = \sum_n |b_n|^2 = \sum_n \sigma_n^2 |a_n|^2, \tag{81}$$

since the orthonormal sequence $\{v_n\}$ can be completed in V [33 Theorem A10]. The general equation of NAH for forward propagation comes directly from Eqs. (79) and (80), and may be written as

$$v(\mathbf{x}) = \sum_n \sigma_n (u, u_n) v_n(\mathbf{x}). \tag{82}$$

Equation (80) also gives $a_n = b_n / \sigma_n$ as the solution of the inverse problem, and thus the general equation of NAH for backward propagation may be written as

$$u(\mathbf{x}) = \sum_n \frac{(\upsilon, \upsilon_n)}{\sigma_n} u_n(\mathbf{x}). \tag{83}$$

However, Eq. (83) is valid only if υ is in the range of **K**. Since Eq. (77) defines a $u \in \mathrm{U}$ if and only if

$$\sum_n |a_n|^2 < \infty \tag{84}$$

by the Riesz-Fischer theorem [33 Theorem A9], it follows that υ is in the range of **K** if and only if Eq. (79) holds with Fourier coefficients b_n satisfying

$$\sum_n \frac{|b_n|^2}{\sigma_n^2} < \infty. \tag{85}$$

Equations (79) and (85) characterize the range of **K**, giving in principle a method for determining if a given $\upsilon \in \mathrm{R}$. However, the problem of inverse diffraction is unstable even if $\upsilon \in \mathrm{R}$. Although Eq. (83) provides the formal solution, it cannot be effectively applied in practice. The reason for this instability and how it is dealt with is taken up in the following section.

C. Evanescent Waves and Regularization

The unstable nature of backward propagation is connected to the behavior of the singular values of **K**. First, since σ_1 is the largest singular value, an alternate version of Eq. (65) can be derived from Eqs. (78) and (81),

$$\|\upsilon\|_{\mathrm{V}} \leq \sigma_1 \|u\|_{\mathrm{U}}. \tag{86}$$

As already mentioned, a bound relation such as Eq. (86) shows the continuous dependence of υ on u, and therefore the stability of direct diffraction. Now, if there were a finite number of singular values, then from Eq. (73) there would be a minimum value σ_{min}. Equation (83) would then give the new bound relation

$$\|u\|_{\mathrm{U}} \leq \frac{1}{\sigma_{min}} \|\upsilon\|_{\mathrm{V}}, \tag{87}$$

and inverse diffraction would also be stable. However, there are an infinite number of singular values. Equation (87) would still remain valid if the σ_n

approached a nonzero limit point σ_∞, only σ_{min} would be replaced by σ_∞. Unfortunately, it can be shown that the singular values satisfy the relation

$$\sum_n \sigma_n^2 = C_K^2, \tag{88}$$

where C_K is given by Eq. (71) [33 Corollary 4.20]. From Eq. (88) it follows that $\sigma_n \to 0$ as $n \to \infty$. For any $\varepsilon > 0$, a value of n can therefore be chosen, say n', such that $\sigma_n < \varepsilon$ for all $n \geq n'$. Thus a bound relation such as Eq. (87) cannot hold. Alternately, if u is the solution of $\mathbf{K}u = \upsilon$ then from Eq. (75) $u' = u + u_{n'}$ is the solution of $\mathbf{K}u' = \upsilon'$ for $\upsilon' = \upsilon + \sigma_{n'} \upsilon_{n'}$. Hence, $\|u' - u\|_{\mathrm{U}} = 1$ even though $\|\upsilon' - \upsilon\|_{\mathrm{V}} = \sigma_{n'} < \varepsilon$ and ε can be made arbitrarily small. A small perturbation in υ, caused for example by the presence of noise, can therefore result in a large alteration in u.

Since the unstable nature of inverse diffraction is due to the decay of the singular values to zero, the rate at which this occurs is of interest. In a sense, the decay rate controls the degree of instability: the more rapid the decay to zero, the more influential are perturbations in the data. For an integral operator such as **K**, the decay rate of the singular values is linked to the smoothness properties of the kernel. In particular, the order of the decay rate is exponential for an analytic kernel [31 Theorem 15.20]. For Eq. (63), the kernel K is analytic when both S_u and S_υ are analytic (that is, when for each surface a parameterization, $\mathbf{r} = \mathbf{X}(u_1, u_2)$, exists such that the components of $\mathbf{X}$ are analytic functions). Planar, cylindrical and spherical surfaces are analytic, and in Section IV the exponential decay of the singular values for these forms of NAH will be examined explicitly. The exponential order of the decay cannot be avoided by simply using surfaces that are not analytic. For the more general situation, we may always select an analytic surface $\tilde{S}_u$ that is exterior to S_u and an analytic surface $\tilde{S}_\upsilon$ that is interior to S_υ. The operator **K** can then be decomposed as $\mathbf{K}_\upsilon \cdot \tilde{\mathbf{K}} \cdot \mathbf{K}_u$ where $\mathbf{K}_u$ maps data from S_u to $\tilde{S}_u$, $\tilde{\mathbf{K}}$ from $\tilde{S}_u$ to $\tilde{S}_\upsilon$, and $\mathbf{K}_\upsilon$ from $\tilde{S}_\upsilon$ to S_υ. The analytic surfaces $\tilde{S}_u$ and $\tilde{S}_\upsilon$ can be arranged to approximate S_u and S_υ so closely that the operator $\tilde{\mathbf{K}}$ essentially performs all of the propagation. Thus, since $\tilde{\mathbf{K}}$ has an analytic kernel, it must be expected that the singular values of **K** will exhibit exponential decay in general.

The exponential nature of the decay refers to the order of the decay rate. Typically, the exponential decay becomes evident only in the asymptotic behavior of the singular values, and a transition generally occurs from a much slower decay [34,35]. Each u_n corresponds to the boundary value on S_u for a propagation mode p_n, and the product $\sigma_n \upsilon_n$ represents the evaluation of this mode on the surface S_υ. The SVD therefore separates forward propagation into modes based on the efficiency of radiating to the surface S_υ. Those modes associated with singular values on the exponential decay side of the transition may be called the evanescent modes. Since the transition in the singular values is generally not well defined, the point where a mode becomes evanescent is somewhat arbitrary, except in certain special cases such as for planar surfaces. This definition of an evanescent mode is based on considering σ_n as a function of index n for fixed S_υ, and is somewhat different than the usual concept of an evanescent wave. The term evanescent wave is typically applied to a propagation mode that undergoes spatial exponential decay, and is equivalent to considering σ_n as a function of propagation distance for fixed index. The definition of an evanescent wave may be extended to include strong spatial decay in general, since a mode may decay rapidly in space without the decay being exponential in form. When it is far enough away from the source, such a mode may eventually lose this strong spatial decay and switch instead to a much less rapid decay associated with cylindrical or spherical spreading [17 I.A.1]. In this sense, a mode may change its spatial behavior from evanescent to nonevanescent decay. Although the mode would still be referred to as an evanescent wave, whether or not it is considered an evanescent mode depends on the transition point selected for the singular values (the σ_n cutoff level). Thus, although they generally do coincide, the evanescent modes and the evanescent waves are not strictly identical for the definitions used here.

It is not important to include the evanescent modes in the reconstruction of υ for forward propagation, since they do not contribute significantly. However, they may still carry a significant amount of information concerning the field on S_u, and can therefore be important in the reconstruction of u for backward propagation. It is apparent that Eq. (83) for backward propagation will converge only if the Fourier coefficients (υ, υ_n) have a more rapid decay than the σ_n. This is certainly true for υ in the range of $\mathbf{K}$, but is unlikely to be true for measured data. Even if Eq. (83) does converge, small errors in the data can still produce large errors in the solution. The problem is therefore one of obtaining a stable solution from Eq. (83) while retaining as much of the evanescent information as possible. In general, the method by which a stable solution is obtained to an ill-posed problem is called a regularization, and a number of approaches are available [31,32,36]. The interest here is on methods that can be based directly on the singular value decomposition, and the obvious approach is to include a weighting factor in Eq. (83) to reduce the effect of the smaller singular values,

$$u^{\alpha}(\mathbf{x}) = \sum_{n} W_{\alpha}(\sigma_n) \frac{(v, v_n)}{\sigma_n} u_n(\mathbf{x}). \tag{89}$$

In Eq. (89), $\alpha \in (0,\infty)$ is called the regularization parameter and u^{α} is the regularized solution. Furthermore, if the weighting factor W_{α} satisfies

$$|W_{\alpha}(\sigma)| \leq C(\alpha)\sigma, \tag{90}$$

then Eqs. (78) and (89) show that

$$\|u^{\alpha}\|_{\mathrm{U}} \leq C(\alpha)\|v\|_{\mathrm{V}} \tag{91}$$

and the regularized solution u^{α} depends continuously on v. In addition, if W_{α} is bounded and satisfies

$$\lim_{\alpha \to 0} W_{\alpha}(\sigma) = 1, \tag{92}$$

then Eq. (89) yields a proper regularization method, in the sense that for any $u \in \mathrm{U}$ and $v = \mathbf{K}u$,

$$\lim_{\alpha \to 0} \|u^{\alpha} - u\|_{\mathrm{U}} = 0. \tag{93}$$

Equation (93) shows that in the absence of data errors, perfect accuracy is obtained by letting $\alpha \to 0$. However, since u does not depend continuously on v, $C(\alpha)$ cannot be bounded as $\alpha \to 0$. With a finite error level ε, α must therefore be chosen to achieve an acceptable tradeoff between accuracy and stability. Particular strategies for selecting α, which would depend on the choice of weighting factor, will not be considered here.

Perhaps the two most common regularization methods are spectral truncation and Tikhonov regularization. Spectral truncation is frequently used with the singular value decomposition, especially for the least squares solution of first kind matrix equations. Such equations are often the result of discretizing an original operator equation like (63). The weighting factor for spectral truncation is given by

$$W_\alpha(\sigma)=\begin{cases}1\ , & \sigma\geq\alpha\\ 0\ , & \sigma<\alpha\end{cases} \tag{94}$$

and involves truncating the series to exclude those singular values smaller than α. This weighting factor satisfies Eq. (90) with

$$C(\alpha)=\frac{1}{\alpha}, \tag{95}$$

showing that $C(\alpha)$ is indeed unbounded as $\alpha\to 0$. If U_α is defined as the subspace spanned by the u_n for $n\leq N(\alpha)$, where $N(\alpha)$ is such that $\sigma_n\geq\alpha$ for $n\leq N(\alpha)$, then the regularized solution based on Eq. (94) clearly satisfies $u^\alpha\in\mathrm{U}_\alpha$. In fact, with the least squares functional $L(u)$ given by

$$L(u)=\|\mathbf{K}u-v\|_{\mathrm{V}}^2, \tag{96}$$

spectral truncation is equivalent to minimizing $L(u)$ for $u\in\mathrm{U}_\alpha$. Because of its simplicity, and because a reasonable truncation level can be estimated readily based on the dynamic range of the measurement system [17 I.A.3, 19 I.C], spectral truncation has been the method of choice for NAH [16-19]. The weighting factor for Tikhonov regularization is given by

$$W_\alpha(\sigma)=\frac{\sigma^2}{\sigma^2+\alpha}, \tag{97}$$

and satisfies Eq. (90) with [31 Theorem 15.23]

$$C(\alpha)=\frac{1}{2\sqrt{\alpha}}. \tag{98}$$

With the Tikhonov functional $T_\alpha(u)$ defined as

$$T_\alpha(u)=L(u)+\alpha\|u\|_{\mathrm{U}}^2, \tag{99}$$

Tikhonov regularization is equivalent to minimizing $T_\alpha(u)$ for $u\in\mathrm{U}$ [III.1, Theorem 16.1]. Equation (99) shows that Tikhonov regularization is also the solution of a least squares problem, but over the full space U and with a penalty term involving the norm of u. Other forms of the penalty term may also be considered [31 Sec. 16.5, 32 Sec. 3.1], but they may not lead to a solution representable in the form of Eq. (89).

Since some form of regularization must be used for inverse diffraction, Eq. (89) rather than Eq. (83) is the basic formula for backward propagation in NAH. Equation (82) for forward propagation remains unaltered. The most basic forms of NAH resulting from these equations are for field propagation between planar, cylindrical or spherical surfaces, and these examples will be discussed in the following section. In all three cases both the operator **K** and an appropriate singular system can be determined analytically. Interest will therefore focus on the properties of the singular values, as discussed above, and on the particular forms of the algorithms arising from Eqs. (82) and (89). The situation for more general surfaces will also be discussed, where the interest is mainly on how to obtain a finite rank or matrix approximation to **K**.

IV. IMPLEMENTATION OF NEARFIELD ACOUSTIC HOLOGRAPHY

In this section the basic forms of NAH for planar, cylindrical and spherical surfaces will be presented, and the methods necessary for arbitrary surfaces will be discussed. Although the results of the previous section formally depend on the boundedness of S_u, the surfaces are infinite for both planar and cylindrical NAH. Nevertheless, for these cases both an operator **K** and an appropriate singular system can be determined such that Eqs. (63), (82) and (89) are valid. However, the resulting **K** in Eq. (63) is not compact and a number of alterations occur in Eqs. (82) and (89). The Neumann operator **N** is even unbounded, and this leads to singularities in the associated singular values. For arbitrary, bounded surfaces the results of the previous section apply directly. Unfortunately, the operator **K** cannot be determined analytically and must therefore be approximated. The case for planar surfaces will be discussed first.

A. NAH for Planar Surfaces

For an infinite planar surface the Green's function for both the Dirichlet and Neumann boundary condition can be determined by the method of images. Since the problem is translationally invariant, without loss of generality we may take the surface $S_0 = S_u$ for Eqs. (51) and (53) as the infinite plane $z = 0$. With the positive z axis taken into the fluid volume, V then corresponds to the half-space $z > 0$. The positive z direction is therefore the direction of forward propagation and the actual sources of the field are in the half-space $z < 0$. For the determination of the Green's function an image point source is placed at the location $\bar{\mathbf{r}}' = \mathbf{x}' - z'\mathbf{e}_z$ in the geometry of Figure 5. This image source position is symmetric with respect to the surface S_0 to the location of the actual point source at $\mathbf{r}' = \mathbf{x}' + z'\mathbf{e}_z$. For a Dirichlet boundary condition, the image source must be out of phase, yielding

$$G_D(\mathbf{r}|\mathbf{r}') = G_0(\mathbf{r}|\mathbf{r}') - G_0(\mathbf{r}|\bar{\mathbf{r}}'), \tag{100}$$

while for a Neumann boundary condition it is in phase,

$$G_N(\mathbf{r}|\mathbf{r}') = G_0(\mathbf{r}|\mathbf{r}') + G_0(\mathbf{r}|\bar{\mathbf{r}}'). \tag{101}$$

From Eqs. (52) and (54) the kernels D and N are thus given by

$$D(\mathbf{r},\mathbf{r}') = -2\frac{\partial}{\partial z}[G_0(\mathbf{r}'|\mathbf{r})] = -\frac{1}{2\pi}\left(\frac{z}{R}\right)\left(ik - \frac{1}{R}\right)\frac{e^{ikR}}{R} \tag{102}$$

and

$$N(\mathbf{r},\mathbf{r}') = -2i\omega\rho_0 G_0(\mathbf{r}'|\mathbf{r}) = -Z_0\frac{ik}{2\pi}\frac{e^{ikR}}{R}, \tag{103}$$

where

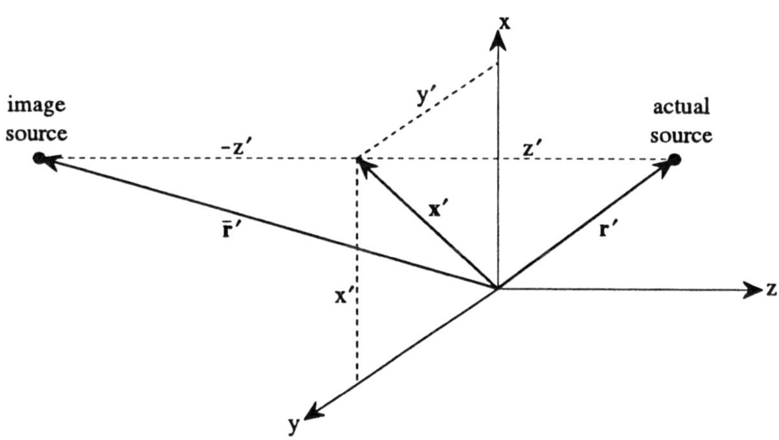

Figure 5. Point source locations for the determination of G for a Dirichlet or Neumann boundary condition imposed over the xy plane.

$$R = \sqrt{(x-x')^2 + (y-y')^2 + z^2} \tag{104}$$

and $Z_0 = \rho_0 c$ is the specific acoustic impedance. Equations (102)-(104) show that the operator **K** is of the form

$$\mathbf{K}u(\mathbf{x}) = \int d\mathbf{x}' \, K(\mathbf{x}-\mathbf{x}'|z) u(\mathbf{x}') = K(\mathbf{x}|z) \otimes u(\mathbf{x}) \tag{105}$$

where the notation has been chosen to emphasize that S_υ can be taken as any constant z plane within the half-space $z > 0$. The operator **K** for plane-to-plane transformation is recognized as a two-dimensional spatial convolution, represented by the symbol ⊗ in Eq. (105).

Since **K** is in the form of a convolution, Eq. (63) could be addressed directly with Fourier transform methods [33 Sec. 9.6]. It is therefore reasonable to assume that suitable singular vectors are given by

$$\upsilon_{\mu\nu}(\mathbf{x}) = u_{\mu\nu}(\mathbf{x}) = \frac{1}{2\pi} e^{i\mu x} e^{i\nu y}, \tag{106}$$

and for this choice it is readily shown that Eq. (75) holds with

$$\sigma_{\mu\nu}(z) = \int d\mathbf{x} \, K(\mathbf{x}|z) e^{-i\mu x} e^{-i\nu y}, \tag{107}$$

provided that this integral exists. From the behavior of the kernels D and N in Eqs. (102) and (103), it is clear that this integral will not converge if either μ or ν is complex. However, Eq. (107) can be evaluated for all real values of μ and ν (excluding certain special values for the Neumann kernel N). The indices on $u_{\mu\nu}$ and $\upsilon_{\mu\nu}$ are therefore real, continuous and in $(-\infty, +\infty)$, and the summation appearing in Eq. (76), and elsewhere in Section III, must be replaced by integration. This is not the only alteration in the equations of Section III, since the $\sigma_{\mu\nu}$ evaluated from Eq. (107) are in general complex. If the singular values are to be real, then it would be necessary to use the magnitude of the $\sigma_{\mu\nu}$ resulting from Eq. (107). The phase would then have to be incorporated into the $\upsilon_{\mu\nu}$, making them dependent on the propagation distance z. However, it is more convenient to let the singular values be complex. It is then also tempting to interpret the $\sigma_{\mu\nu}$ as the eigenvalues of **K**, since U and V are equivalent here and $\upsilon_{\mu\nu} = u_{\mu\nu}$. The difficulty in interpreting the exponentials in Eq. (106) as eigenvectors of **K** is that they are not square-integrable over the infinite xy plane.

In fact, the operator **K** is not compact and the formulation upon which Eqs. (82) and (89) are based is not strictly valid. Even Eq. (35), upon which Eq. (63) is based, is not valid in principle since it was developed only for finite $\overline{S}$. However, when the equations are properly interpreted, all the results of the previous section can be justified, for example by using a separation of variables approach directly on the Helmholtz equation [25 Chapt. 5,6,11].

From the previous discussion it is reasonable to expect that the results for planar NAH will be expressible in terms of the Fourier transform. Indeed, from Eqs. (68) and (106) the coefficients for the expansion in Eq. (76) are given by

$$(u, u_{\mu\nu}) \equiv \tilde{u}(\mu, \nu) = \frac{1}{2\pi}\int d\mathbf{x}\, e^{-i\mu x} e^{-i\nu y} u(\mathbf{x}) = \mathcal{F}\{u(\mathbf{x})|\mu, \nu\}, \tag{108}$$

which is recognized as the two-dimensional Fourier transform of u. The expansion formula itself then becomes

$$u(\mathbf{x}) = \frac{1}{2\pi}\iint d\mu d\nu\, e^{i\mu x} e^{i\nu y} \tilde{u}(\mu, \nu) = \mathcal{F}^{-1}\{\tilde{u}(\mu, \nu)|\mathbf{x}\}, \tag{109}$$

which is just the two-dimensional inverse Fourier transform of $\tilde{u}$. Equation (76) therefore contains the Fourier integral theorem. Of course the assumption has been made that the $u_{\mu\nu}$ are orthonormal as defined, a result which follows from Fourier transform theory in the sense that

$$\frac{1}{(2\pi)^2}\int d\mathbf{x}\, e^{i(\mu-\mu')x} e^{i(\nu-\nu')y} = \delta(\mu-\mu')\delta(\nu-\nu') \tag{110}$$

where δ is again the Dirac distribution. In fact, from Fourier transform theory the exponentials in Eq. (106) with μ and ν in $(-\infty, +\infty)$ form a basis for the expansion of any function square-integrable over the infinite xy plane. Furthermore, the $\sigma_{\mu\nu}$ in Eq. (107) give the spectrum of **K** [33 Sec. 9.6], and thus Eq. (72) provides a spectral decomposition of the operator. In the notation of Eqs. (108) and (109), and using similar results for the expansion of υ, Eqs. (82) and (89) can now be written as

$$\upsilon(\mathbf{x}) = \mathcal{F}^{-1}\{\sigma_{\mu\nu}(z)\tilde{u}(\mu, \nu)|\mathbf{x}\} \tag{111}$$

and

$$u^{\alpha}(\mathbf{x}) = \mathcal{F}^{-1}\left\{W_{\alpha}(|\sigma_{\mu\nu}|)\sigma_{\mu\nu}^{-1}(z)\tilde{v}(\mu,\nu)|\mathbf{x}\right\}. \tag{112}$$

Since $\sigma_{\mu\nu}$ is complex, its magnitude rather than the value itself is used as the argument for the weighting factor. From Eq. (111), forward propagation is accomplished by first Fourier transforming u to obtain $\tilde{u}$, multiplying this result by the singular values $\sigma_{\mu\nu}$ for the appropriate propagation distance z and then inverse transforming to obtain v. Backward propagation follows from Eq. (112) in reverse order: v is first Fourier transformed to obtain $\tilde{v}$, this is then divided by the singular values, filtered for regularization and inverse transformed to obtain u^{α}. It is a distinct advantage that the processing is in terms of the Fourier transform, since the computationally efficient FFT algorithm can be used for discrete measurement data [16].

The formula for the singular values in Eq. (107) is also a Fourier transform, and one that can be evaluated analytically. The evaluation of Eq. (107) follows from the integral representation of the free-space Green's function

$$G_{0}(\mathbf{r}|\mathbf{r}') = \frac{i}{8\pi^{2}}\iint d\mu d\nu \frac{e^{i\kappa(z-z')}}{\kappa} e^{i\mu(x-x')} e^{i\nu(y-y')} \tag{113}$$

for $z > z'$ [37], where

$$\kappa = \begin{cases} \sqrt{k^{2}-\lambda^{2}} & , \ \lambda \le k \\ i\sqrt{\lambda^{2}-k^{2}} & , \ \lambda > k \end{cases} \tag{114}$$

and

$$\lambda = \sqrt{\mu^{2}+\nu^{2}}. \tag{115}$$

From Eqs. (103) and (113), and Eq. (110) or the Fourier inversion theorem, the singular values for the Neumann operator $\mathbf{N}$ are found to be

$$\sigma_{\mu\nu}^{N}(z) = Z_{0}\frac{k}{\kappa}e^{i\kappa z}. \tag{116}$$

By differentiating with respect to z under the integral sign in Eq. (113), which is valid since the integral is uniformly convergent for $z > 0$ with $z' = 0$, the singular values for the Dirichlet operator $\mathbf{D}$ are found to be

$$\sigma_{\mu\nu}^{D}(z) = e^{i\kappa z}. \tag{117}$$

Both sets of singular values are seen to have exponential decay for $\lambda > k$, with the asymptotic forms

$$\sigma_{\mu\nu} \sim e^{-s\sqrt{p^2+q^2}} \begin{cases} -iZ_o \big/ \sqrt{p^2+q^2} & , \; K=N \\ 1 & , \; K=D \end{cases} \tag{118}$$

for $\sqrt{p^2+q^2} >> 1$, where the normalized indices p and q are given by $p = \mu/k$ and $q = \nu/k$, and $s = kz$. Even though **K** is not compact, the singular values are still seen to approach zero for the higher values of the indices μ and ν. A contour plot of $\log|\sigma^D_{\mu\nu}|$ is shown in Figure 6 for $s = 2\pi$. Only positive values of p and q are considered since $\sigma^D_{\mu\nu}$ does not depend on the sign of the indices. In fact, although the two indices μ and ν cannot be combined into a single index n, as used in Section III, Eqs (114)-(117) show that the singular values depend only on λ. This is evident in Figure 6, as is the fact that there is no decay in amplitude, only a change in phase, for $\sqrt{p^2+q^2} < 1$. The transition to an exponential decay occurs at $\sqrt{p^2+q^2} = 1$, which is referred to as the radiation circle.

Since the transition to an exponential decay is well defined, the definition

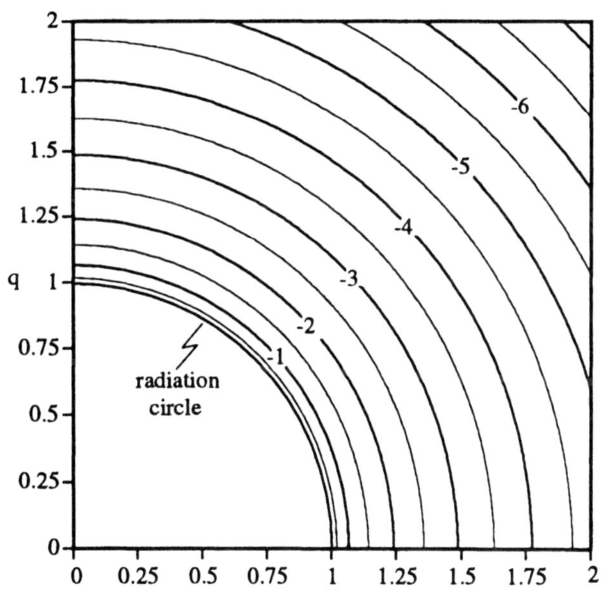

Figure 6. Contour plot of $\log|\sigma^D_{\mu\nu}|$ for planar NAH with $s = 2\pi$.

of an evanescent mode is unambiguous, and coincides with the standard concept of an evanescent wave. Indeed, Eq. (113) is a plane wave expansion of the free-space Green's function, and Eq. (111) yields a plane wave expansion of the pressure field. Each of the plane wave components has the form $\exp(i\mathbf{k}\cdot\mathbf{r})$, where the wave number vector $\mathbf{k} = \mu\mathbf{e}_x + \nu\mathbf{e}_y + \kappa\mathbf{e}_z$ is required by the Helmholtz equation to have a magnitude equal to k. Thus Eq. (114) for κ results, and the plane wave components with $\lambda > k$ are in the form of decaying exponentials for all $z > 0$. At the transition point $\lambda = k$, the plane wave components are of the form

$$p(\mathbf{r}) = e^{ik(x\cos\varphi + y\sin\varphi)}, \tag{119}$$

for which $v_n \equiv 0$ on the surface $z = 0$. For this reason the operator $\mathbf{N}$ is unbounded and the singularity

$$|\sigma_{\mu\nu}^N| \sim \frac{1}{\sqrt{|\lambda - k|}} \tag{120}$$

for $\lambda \to k$ occurs in the associated singular values. The propagation distance for $s = 2\pi$ is only one acoustic wavelength, yet Figure 6 shows that the singular values have already decayed by a factor of more than 100 for $\sqrt{\mathrm{p}^2 + \mathrm{q}^2} = 1.25$. Figure 7 shows plots of $\log|\sigma_{\mu\nu}^D|$ for several other values of s. From Figure 7, and Eq. (118), it would seem that the degree of instability for backward propagation in planar NAH continues to increase with increasing z. This is somewhat misleading, however, since there is no decay of the singular values at all for $\lambda < k$, while the decay for $\lambda > k$ is so rapid once z is beyond several wavelengths that little is lost by further increases in propagation distance. That is, once z is many wavelengths, $\mathbf{K}$ becomes essentially an ideal low pass filter with cutoff frequency $\lambda = k$, as illustrated by the $s \to \infty$ curve in Figure 7. It is therefore important that the measurement surface be within several wavelengths of the source if at least some information concerning the evanescent plane wave components is to be retained. It is not difficult to show that the field from a finite source must contain such evanescent components, since otherwise, as well as being singular, it could also be extended to all of space as an entire field and would therefore have to vanish identically.

The $s \to \infty$ curve in Figure 7 acts as an ideal low pass filter because the plane wave components outside the radiation circle correspond to the higher

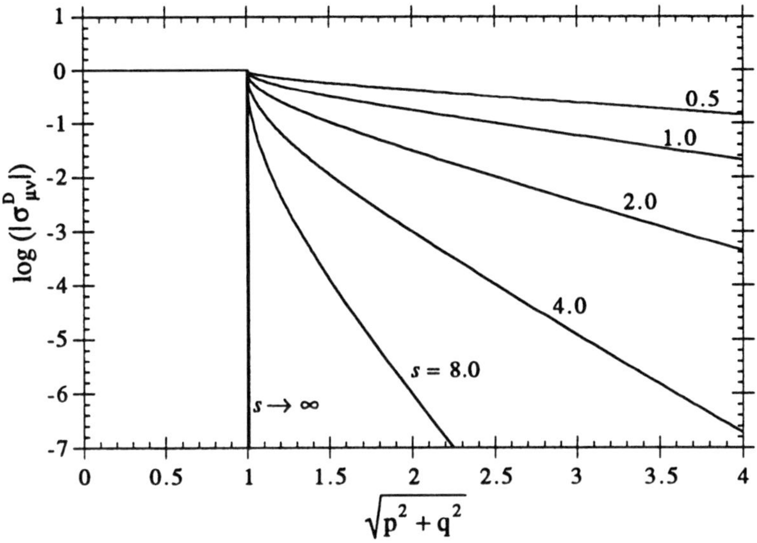

Figure 7. Log$|\sigma^D_{\mu\nu}|$ vs. $\sqrt{p^2+q^2}$ for planar NAH and various values of s, including the limit $s \to \infty$.

spatial frequencies. Forward transformation from plane-to-plane in general behaves as a filter, since it smooths the spatial variations in the data. That is, an abrupt transition in the data at $z=0$, which requires arbitrarily high spatial frequency components, will become smooth on any plane $z>0$. Since regularization must also damp out the higher spatial frequencies, backward propagation with Eq. (112) also behaves like a filter, at least in the absence of data errors. For exact data, the regularized solution u^α may be written as

$$u^\alpha(\mathbf{x}) = \frac{1}{2\pi} w_\alpha(\mathbf{x}) \otimes u(\mathbf{x}) = \frac{1}{2\pi}\int d\mathbf{x}'\, w_\alpha(\mathbf{x}-\mathbf{x}')u(\mathbf{x}'), \tag{121}$$

where u is the exact solution (such that $\mathbf{K}u = v$) and the convolution filter w_α is the inverse Fourier transform of the weighting factor W_α. For spectral truncation, this filter can be evalutated as

$$w_\alpha(\mathbf{x}) = k_\alpha^2 \frac{J_1(k_\alpha|\mathbf{x}|)}{k_\alpha|\mathbf{x}|} \tag{122}$$

where k_α is the cutoff frequency for the value of α. The filter in Eq. (122) has a dominant central peak with a diameter of approximately $2\pi / k_\alpha$, which essentially gives the spatial resolution of spectral truncation. The resolution therefore improves with increasing k_α. However, increasing k_α corresponds to decreasing α, which also decreases the stability of the method. The need for a compromise between resolution/accuracy and stability is thus highlighted once again. Backward propagation within conventional holography may also be put in the form of Eq. (112). In conventional holography, the conjugate image Ψ is formed by forward propagating the complex conjugate of the data,

$$\Psi(\mathbf{x}) = \int d\mathbf{x}' \, K(\mathbf{x}-\mathbf{x}'|z)\, v^*(\mathbf{x}') = K(\mathbf{x}|z) \otimes v^*(\mathbf{x}) . \tag{123}$$

The backward propagated field may be considered to be $\hat{u} = \Psi^*$, yielding

$$\hat{u}(\mathbf{x}) = \mathcal{F}^{-1}\left\{\sigma^*_{\mu\nu}(z)\,\tilde{v}(\mu,\nu)|\mathbf{x}\right\}, \tag{124}$$

which can be put in the form of Eq. (112) with

$$W_\alpha(|\sigma_{\mu\nu}|) = |\sigma_{\mu\nu}|^2 . \tag{125}$$

A superscript α has not been used in Eq. (124) since it is not a regularization method, but a circumflex has been used to indicate approximation. For the Dirichlet operator, $|\sigma_{\mu\nu}|=1$ for $\lambda \leq k$, while for $\lambda > k$ it is essentially negligible once z is several wavelengths or more, as already discussed and shown in Figure 7. For such z, Eq. (125) is thus equivalent to spectral truncation with $k_\alpha = k$, which gives the wavelength resolution limit of conventional holography. This conjugation technique has been extended in the theory of generalized holography to arbitrarily shaped surfaces [10]. Such an approach does not take advantage of evanescent wave information in general, as is possible with NAH.

B. NAH for Cylindrical Surfaces

The transformation of fields between cylindrical surfaces provides another example that can be formulated using Eqs. (82) and (89) even though the resulting **K** is not a compact operator. In this case, simple analytical forms for D and N, such as Eqs. (102) and (103), are not available. However, the method of separation of variables can be used for the circular cylindrical coordinate system

(ρ,ϕ,z) [25 Chapt. 7, 38 Chapt. 16]. The results for the Dirichlet and Neumann Green's functions are given by

$$G_\beta(\mathbf{r}|\mathbf{r}')=\frac{i}{8\pi}\sum_m\int d\lambda C^\beta_{\lambda m}(\rho_<|\rho_s)H_m(\kappa\rho_>)e^{im(\phi'-\phi)}e^{i\lambda(z'-z)} \tag{126}$$

where

$$C^\beta_{\lambda m}(\rho|\rho_s)=\begin{cases}J_m(\kappa\rho)-\{J_m(\kappa\rho_s)/H_m(\kappa\rho_s)\}H_m(\kappa\rho) & ,\ \beta=D\\ J_m(\kappa\rho)-\{J'_m(\kappa\rho_s)/H'_m(\kappa\rho_s)\}H_m(\kappa\rho) & ,\ \beta=N\end{cases} \tag{127}$$

and both the discrete index m and the continuous index λ run from $-\infty$ to $+\infty$. In Eqs. (126) and (127), J_m is the Bessel function of the first kind while H_m is the Hankel function of the first kind, $\rho_>$ is the larger of ρ and ρ' while $\rho_<$ is the lesser of them, and ρ_s is the location of the surface over which the boundary condition holds. By letting $\rho_s\to\rho'$, the Neumann kernel N in Eq. (52) is found to be

$$N(\mathbf{r},\mathbf{r}')=\frac{iZ_0}{\rho'}\sum_m\int d\lambda\,\frac{kH_m(\kappa\rho)}{\kappa H'_m(\kappa\rho')}e^{im(\phi-\phi')}e^{i\lambda(z-z')} \tag{128}$$

while the Dirichlet kernel in Eq. (54) is

$$D(\mathbf{r},\mathbf{r}')=\frac{1}{\rho'}\sum_m\int d\lambda\,\frac{H_m(\kappa\rho)}{H_m(\kappa\rho')}e^{im(\phi-\phi')}e^{i\lambda(z-z')}. \tag{129}$$

Using $dS=\rho d\phi dz=\rho dA$, both Eq. (128) and (129) lead to Eq. (63) with K in the form

$$K(\mathbf{x},\mathbf{x}')=\sum_m\int d\lambda\,\sigma_{\lambda m}(\rho,\rho')u^*_{\lambda m}(\mathbf{x}')v_{\lambda m}(\mathbf{x}). \tag{130}$$

The method of separation of variables is therefore seen to directly yield a spectral decomposition of **K**, from which a singular system can be determined by inspection. As in the planar case, it is convenient to let the singular values be complex, and thus from Eqs. (128) and (129) the singular vectors are taken as

$$u_{\lambda m}(\mathbf{x})=v_{\lambda m}(\mathbf{x})=\frac{1}{2\pi}e^{im\phi}e^{i\lambda z} \tag{131}$$

for cylindrical NAH. The singular values are then

$$\sigma^D_{\lambda m}(\rho,\rho')=\frac{H_m(\kappa\rho)}{H_m(\kappa\rho')} \tag{132}$$

and

$$\sigma^N_{\lambda m}(\rho,\rho')=iZ_0\frac{kH_m(\kappa\rho)}{\kappa H'_m(\kappa\rho')}. \tag{133}$$

The singular vectors $u_{\lambda m}$ as defined in Eq. (131) are orthonormal, since

$$\frac{1}{(2\pi)^2}\int d\mathbf{x}\, e^{i(m-m')\phi}e^{i(\lambda-\lambda')z}=\delta(\lambda-\lambda')\delta_{mm'} \tag{134}$$

where the integration is over $-\infty<z<+\infty$ and $-\pi\le\phi<\pi$.

Equations (82) and (89) for forward and backward propagation can again be interpreted in terms of the Fourier transform, although the finite range of ϕ strictly results in a Fourier series for the expansion with respect to that coordinate. From Eqs. (68) and (131), the coefficients for the expansion in Eq. (76) are given by

$$(u,u_{\lambda m})\equiv\tilde{u}(\lambda,m)=\frac{1}{2\pi}\int d\mathbf{x}\, e^{-im\phi}e^{-i\lambda z}u(\mathbf{x})=\mathcal{F}\{u(\mathbf{x})|\lambda,m\}. \tag{135}$$

The expansion formula itself becomes

$$u(\mathbf{x})=\frac{1}{2\pi}\sum_m\int d\lambda\, e^{im\phi}e^{i\lambda z}\,\tilde{u}(\lambda,m)=\mathcal{F}^{-1}\{\tilde{u}(\lambda,m)|\mathbf{x}\}, \tag{136}$$

where the inverse Fourier transform over a discrete index is to be interpreted as a Fourier series. Using the notation of Eqs. (135) and (136), Eqs. (82) and (89) may be written as

$$v(\mathbf{x})=\mathcal{F}^{-1}\{\sigma_{\lambda m}(\rho,\rho')\tilde{u}(\lambda,m)|\mathbf{x}\} \tag{137}$$

and

$$u^\alpha(\mathbf{x})=\mathcal{F}^{-1}\{W_\alpha(|\sigma_{\lambda m}|)\sigma^{-1}_{\lambda m}(\rho,\rho')\tilde{v}(\lambda,m)|\mathbf{x}\}. \tag{138}$$

Thus the processing for the transformation between cylindrical surfaces is also in terms of the Fourier transform, and the computationally efficient FFT algorithm is again available for discrete measurement data.

The qualitative behavior of the singular values is also very similar to that for planar NAH, although the quantitative details are more complicated. Figure 8 shows a contour plot of $\log|\sigma^D_{\lambda m}|$ for $s = 2\pi$ and $t = 20$. The normalized propagation distance is now given by $s = k(\rho' - \rho)$, and the parameter $t = k\rho'$ is the normalized radius of the surface $S_0 = S_u$. An alternate propagation parameter that will also be used is $\zeta = s/t = (\rho - \rho')/\rho'$. The normalized indices p and q are now given by $\mathrm{p} = m/t$ and $\mathrm{q} = \lambda/k$. Although p is therefore discrete, t has been chosen large enough to enable smooth curves to be drawn in Figure 8, and this will also be true for other plots to follow. Again only positive values of p and q are considered since $\sigma^D_{\lambda m}$ does not depend on the sign of the indices. However, unlike planar NAH, it is clear from Figure 8 that the $\sigma^D_{\lambda m}$ depend not only on $\sqrt{\mathrm{p}^2 + \mathrm{q}^2}$ but also on the relative angle φ between p and q $(\tan\varphi = \mathrm{q}/\mathrm{p})$. This dependence naturally shows up in the asymptotic exponential decay of the singular values,

$$\sigma_{\lambda m} \sim \frac{1}{\sqrt{1+\eta(\varphi)}} e^{-\tau(\varphi)\sqrt{\mathrm{p}^2+\mathrm{q}^2}} \begin{cases} -iZ_0 \Big/ \sqrt{\mathrm{p}^2+\mathrm{q}^2} & , \ K = N \\ 1 & , \ K = D \end{cases} \tag{139}$$

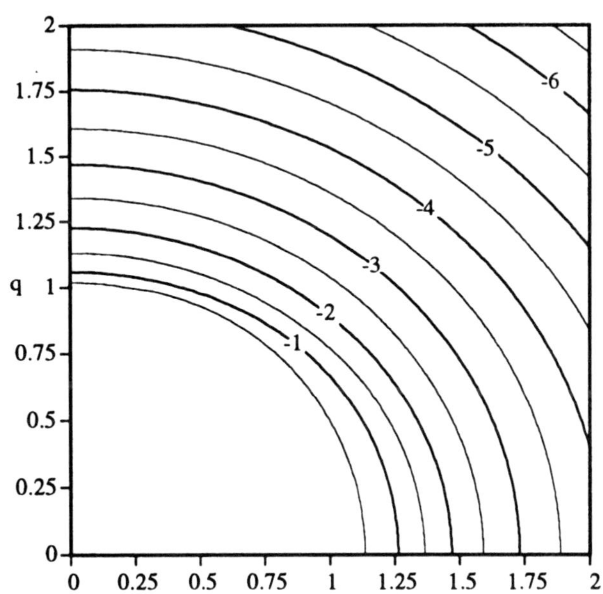

Figure 8. Contour plot of $\log|\sigma^D_{\lambda m}|$ for cylindrical NAH with $s = 2\pi$ and $t = 20$.

for $\sqrt{p^2+q^2} >> 1$. The decay rate τ is given by

$$\frac{\tau(\varphi)}{t} = \eta(\varphi) + \left\{ \ln(1+\zeta) - \ln\left[1 + \frac{\eta(\varphi)}{1+\cos\varphi}\right] \right\} \cos\varphi \tag{140}$$

where

$$\eta(\varphi) = \sqrt{1+\zeta(\zeta+2)\sin^2\varphi} - 1, \tag{141}$$

and is plotted in Figure 9 as a function of ζ for various φ values. The limiting cases in angle are $\tau(90°) = t\zeta = s$, which is the asymptotic decay rate for planar NAH, and $\tau(0°) = t\ln(1+\zeta)$. For any finite ζ value the decay rate varies monotonically with φ between these limiting cases. Only when $\zeta \to 0$, for which $\tau(\varphi) \sim s$, is τ essentially independent of φ.

From Figure 9, the nonsymmetry evident in Figure 8 should become increasingly more pronounced with increasing propagation distance. This is verified by Figures 10-12, which show plots of $\log(\sqrt{1+\zeta}|\sigma^D_{\lambda m}|)/t$ for $\varphi = 90°$, 45° and 0°, respectively, and for various values of ζ. The factor of $\sqrt{1+\zeta}$ is included to account for the geometric decay due to cylindrical spreading, which makes $|\sigma^D_{\lambda m}| \to 0$ for all λ and m as $\zeta \to \infty$, while the division by t makes the

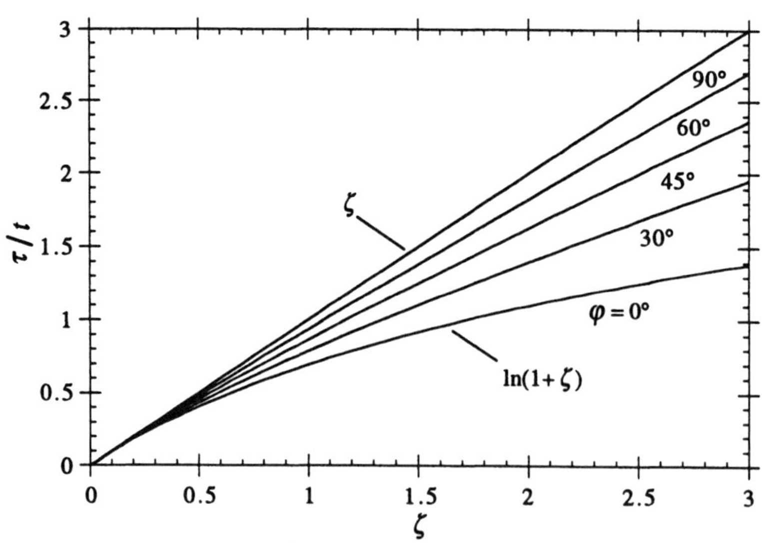

Figure 9. Asymptotic exponential decay rate for cylindrical NAH from Eq. (140).

curves essentially independent of this parameter (although only qualitatively for small t). The vertical line in Figure 10 is correct for the $\zeta \to \infty$ limit, since $\sigma^D_{\lambda m} \to (1+\zeta)^{|m|}$ as $q \to 1$, and thus when $q = 1$ $\sqrt{1+\zeta}|\sigma^D_{\lambda 0}| \to \infty$ for $\zeta \to \infty$. Yet, for any $q > 1$ $|\sigma^D_{\lambda m}| \to 0$ exponentially as $\zeta \to \infty$, giving the indicated sharp cutoff as in planar NAH. Although the results in Figure 10 for finite ζ and $q < 1$ appear to be identically zero, as in planar NAH, this is due to scaling. The exceptional behavior of $\sigma_{\lambda 0}$ near $q = 1$ is linked to the fact that, from Eq. (131), the singular vector for $p = 0$ and $q = 1$ is equivalent to a plane wave traveling along the z axis,

$$p(\mathbf{r}) = e^{ikz}. \tag{142}$$

Equation (142) is by itself a solution of the Helmholtz equation in cylindrical coordinates, and one that does not satisfy the radiation condition, in the sense that it propagates along rather than away from any cylindrical, constant ρ surface. Furthermore, the normal velocity $v_n \equiv 0$ on any such surface. As a result, the operator **N** is again unbounded and the singularity

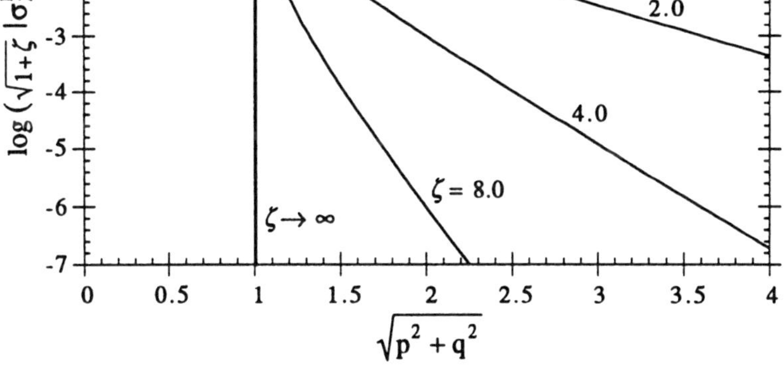

Figure 10. $\text{Log}(\sqrt{1+\zeta}|\sigma^D_{\lambda m}|)/t$ vs. $\sqrt{p^2+q^2}$ for cylindrical NAH with $\varphi = 90°$ and various values of ζ (evaluated for $t = 20$).

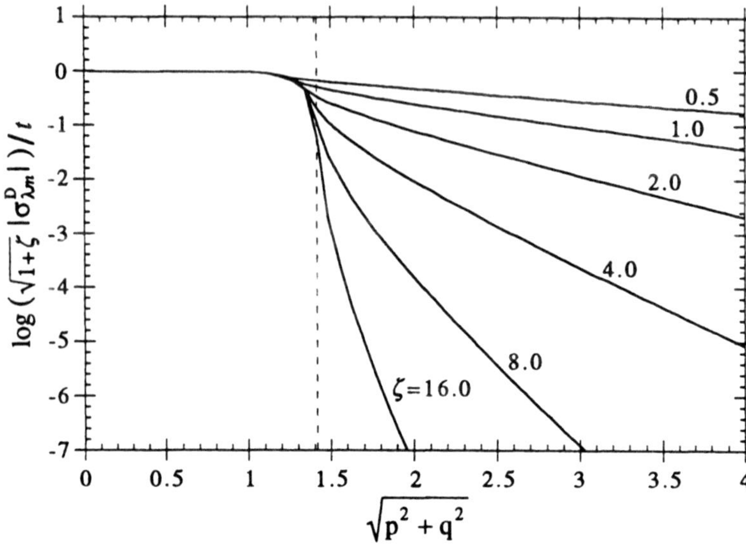

Figure 11. Log$(\sqrt{1+\zeta}|\sigma^{D}_{\lambda m}|)/t$ vs. $\sqrt{p^2+q^2}$ for cylindrical NAH with $\varphi = 45°$ and various values of ζ (evaluated for $t = 20$).

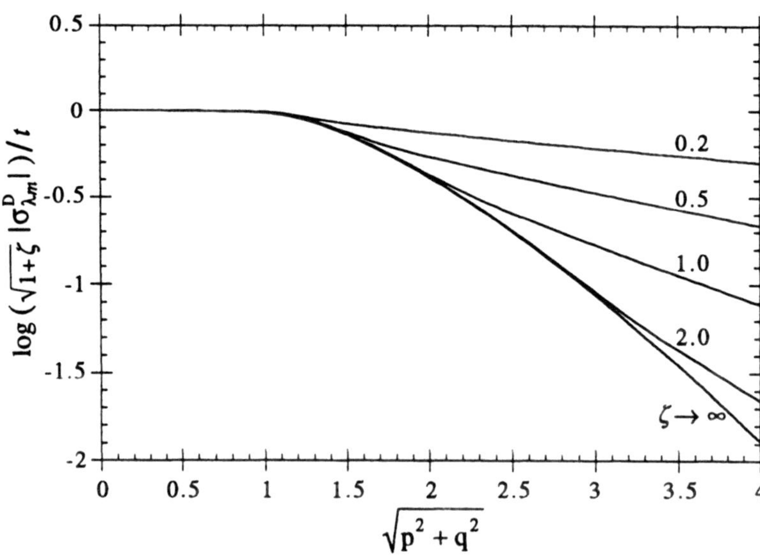

Figure 12. Log$(\sqrt{1+\zeta}|\sigma^{D}_{\lambda m}|)/t$ vs. $\sqrt{p^2+q^2}$ for cylindrical NAH with $\varphi = 0°$ and various values of ζ (evaluated for $t = 20$).

$$|\sigma^N_{\lambda 0}| \sim -\ln|\lambda - k| \tag{143}$$

for $\lambda \to k$ occurs in the associated singular values. From Figure 10, it is evident that for $\varphi = 90°$ all the contours in a plot such as Figure 8 will accumulate at the point $q = 1$ as $\zeta \to \infty$. However, from Figure 12, the contours for $\varphi = 0°$ will reach separate, fixed positions without such an accumulation point (indicating that the modes experience a transition from evanescent wave to nonevanescent wave behavior). These are the limiting cases. For any other φ, of which Figure 11 for $\varphi = 45°$ is an example, the contours will reach fixed positions but with the finite accumulation point $q = 1$, or $\sqrt{p^2 + q^2} = 1/\sin\varphi$. This arises because the $\sigma^D_{\lambda m}$ have spatial exponential decay for all ζ when $q > 1$. The accumulation point for $\varphi = 45°$ is marked by the dashed line in Figure 11.

The accumulation of the contours at $q = 1$ is evident in Figure 13, which shows a contour plot of $\log(\sqrt{1+\zeta}|\sigma^D_{\lambda m}|)$ in the $\zeta \to \infty$ limit. The difficulty in representing the behavior of the singular values near $q = 1$ with a finite spacing between data points is also evident. Clearly there are modes that propagate to the farfield with little or no decay, excluding that due to cylindrical spreading. A radiation circle therefore exists for cylindrical NAH, and to obtain a definition independent of t, propagation to the farfield in the limit $t \to \infty$ is considered,

$$\lim_{\zeta, t \to \infty} (\sqrt{1+\zeta}|\sigma^D_{\lambda m}|) = \begin{cases} \left[\left(1 - p^2 - q^2\right)/\left(1 - q^2\right)\right]^{1/4} & , \sqrt{p^2 - q^2} < 1 \\ 0 & , \sqrt{p^2 - q^2} > 1 \end{cases} \tag{144}$$

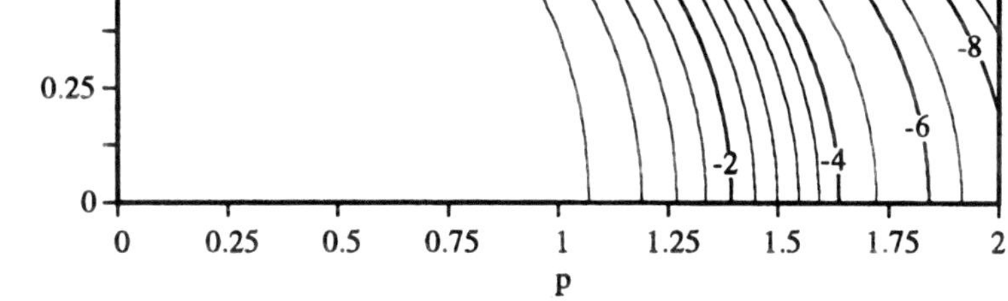

Figure 13. Contour plot of $\log(\sqrt{1+\zeta}|\sigma^D_{\lambda m}|)$ in the farfield ($\zeta \to \infty$) for cylindrical NAH with $t = 20$.

Letting $t \to \infty$ for fixed ρ' corresponds to the high frequency limit. The result in Eq. (144) is plotted in Figure 14 for various φ values, from which it is evident that $\sqrt{p^2+q^2}=1$ may also be defined as the radiation circle for cylindrical NAH. From Eq. (131), the singular vectors that fall outside this radiation circle once again correspond to the higher spatial frequencies, and the behavior of forward and backward propagation with Eqs. (137) and (138) is essentially the same as for planar NAH.

C. NAH for Spherical Surfaces

The separation of variables approach used for cylindrical NAH is also applicable for the transformation of fields between other conformal surfaces. However, in order for the procedure to work, it is necessary that the surfaces correspond to a fixed radial or propagation coordinate in a coordinate system for which the Helmholtz equation is separable. There are a number of such coordinate systems available [25 Chapt. 5]. The simplest example for the transformation of fields between bounded surfaces is provided by the spherical coordinates (r, θ, ϕ). Following the procedure of the previous section, the first

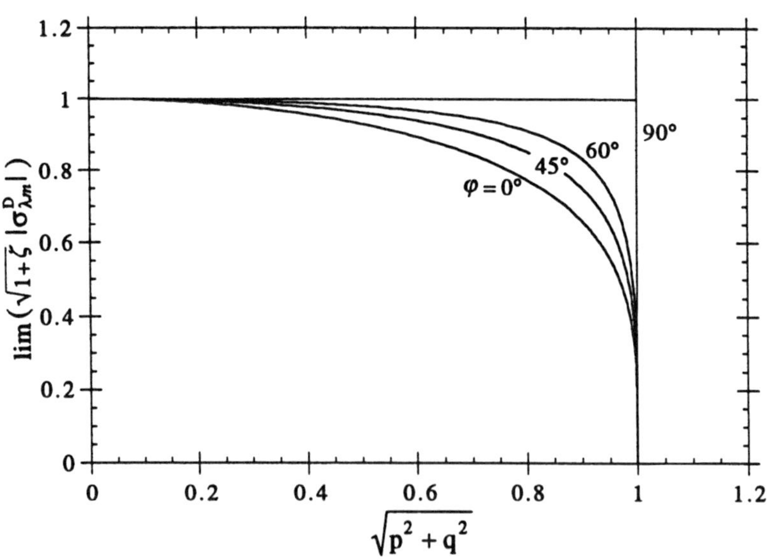

Figure 14. The farfield ($\zeta \to \infty$) and high frequency ($t \to \infty$) limit of $\sqrt{1+\zeta}\,|\sigma_{\lambda m}^{D}|$ for cylindrical NAH and various φ values.

step is to use the method of separation of variables to obtain the Green's function for Dirichlet and Neumann boundary conditions. The result for spherical coordinates is given by

$$G_\beta(\mathbf{r}|\mathbf{r}') = ik\sum_l\sum_m C_{lm}^\beta(r_<|r_s)h_l(kr_>)Y_{lm}(\theta',\phi')Y_{lm}^*(\theta,\phi) \tag{145}$$

where

$$C_{lm}^\beta(r|r_s) = \begin{cases} j_l(kr) - \{j_l(kr_s)/h_l(kr_s)\}h_l(kr) \ , \ \beta = D \\ j_l(kr) - \{j_l'(kr_s)/h_l'(kr_s)\}h_l(kr) \ , \ \beta = N \end{cases} \tag{146}$$

and the indices $l \in [0,\infty)$ and $m \in [-l,+l]$ are both discrete. In Eqs. (145) and (146), j_l is the spherical Bessel function of the first kind while h_l is the spherical Hankel function of the first kind, $r_>$ is the larger of r and r' while $r_<$ is the lesser of them, and r_s is the location of the surface over which the boundary condition holds. Furthermore, the Y_{lm} are the scalar spherical harmonics

$$Y_{lm}(\theta,\phi) = \left[\frac{2l+1}{4\pi}\frac{(l-|m|)!}{(l+|m|)!}\right]^{1/2} P_l^{|m|}(\cos\theta)e^{im\phi} \tag{147}$$

where the P_l^m are the associated Legendre functions. The spherical harmonics are orthonormal over the unit sphere

$$\int d\Omega Y_{l'm'}^*(\theta,\phi)Y_{lm}(\theta,\phi) = \delta_{ll'}\delta_{mm'} \tag{148}$$

where $d\Omega = \sin\theta\, d\theta d\phi$ and the integration is over $0 \le \theta < \pi$ and $-\pi \le \phi < \pi$.

With the Green's function determined, the Neumann and Dirichlet kernels N and D can be evaluated from Eqs. (52) and (54) by letting $r_s \to r'$ in Eqs. (145) and (146). This yields for the Neumann kernel N,

$$N(\mathbf{r},\mathbf{r}') = \frac{iZ_0}{r'^2}\sum_l\sum_m \frac{h_l(kr)}{h_l'(kr')}Y_{lm}(\theta,\phi)Y_{lm}^*(\theta',\phi'), \tag{149}$$

and for the Dirichlet kernel D,

$$D(\mathbf{r},\mathbf{r}') = \frac{1}{r'^2}\sum_l\sum_m \frac{h_l(kr)}{h_l(kr')}Y_{lm}(\theta,\phi)Y_{lm}^*(\theta',\phi'). \tag{150}$$

Using $dS = r^2 d\Omega = r^2 dA$, Eq.(63) follows from both Eq. (149) and (150), and an appropriate singular system can again be selected by inspection. For spherical NAH the singular vectors are thus taken as

$$u_{lm}(\mathbf{x}) = v_{lm}(\mathbf{x}) = Y_{lm}(\theta,\phi). \tag{151}$$

The singular values are then

$$\sigma_{lm}^{D}(r,r') = \frac{h_l(kr)}{h_l(kr')} \tag{152}$$

and

$$\sigma_{lm}^{N}(r,r') = iZ_0 \frac{h_l(kr)}{h_l'(kr')}. \tag{153}$$

Once again it is more convenient to let the singular values be complex, even though this case strictly falls within the formulation under which Eqs. (82) and (89) were developed. It is also more convenient to use the two indices l and m, rather than combining them into a single index n as in Section III. What is different from before is that the singular values and singular vectors are the eigenvalues and eigenvectors of **K**, since U and V are equivalent and **K** is now a compact operator.

For spherical NAH, forward and backward propagation based on Eqs. (82) and (89) is not expressed in terms of a two-dimensional Fourier transform. Instead, from Eqs. (68) and (151) the coefficients for the expansion in Eq. (76) are given by

$$(u, u_{lm}) \equiv \tilde{u}(l,m) = \int d\Omega\, Y_{lm}^{*}(\theta,\phi) u(\mathbf{x}), \tag{154}$$

and the expansion formula itself is

$$u(\mathbf{x}) = \sum_{l}\sum_{m} \tilde{u}(l,m) Y_{lm}(\theta,\phi). \tag{155}$$

For convenience the same notation is used for the coefficients as before, and thus Eqs. (82) and (89) may be written as

$$v(\mathbf{x}) = \sum_{l}\sum_{m} \{\sigma_{lm}(r,r')\tilde{u}(l,m)\} Y_{lm}(\theta,\phi) \tag{156}$$

and

$$u^{\alpha}(\mathbf{x}) = \sum_{l}\sum_{m}\left\{W_{\alpha}(|\sigma_{lm}|)\sigma_{lm}^{-1}(r,r')\tilde{v}(l,m)\right\}Y_{lm}(\theta,\phi). \tag{157}$$

The processing for the transformation between spherical surfaces is thus in terms of the projection onto and summation over the spherical harmonics Y_{lm}. However, from Eq. (147) the expansion with respect to ϕ is still a Fourier series.

Since h_l is simply related to $H_{l+1/2}$, the behavior of the singular values for spherical NAH is qualitatively the same as for cylindrical NAH with $\lambda = 0$ ($\varphi = 0°$). The major difference is that the decay due to cylindrical spreading is now replaced by that for spherical spreading. In fact, $(1+\zeta)\sigma_{lm}^{D}$ for spherical NAH is exactly the same as $\sqrt{1+\zeta}\,\sigma_{0,l+1/2}^{D}$ for cylindrical NAH. Figure 12 and the $\varphi = 0°$ curve in Figure 14 thus hold with $\mathrm{p} = (l+1/2)/t$ ($\mathrm{q} = 0$). The singular values therefore decay rapidly for large l. This is also true for large m, since l must be greater than or equal to $|m|$. It therefore follows that the higher spatial frequencies are again linked to the smaller singular values, since the singular vectors become more oscillatory with increasing l and m. This connection between the decay of the singular values and the increasingly oscillatory behavior of the singular vectors holds in general, and is a result of the smoothing properties of forward propagation. The resolution obtainable for backward propagation is therefore related to the oscillation period of the singular vector associated with the smallest measureable singular value.

D. NAH for General Surfaces

The separation of variables technique works for only a limited number of surface shapes. For general surfaces, numerical techniques must be used to obtain an approximation $\tilde{\mathbf{K}}$ to the operator $\mathbf{K}$. Usually this approximation is of finite rank, and may therefore be represented by a finite dimensional matrix. Both u and v are then finite dimensional column vectors, and Eq. (63) becomes a first kind matrix equation. The singular value decomposition is then applied to the matrix $\tilde{\mathbf{K}}$, and the actual evaluation of Eqs. (82) and (89) is straightforward. Since routines are readily available for computing the singular value decomposition of a matrix, the main interest is therefore on how to obtain $\tilde{\mathbf{K}}$. Although there are a number of approaches, only those that work directly in terms of the actual field variables will be considered here. Indirect methods that use an intermediary such as a surface or volume source distribution will not be described (single and double layer approaches being examples). Even so, the discussion cannot be comprehensive, and thus only a general outline is presented.

The Kirchhoff-Helmholtz integral provides a representation for the acoustic field given both of the boundary values p_{n} and v_{n}. However, since only one of these may be specified as a boundary condition, the other must first be determined in order to evaluate the integral. Colton and Kress discuss in [29] the existence of the Dirichlet to Neumann map,

$$v_{\mathrm{n}}(\mathbf{r}) = \mathbf{Y}p_{\mathrm{n}}(\mathbf{r}) = \int_{S_{\mathrm{o}}} dS(\mathbf{r}')Y(\mathbf{r},\mathbf{r}')p_{\mathrm{n}}(\mathbf{r}'), \tag{158}$$

and its inverse,

$$p_{\mathrm{n}}(\mathbf{r}) = \mathbf{Z}v_{\mathrm{n}}(\mathbf{r}) = \int_{S_{\mathrm{o}}} dS(\mathbf{r}')Z(\mathbf{r},\mathbf{r}')v_{\mathrm{n}}(\mathbf{r}'). \tag{159}$$

Here S_{o} should be interpreted as S_m for forward propagation and as S_r for backward propagation. The integral expressions for **Y** and **Z** in these equations would seem to be a result of Eqs. (51) and (53), with

$$Z(\mathbf{r},\mathbf{r}') = \lim_{\bar{\mathbf{r}}\to\mathbf{r}} N(\bar{\mathbf{r}},\mathbf{r}') \tag{160}$$

and

$$Y(\mathbf{r},\mathbf{r}') = \frac{1}{i\omega\rho_{\mathrm{o}}} \lim_{\bar{\mathbf{r}}\to\mathbf{r}} \mathbf{n}_{\mathrm{o}}(\mathbf{r})\cdot\nabla D(\bar{\mathbf{r}},\mathbf{r}') \tag{161}$$

for $\mathbf{r},\mathbf{r}' \in S_{\mathrm{o}}$ and $\bar{\mathbf{r}} \in V$. Since D and N are Green's functions, the limits in Eqs. (160) and (161) will define functions that are singular for $\mathbf{r} = \mathbf{r}'$. For Eqs. (104), (128) and (149) it can be shown that the kernel Z, generally referred to as the radiation impedance, is weakly singular, and thus the integral form in Eq. (159) is valid when interpreted as an improper integral. The kernel Y will clearly be more singular and must be treated in general as a distribution for a proper interpretation of the integral form in Eq. (158), or **Y** may be represented instead as an integro-differential operator [39 Eq. (A5)]. Combining Eq. (159) with the Kirchhoff-Helmholtz integral (37) for the geometry of Figure 2 yields

$$p(\mathbf{r}) = \mathbf{N}_{\mathrm{o}}v_{\mathrm{n}}(\mathbf{r}) + \mathbf{D}_{\mathrm{o}}\mathbf{Z}v_{\mathrm{n}}(\mathbf{r}) = \mathbf{N}v_{\mathrm{n}}(\mathbf{r}), \tag{162}$$

where the operator $\mathbf{N}_0$ is given by

$$\mathbf{N}_0 v_n(\mathbf{r}) = -i\omega\rho_0 \int_{S_0} dS(\mathbf{r}') G_0(\mathbf{r}'|\mathbf{r}) v_n(\mathbf{r}'), \tag{163}$$

and the operator $\mathbf{D}_0$ by

$$\mathbf{D}_0 p_n(\mathbf{r}) = \int_{S_0} dS(\mathbf{r}') \mathbf{n}_0(\mathbf{r}') \cdot \nabla' G_0(\mathbf{r}'|\mathbf{r}) p_n(\mathbf{r}'). \tag{164}$$

The properties of the operators $\mathbf{N}_0$ and $\mathbf{D}_0$, which correspond respectively to acoustic single and double layer potentials, are well-known [29 Sec. 3.1]. Using Eqs. (162)-(164), an approximation to the operator $\mathbf{N}$ will follow from an appropriate approximation to $\mathbf{Z}$. Although similar results hold for $\mathbf{D}$, only the case for the operator $\mathbf{N}$ will be considered for convenience.

In order to obtain an approximation to $\mathbf{Z}$, it is first necessary to link the pressure and normal derivative on the boundary. One such relation comes from the second part of Green's representation integral in Eq. (33) with G taken as the free-space Green's function G_0. Since G_0 is defined for all $\mathbf{r} \neq \mathbf{r}'$, the result is nontrivial and for the geometry in Figure 2 may be written as

$$\mathbf{D}_0 p_n(\mathbf{r}) + \mathbf{N}_0 v_n(\mathbf{r}) = 0 \tag{165}$$

for $\mathbf{r} \in V_0$ (for the interior problem it holds for $\mathbf{r} \in V$). Equation (165) is often referred to as the extended boundary condition or the extinction theorem. In the null-field method, it is reduced to a system of equations yielding a generalized moment problem. These equations are uniquely solvable for p_n given v_n or vice versa [40,41]. Another common approach is to take the limit of the Kirchhoff-Helmholtz integral as $\mathbf{r} \in V$ approaches a point on the surface S_0, or equivalently to consider Eq. (165) in a similar limit. Using the properties of the single and double layer operators $\mathbf{N}_0$ and $\mathbf{D}_0$, the result in either case is the boundary integral equation

$$p_n(\mathbf{r}) = 2\mathbf{D}_0 p_n(\mathbf{r}) + 2\mathbf{N}_0 v_n(\mathbf{r}) \tag{166}$$

for $\mathbf{r} \in S_0$. A distinction between the operators $\mathbf{N}_0$ and $\mathbf{D}_0$ on and off the surface S_0 has not been made, even though the kernels are singular for Eq. (166) but not for Eq. (162) or (165).

A number of methods have been developed for obtaining approximate solutions to integral equations such as (166) [31]. Many of these methods are based on representing the boundary values as

$$p_{\mathrm{n}}(\mathbf{r}) = \sum_i c_i f_i(\mathbf{r}) = \sum_i f_i(\mathbf{r}) \mathbf{F}_i\{p_{\mathrm{n}}\} \tag{167}$$

and

$$\upsilon_{\mathrm{n}}(\mathbf{r}) = \sum_i d_i h_i(\mathbf{r}) = \sum_i h_i(\mathbf{r}) \mathbf{H}_i\{\upsilon_{\mathrm{n}}\}. \tag{168}$$

Even if the functions $\{f_i\}$ and $\{h_i\}$ are complete in $C(S_o)$, Eqs. (167) and (168) are actually approximations since the summations are in general finite. The $\mathbf{F}_i$ and $\mathbf{H}_i$ are linear functionals which yield the correct coefficients c_i and d_i for the chosen representations. For example, in interpolation methods $\mathbf{F}_i\{p_{\mathrm{n}}\} = p_{\mathrm{n}}(\mathbf{r}_i)$ where the $\mathbf{r}_i$ are the nodal points. The f_i are then global interpolation functions such that $f_i(\mathbf{r}_j) = \delta_{ij}$. Often the interpolation is considered only locally in the actual implementation, as in the boundary element method (BEM). In the BEM the surface is partitioned into a number of surface elements over which the boundary values are approximated by local shape functions [42]. These shape functions deal only with the nodal points that are on the surface element to which they apply. Although it is not done in practice, the approximation of the boundary values could be written as in Eqs. (167) and (168). If Eqs. (167) and (168) are substituted into the right hand side of Eq. (166) and the linear functional $\mathbf{F}_i$ is then applied to both sides, the result may be written as

$$c_i = \mathbf{F}_i\{p_{\mathrm{n}}\} = \sum_j A_{ij} c_j + \sum_j B_{ij} d_j \tag{169}$$

where

$$A_{ij} = 2\mathbf{F}_i\{\mathbf{D}_o f_j\} \text{ and } B_{ij} = 2\mathbf{F}_i\{\mathbf{N}_o h_j\}. \tag{170}$$

The elements of the global matrices **A** and **B** need not be evaluated directly from Eq. (170). For example, in the BEM they are assembled from similar local matrices defined in terms of the shape functions and evaluated for each surface element.

Equation (169) can presumably be solved for the coefficients c_i. This will yield values for the c_i such that the approximation of both sides of Eq. (166) via Eq. (167) are consistent. The procedure does not yield a boundary value p_n such that Eq. (166) is satisfied for all $\mathbf{r}$. How well the boundary integral equation is satisfied on a point by point basis will depend on the quality of the approximation via Eq. (167). The weighted residual procedures provide alternate approaches to obtaining a system of equations for evaluating the c_i [43]. With the residual defined as

$$E(\mathbf{r}) = p_n(\mathbf{r}) - 2\mathbf{D}_o p_n(\mathbf{r}) - 2\mathbf{N}_o v_n(\mathbf{r}) = \sum_j c_j \left[f_j - 2\mathbf{D}_o f_j \right] - 2\sum_j d_j \mathbf{N}_o h_j, \quad (171)$$

the collocation procedure forces $E(\mathbf{r}_n) = 0$ at enough collocation points $\mathbf{r}_n$ for a solution. If these collocation points are the nodal points for an interpolation, then the result is the same as Eq. (169). In the Galerkin procedure, the residual is forced to be orthogonal to the subspace spanned by the f_i, or

$$\int_{S_o} dS(\mathbf{r}) E(\mathbf{r}) f_i(\mathbf{r}) = 0 \quad (172)$$

for all i. The least-squares procedure minimizes the integrated, squared error over the surface, or

$$\frac{\partial}{\partial c_i^*} \int_{S_o} dS(\mathbf{r}) |E(\mathbf{r})|^2 = 0 \quad (173)$$

for all i. Both the least-squares and the Galerkin procedures require the evaluation of a double integration over S_o. Since the collocation procedure requires only a single integration, it is more commonly used for integral equations such as (166).

The boundary integral equation (166) was derived from the representation integral, not directly from the boundary condition. It does not therefore guarantee that the boundary condition is satisfied. In fact, the boundary integral equation suffers from the existence of fictitious eigenfrequencies. At these frequencies the boundary integral equation does not have a unique solution, even though a unique solution does exist for the boundary value problem. This is discussed by Kleinman and Roach [44] who show that a compatibility equation can be used to ensure uniqueness. It is also possible to make use of the extinction theorem (165) [45], or to use a modified Green's function in place of

G_0 [22 Theorem 3.35 and 3.36]. Provided one of these techniques is used to ensure a unique solution, then the coefficients c_i may be written in terms of the d_i as

$$c_i = \sum_j \tilde{Z}_{ij} d_j . \tag{174}$$

Using Eq. (174) in Eqs. (167) and (168) yields the finite rank approximation

$$\tilde{\mathbf{Z}} v_n(\mathbf{r}) = \sum_i \sum_j \tilde{Z}_{ij} f_i(\mathbf{r}) \mathbf{H}_j \{ v_n \} \tag{175}$$

for the operator $\mathbf{Z}$. The use of Eq. (175) in Eq. (162) then results in the $\mathbf{D}_0$ term being of finite rank. The $\mathbf{N}_0$ term is also of finite rank if Eq. (168) is used, and together this yields the finite rank approximation

$$\tilde{\mathbf{N}} v_n = \sum_j \tilde{N}_j(\mathbf{r}) \mathbf{H}_j \{ v_n \} \tag{176}$$

for the operator $\mathbf{N}$, where

$$\tilde{\mathrm{N}}_j(\mathbf{r}) = \mathrm{N}_0 \mathrm{h}_j(\mathbf{r}) + \sum_i \mathbf{D}_0 \mathrm{f}_i(\mathbf{r}) \tilde{\mathrm{Z}}_{ij} . \tag{177}$$

Equation (176) is the required approximation, and it can be used directly for forward propagation if desired.

More commonly, Eq. (176) is converted into a matrix equation by setting it equal to the pressure at the locations for which it is either desired (for forward propagation) or measured (for backward propagation). This procedure results in the system of equations

$$p(\mathbf{r}_i) = \sum_j \tilde{N}_j(\mathbf{r}_i) d_j = \sum_j \tilde{N}_{ij} d_j , \tag{178}$$

and in this case the matrix $\tilde{\mathbf{N}}$, with elements $\tilde{N}_{ij}$, need not be square. In fact, because of the presence of noise, it may be desirable for backward propagation to have the number of measurement locations $\mathbf{r}_i$ larger than the number of coefficients d_i sought. This does not cause any difficulties for the singular value decomposition formulation of NAH. Since $\tilde{\mathbf{N}}$ is of finite rank, perhaps a more correct approach for backward propagation would be to use the Gramm-Schmidt orthogonalization procedure on the functions $\tilde{N}_j(\mathbf{r})$ to obtain an orthonormal basis spanning the range of $\tilde{\mathbf{N}}$. This basis could then be used to expand p over the measurement surface, and the matrix elements relating these coefficients to the d_i could be determined from Eq. (176). Of course this requires more computational effort than the simple point evaluation approach. In any case, a matrix approximation is obtained for the operator **N**. Similar techniques can be used for the Dirichlet operator **D** and the more general operator **K**.

V. DISCUSSION

The development of NAH as presented here, although complete with regard to the analytical formulation, discussed only briefly, or omitted entirely, a number of important implementation aspects. For example, only two methods of regularization were discussed, that of spectral truncation and Tikhonov regularization, while strategies for selecting an appropriate, preferably optimal, value of the regularization parameter were completely neglected. Clearly the success of backward propagation in any implementation will depend critically on the choice of both the regularization method and the associated regularization parameter, the aim being to retain as much of the evanescent information as possible without amplifying the noise level. The second major area not discussed involves the measurement aspects of sampling and windowing. Sampling refers to the measurement of the data at a set of discrete points, with the location and spacing selected to ensure an adequate representation of the information content. For a planar measurement surface the details can be worked out explicitly based on the sampling theorem and the highest spatial frequency present in the data [8 Chapt. 5]. Windowing refers to the measurement of the data over a finite segment or aperture of the full measurement surface, so that only partial information is retained. This does not strictly include situations where the data over the remaining part of the measurement surface is known to be negligible. The finite aperture problem, that of forward or backward propagating from measurements over an open surface, is ill-posed. The direct approach to overcome this is to add appropriate zero data points to the actual measured data in order to fill out or close the measurement surface. Nearfield acoustic holography can then be applied to the expanded data set, and if the distance of propagation from the measurement surface is small then it may be reasonable to expect that the error incurred will also be small. Such a procedure has recently been applied with reasonable success [46].

Nearfield acoustic holography is based on an exact approach to the problems of direct and inverse diffraction. The method utilizes the singular value decomposition of the forward propagator **K**, an operator representing the exact solution to direct diffraction. For general surface shapes it is usually possible to obtain only an approximation to **K**. This is a numerical approximation and differs from the asymptotic approximations to direct diffraction used in Fresnel and Fourier optics. In those theories, **K** is replaced by a new operator that is strictly equivalent only in appropriate asymptotic situations, such as paraxial, farfield or high frequency propagation. This may be sufficient for forward propagation, but is generally not a satisfactory method upon which to base backward propagation [47]. Although NAH attempts to deal with inverse diffraction in an exact manner, the problem is ill-posed and requires regularization. In practice, backward propagation in NAH is therefore an approximation, even in a strictly analytical formulation. However, it can still provide enhanced resolution over direct diffraction imaging, as extended to arbitrary surfaces in the theory of generalized holography [10], since at least some of the evanescent wave information can be correctly included in the reconstruction [15]. This does require the use of a priori information concerning the field source, at least to the extent that the space between the measurement and reconstruction surface should strictly be free of sources. Generalized holography, on the other hand, can be applied without any concern for the size and location of the field source. With regularization, it is possible to back propagate through the source in NAH. However, the effect this would have on the reconstruction is unclear and would certainly depend on the regularization method, as would the possibility of detecting it. The full potential of NAH is therefore best exploited when the source volume is known a priori, such as in reconstructing the surface motion of a vibrating body.

VI. REFERENCES

[1] D. Gabor, "A new microscope principle," *Nature* **161**, 777 (1948).

[2] D. Gabor, "Microscopy by reconstructed wave front," *Proc. Roy. Soc.* **A197**, 454 (1949).

[3] R. Mittra and P. L. Ransom, "Imaging with coherent fields," in *Proceedings of the Symposium on Modern Optics,* Microwave Research Institute Symposia Series, Vol. 17, Polytechnic Press, Polytechnic Institute of Brooklyn, New York (1967).

[4] E. N. Leith and J. Upatnieks, "Reconstructed wave fronts and communication theory," *J. Opt. Soc. Am.* **52**, 1123 (1962).

[5] E. N. Leith and J. Upatnieks, "Wavefront reconstructions with continuous-tone objects," *J. Opt. Soc. Am.* **53**, 1377 (1963).

[6] R. K. Mueller, "Acoustical holography survey," in *Advances In Holography*, Vol. 1 (N. H. Farhat, ed.), Marcel Dekker, New York (1975).

[7] J. R. Shewell and E. Wolf, "Inverse diffraction and a new reciprocity theorem," *J. Opt. Soc. Am.* **58**, 1596 (1968).

[8] B. P. Hildebrand and B. B. Brenden, *An Introduction To Acoustical Holography*, Plenum Press, New York (1972).

[9] J. W. Goodman, "Digital image formation from detected holographic data," in *Acoustical Holography*, Vol. 1 (A. F. Metheral, et. al., ed.), Plenum Press, New York (1969).

[10] R. P. Porter, "Generalized holography with application to inverse scattering and inverse source problems," in *Progress In Optics*, Vol. 27 (E. Wolf, ed.), North-Holland, Amsterdam (1989).

[11] M. M. Sondhi, "Reconstruction of objects from their sound-diffraction patterns," *J. Acoust. Soc. Am.* **46**, 1158 (1969).

[12] A. L. Boyer, et. al., "Reconstruction of ultrasonic images by backward propagation," in *Acoustical Holography*, Vol. 3 (A. F. Metheral, ed.), Plenum Press, New York (1971).

[13] P. R. Stepanishen and K. C. Benjamin, "Forward and backward projection of acoustic fields using FFT methods," *J. Acoust. Soc. Am.* **71**, 803 (1982).

[14] E. G. Williams and J. D. Maynard, "Numerical evaluation of the Rayleigh integral for planar radiators using the FFT," *J. Acoust. Soc. Am.* **72**, 2020 (1982).

[15] E. G. Williams and J. D. Maynard, "Holographic imaging without the wavelength resolution limit," *Phys. Rev. Lett.* **45**, 554 (1980).

[16] J. D. Maynard, et. al., "Nearfield acoustic holography: I. Theory of generalized holography and the development of NAH," *J. Acoust. Soc. Am.* **78**, 1395 (1985).

[17] E. G. Williams, et. al., "Generalized nearfield acoustic holography for cylindrical geometry: Theory and experiment," *J. Acoust. Soc. Am.* **81**, 389 (1987).

[18] W. A. Veronesi and J. D. Maynard "Digital holographic reconstruction of sources with arbitrarily shaped surfaces," *J. Acoust. Soc. Am.* **85**, 588 (1989).

[19] G. V. Borgiotti, et. al., "Conformal generalized nearfield acoustic holography for axisymmetric geometries," *J. Acoust. Soc. Am.* **88**, 199 (1990).

[20] R. F. Millar, "The Rayleigh hypothesis and a related least-squares solution to scattering problems for periodic surfaces and other scatterers," *Radio Sci.* **8**, 785 (1973).

[21] D. S. Jones, *Acoustic and Electromagnetic Waves*, Clarendon Press, Oxford (1986).

[22] D. Coltan and R. Kress, *Integral Equation Methods in Scattering Theory*, Wiley, New York (1983).

[23] P. Grisvard, *Elliptic Problems in Nonsmooth Domains*, Pitman, Boston (1985).

[24] C. H. Wilcox, "A generalization of theorems of Rellich and Atkinson," *Proc. Amer. Math. Soc.* **7**, 271-276 (1956).

[25] P. M. Morse and H. Feshbach, *Methods of Theoretical Physics*, McGraw-Hill, New York (1953).

[26] I. Stakgold, *Boundary Value Problems of Mathematical Physics*, Macmillan, New York, Vol. II, Chapt. 5 (1968).

[27] W. Rudin, *Functional Analysis*, McGraw-Hill, New York, Theorem 8.12, p. 219 (1991).

[28] W. Kaplan, *Advanced Calculus*, Addison-Wesley, Reading, MA, Chapt. 6 (1984).

[29] D. Coltan and R. Kress, *Inverse Acoustic and Electromagnetic Scattering Theory*, Springer-Verlag, New York (1992).

[30] P. R. Garabedian, *Partial Differential Equations*, Wiley, New York (1964).

[31] R. Kress, *Linear Integral Equations* Springer-Verlag, New York (1989).

[32] J. Baumeister, *Stable solution of inverse problems*, Friedr. Vieweg, Braunschweig (1986).

[33] D. Porter and D. S. G. Stirling, *Integral Equations: a practical treatment, from spectral theory to applications*, Cambridge University Press, Cambridge (1990).

[34] G. V. Borgiotti, "The power radiated by a vibrating body in an acoustic fluid and its determination from boundary measurements," *J. Acoust. Soc. Am.* **88**, 1884 (1990).

[35] G. V. Borgiotti and K. E. Jones, "The determination of the acoustic far field of a radiating body in an acoustic fluid from boundary measurements," *J. Acoust. Soc. Am.* **93**, 2788 (1993).

[36] A. N. Tikhonov and V. Y. Arsenin, *Solutions of Ill-Posed Problems*, Winston, Washington (1977).

[37] A. Baños, *Dipole Radiation in the Presence of a Conducting Half-Space*, Pergamon Press, New York, Eq. (2.19), p.18 (1966).

[38] G. Arfken, *Mathematical Methods for Physicists*, Academic Press, New York (1985).

[39] E. G. Williams, "Numerical evaluation of the radiation from unbaffled, finite plates using the FFT," *J. Acoust. Soc. Am.* **74**, 343 (1983).

[40] P. A. Martin, "On the null-field equations for the exterior problems of acoustics," *Q. J. Mech. Appl. Math.* **33**, 385-396 (1980).

[41] D. Coltan and R. Kress, "The unique solvability of the null-field equations of acoustics," *Q. J. Mech. Appl. Math.* **36**, 87-95 (1983).

[42] T. W. Wu, et. al., "An efficient boundary element algorithm for multi-frequency acoustical analysis," *J. Acoust. Soc. Am.* **94**, 447 (1993).

[43] K.-J. Bathe, *Finite Element Procedures in Engineering Analysis*, Prentice-Hall, Englewood Cliffs, NJ, Section 3.3.3 (1982).

[44] R. E. Kleinman and G. F. Roach, "Boundary integral equations for the three-dimensional Helmholtz equation," *SIAM Review* **16**, 214-236 (1974).

[45] H. A. Schenck, "Improved integral formulation for acoustic radiation problems," *J. Acoust. Soc. Am.* **44**, 41 (1968).

[46] A. Sarkissian, et. al., "Reconstruction of the acoustic field over a limited surface area on a vibrating cylinder," *J. Acoust. Soc. Am.* **93**, 48 (1993).

[47] G. Crosta, "On approximations of Helmholtz equation in the the halfspace: their relevance to inverse diffraction," *Wave Motion* **6**, 237 (1984).

A Design Technique for 2-D Linear Phase Frequency Sampling Filters with Fourfold Symmetry

Peter A. Stubberud
University of Nevada, Las Vegas

Abstract

In this chapter, system functions are developed for two dimensional (2-D) frequency sampling filters that have real impulse responses and linear phase and for 2-D frequency sampling filters that have real impulse responses, linear phase and fourfold symmetry. Under certain conditions, these frequency sampling filters can implement narrowband 2-D linear phase filters and narrowband 2-D linear phase filters with fourfold symmetry much more efficiently than direct convolution implementations. Also, a technique for determining optimal frequency sampling filter coefficients is developed for frequency sampling filters that have real impulse responses, linear phase and fourfold symmetry. This technique approximates a desired frequency response by minimizing a weighted mean square error over the passbands and stopbands subject to constraints on the filter's amplitude response.

I. Introduction

Some two dimensional (2-D) signal processing systems, including image processing systems, require linear phase or zero phase filters. A 2-D linear phase or zero phase filter implemented by direct convolution uses the filter's impulse response as coefficients. If a 2-D linear phase filter has a region of support, R_N, where

CONTROL AND DYNAMIC SYSTEMS, VOL. 77

$$R_N = \{(n_1, n_2) : 0 \le n_1 \le N_1 - 1, 0 \le n_2 \le N_2 - 1\},$$

and N_1, N_2, n_1 and n_2 are members of the set of integers ($N_1, N_2, n_1, n_2 \in I$), then the filter's impulse response has the form

$$h(n_1, n_2) = h(N_1 - 1 - n_1, N_2 - 1 - n_2)$$

[1]. If the filter is implemented using direct convolution, approximately $N_1N_2/2$ multiplies are required to compute each output sample. To reduce the number of multiplies required by a direct convolution implementation of a linear phase filter, fourfold symmetry conditions are often imposed on linear phase filters. If a 2-D fourfold symmetric linear phase filter has support over the region, R_N, then the filter's impulse response has the form

$$\begin{aligned} h(n_1, n_2) &= h(n_1, N_2 - 1 - n_2) \\ &= h(N_1 - 1 - n_1, n_2) \\ &= h(N_1 - 1 - n_1, N_2 - 1 - n_2) \end{aligned}$$

[2], and a direct convolution implementation of the filter requires approximately $N_1N_2/4$ multiplies per output sample.

Unlike direct convolution implementations which use the filter's impulse response as coefficients in the filter's implementation, frequency sampling filters use frequency samples, which are specific values from the filter's frequency response, as coefficients in the filter's implementation. The frequency sampling filters discussed in this chapter interpolate a frequency response through a set of (N_1N_2) frequency samples. When using a frequency sampling filter to implement a frequency selective filter, the frequency samples that lie in the filter's stopbands can be set to zero and do not have to be implemented. Thus, only the non-zero frequency samples which lie in the filter's passbands and transition bands are required in the filter's implementation.

As a filter's passbands or transition bands narrow or its stopband specifications become more stringent, the filter's region of support increases which implies that the values of N_1 and N_2 increase. Because the number of multiplies per output sample of a direct convolution filter is approximately proportional to (N_1N_2), even a small increase in the filter's specifications can substantially increase a direct convolution filter's computational requirements. However, as a frequency sampling filter's passbands and transition bands narrow and its stopbands increase, the number of nonzero frequency samples decrease; and thus, the number of coefficients required in the filter's implementation decreases. Thus, frequency sampling filters have the potential to implement narrowband frequency selective filters much more efficiently than direct convolution filters.

In this chapter, the 2-D Type 1-1 frequency sampling filter system function described in reference [3] is further developed for frequency sampling filters that have real impulse responses and linear phase and for frequency sampling filters that have real impulse responses, linear phase and fourfold symmetry. This chapter also introduces three new types, Type 1-2, Type 2-1 and Type 2-2, of frequency sampling filter system functions. These system functions are also developed for frequency sampling filters that have real impulse responses and linear phase and for frequency sampling filters that have real impulse responses, linear phase and fourfold symmetry. The resulting system functions for frequency sampling filters that have a real impulse responses and linear phase interpolate a frequency response through a set of approximately $N_1N_2/2$ independent samples, and the resulting system functions for frequency sampling filters that have a real impulse responses, linear phase and fourfold symmetry interpolate a frequency response through a set of approximately $N_1N_2/4$ independent samples. As a result these system functions are computationally more efficient for frequency sampling filters that have real impulse responses, linear phase and for frequency sampling filters that have real impulse responses, linear phase and fourfold symmetry than the system function described in reference [3]. Although frequency sampling filters interpolate a frequency response through a set of frequency samples, the frequency response may not be well behaved between samples. References [3; 4] describe two of the design methods currently used to control the interpolation errors between frequency samples and approximate a desired frequency response. In this chapter, a design technique is developed that controls the interpolation errors and approximates a desired frequency response by minimizing a weighted mean square error over the passbands and stopbands subject to constraints on the filter's amplitude response. This results in a constrained optimization problem which can be solved by using the Lagrange multiplier optimization method.

II. 2-D Frequency Sampling Filters

A. Type 1-1 Frequency Sampling Filters

Consider a filter that has an impulse response $h(\mathbf{n})$ where $\mathbf{n} = (n_1,n_2)$, a region of support R_N where

$$R_N = \{(n_1,n_2): 0 \le n_1 \le N_1 - 1, 0 \le n_2 \le N_2 - 1\} \quad n_1,n_2 \in I,$$

and a frequency response, $H_\omega(\boldsymbol{\omega})$, where

$$\boldsymbol{\omega} = (\omega_1,\omega_2) \in R_\Omega = \{(\omega_1,\omega_2): 0 \le \omega_1 < 2\pi, 0 \le \omega_2 < 2\pi\}$$

ω_1, ω_2 $\in$ the set of real numbers ($\Re$). Let $H_k(\mathbf{k})$ where

$$\mathbf{k} = (k_1,k_2) \in R_K = \{(k_1,k_2): 0 \le k_1 \le N_1 - 1, 0 \le k_2 \le N_2 - 1\} \quad k_1,k_2 \in I$$

represent a set of (N_1N_2) frequency samples taken from the filter's frequency response, $H_\omega(\boldsymbol{\omega})$ for $\boldsymbol{\omega} \in R_\Omega$. For a Type 1-1 frequency sampling filter, the set, $H_k(\mathbf{k})$ for $\mathbf{k} \in R_K$, of frequency samples are selected such that

$$H_k(\mathbf{k}) = H_\omega(\omega_1,\omega_2)\Big|_{\omega_1 = \frac{2\pi}{N_1}k_1, \omega_2 = \frac{2\pi}{N_2}k_2}$$

where $H_k(\mathbf{k})$ can be written as

$$H_k(\mathbf{k}) = |H_k(\mathbf{k})| e^{j\theta(\mathbf{k})}$$

where $\mathbf{k} \in R_K$, $j = \sqrt{-1}$ and $\theta(\mathbf{k})$ is the phase of $H_k(\mathbf{k})$. Thus, for a Type 1-1 frequency sampling filter, the set, $H_k(\mathbf{k})$ for $\mathbf{k} \in R_K$, of frequency samples corresponds to the Discrete Fourier Transform (DFT) coefficients. Also, the impulse response, $h(\mathbf{n})$ for $\mathbf{n} \in R_N$, which interpolates a frequency response through the set of frequency samples, $H_k(\mathbf{k})$ for $\mathbf{k} \in R_K$, can be determined from the inverse discrete Fourier transform (IDFT),

$$h(\mathbf{n}) = \frac{1}{N_1N_2} \sum_{\mathbf{k} \in R_K} H_k(\mathbf{k}) e^{j(2\pi/N_1)n_1k_1} e^{j(2\pi/N_2)n_2k_2}, \tag{1}$$

and a set, $H_k(\mathbf{k})$ for $\mathbf{k} \in R_K$, of frequency samples can be determined from the filter's impulse response by the discrete Fourier transform (DFT),

$$H_k(\mathbf{k}) = \sum_{\mathbf{n} \in R_N} h(\mathbf{n}) e^{-j(2\pi / N_1) n_1 k_1} e^{-j(2\pi / N_2) n_2 k_2} .$$

If we let $H(z_1,z_2)$ represent the z transform of $h(\mathbf{n})$, then the system function, $H(z_1,z_2)$, of the filter that interpolates a frequency response through the set, $H_k(\mathbf{k})$ for $\mathbf{k} \in R_K$, of frequency samples is

$$H(z_1, z_2) = \sum_{n_1=0}^{N_1-1} \sum_{n_2=0}^{N_2-1} h(n_1, n_2) z_1^{-n_1} z_2^{-n_2} . \tag{2}$$

By substituting Equation (1) into Equation (2), interchanging the order of the summations and performing the summation over the n_1 and n_2 indices, $H(z_1,z_2)$ can be written as

$$H(z_1, z_2) = \frac{1 - z_1^{-N_1}}{N_1} \frac{1 - z_2^{-N_2}}{N_2} \times$$
$$\sum_{k_1=0}^{N_1-1} \sum_{k_2=0}^{N_2-1} \frac{H_k(k_1, k_2)}{\left(1 - e^{j(2\pi / N_1) k_1} z_1^{-1}\right)\left(1 - e^{j(2\pi / N_2) k_2} z_2^{-1}\right)} \tag{3}$$

Equation (3), originally developed in [3], describes the general form of a Type 1-1 frequency sampling filter, and has the form of an interpolation formula. The complex function, $H(z_1,z_2)$, interpolates a polynomial through the frequency samples in the set, $H_k(\mathbf{k})$ for $\mathbf{k} \in R_K$, so that

$$H(z_1, z_2)\big|_{z_1 = e^{j(2\pi / N_1) k_1}, z_2 = e^{j(2\pi / N_2) k_2}} = H_k(k_1, k_2) .$$

Thus, as desired, the frequency sampling filter's frequency response passes through the $(N_1 N_2)$ frequency samples in the set $H_k(\mathbf{k})$ for $\mathbf{k} \in R_K$.

The frequency sampling filter in Equation (3) can be expressed in a computationally more efficient form by constraining the filter to have a real impulse response and linear phase. A filter with a real impulse response has a DFT of the form

$$|H_k(k_1, k_2)| = |H_k(N_1 - k_1, N_2 - k_2)|$$

$$\theta(k_1, k_2) = -\theta(N_1 - k_1, N_2 - k_2).$$

It can be shown[1] that the phase of a linear phase FIR filter with a region of support R_N is

$$\arg\left[H_\omega(\omega_1,\omega_2)\right] = -\omega_1\left(\frac{N_1-1}{2}\right) - \omega_2\left(\frac{N_2-1}{2}\right). \tag{4}$$

Therefore, the phase of the frequency samples is

$$\theta(k_1,k_2) = -\frac{2\pi}{N_1}k_1\left(\frac{N_1-1}{2}\right) - \frac{2\pi}{N_2}k_2\left(\frac{N_2-1}{2}\right).$$

Substituting these constraints into Equation (3), the filter's system function can be written as

$$\begin{aligned}
H_z(z_1,z_2) = {} & \frac{1-z_1^{-N_1}}{N_1}\frac{1-z_2^{-N_2}}{N_2}\left\{\frac{H_k(0,0)}{\left(1-z_1^{-1}\right)\left(1-z_2^{-1}\right)}\right. \\
& + \sum_{k_1=1}^{M_1} \frac{(-1)^{k_1}2\left|H_k(k_1,0)\right|\cos\left(\frac{\pi k_1}{N_1}\right)\left(1-z_1^{-1}\right)}{\left[1-2\cos\left(\frac{2\pi}{N_1}k_1\right)z_1^{-1}+z_1^{-2}\right]\left(1-z_2^{-1}\right)} \\
& + \sum_{k_2=1}^{M_2} \frac{(-1)^{k_2}2\left|H_k(0,k_2)\right|\cos\left(\frac{\pi k_2}{N_2}\right)\left(1-z_2^{-1}\right)}{\left(1-z_1^{-1}\right)\left[1-2\cos\left(\frac{2\pi}{N_2}k_2\right)z_2^{-1}+z_2^{-2}\right]} \\
& + \sum_{k_1=1}^{N_1-1}\sum_{k_2=1}^{M_2}(-1)^{k_1+k_2}2\left|H_k(k_1,k_2)\right| \times \\
& \left.\frac{\left[\cos\left(\frac{\pi k_1}{N_1}+\frac{\pi k_2}{N_2}\right)\left(1+z_1^{-1}z_2^{-1}\right)-\cos\left(\frac{\pi k_1}{N_1}-\frac{\pi k_2}{N_2}\right)\left(z_1^{-1}+z_2^{-1}\right)\right]}{\left[1-2\cos\left(\frac{2\pi}{N_1}k_1\right)z_1^{-1}+z_1^{-2}\right]\left[1-2\cos\left(\frac{2\pi}{N_2}k_2\right)z_2^{-1}+z_2^{-2}\right]}\right\}
\end{aligned} \tag{5}$$

where $M_1 = (N_1\text{-}1)/2$, $M_2 = (N_2\text{-}1)/2$, and N_1 and N_2 are odd.

The frequency sampling filter in Equation (5) can be expressed in a computationally more efficient form if we also constrain the filter to have fourfold symmetry. A Type 1-1 frequency sampling filter with fourfold symmetry has a DFT of the form

$$\left|H_k(k_1,k_2)\right| = \left|H_k(k_1,N_2-k_2)\right| = \left|H_k(N_1-k_1,k_2)\right| = \left|H_k(N_1-k_1,N_2-k_2)\right|.$$

Substituting these constraints into Equation (5), the filter's system function can be written as

$$
\begin{aligned}
H_z(z_1,z_2) = \frac{1-z_1^{-N_1}}{N_1}\frac{1-z_2^{-N_2}}{N_2}\Bigg\{ & \frac{H_k(0,0)}{\left(1-z_1^{-1}\right)\left(1-z_2^{-1}\right)} \\
& + \sum_{k_1=1}^{M_1} \frac{(-1)^{k_1}\, 2\left|H_k(k_1,0)\right| \cos\left(\frac{\pi k_1}{N_1}\right)\left(1-z_1^{-1}\right)}{\left[1-2\cos\left(\frac{2\pi}{N_1}k_1\right)z_1^{-1}+z_1^{-2}\right]\left(1-z_2^{-1}\right)} \\
& + \sum_{k_2=1}^{M_2} \frac{(-1)^{k_2}\, 2\left|H_k(0,k_2)\right| \cos\left(\frac{\pi k_2}{N_2}\right)\left(1-z_2^{-1}\right)}{\left(1-z_1^{-1}\right)\left[1-2\cos\left(\frac{2\pi}{N_2}k_2\right)z_2^{-1}+z_2^{-2}\right]} \\
& + \sum_{k_1=1}^{M_1}\sum_{k_2=1}^{M_2} (-1)^{k_1+k_2}\, 4\left|H_k(k_1,k_2)\right| \times \\
& \qquad \frac{\cos\left(\frac{\pi k_1}{N_1}\right)\cos\left(\frac{\pi k_2}{N_2}\right)\left(1-z_1^{-1}\right)\left(1-z_2^{-1}\right)}{\left[1-2\cos\left(\frac{2\pi}{N_1}k_1\right)z_1^{-1}+z_1^{-2}\right]\left[1-2\cos\left(\frac{2\pi}{N_2}k_2\right)z_2^{-1}+z_2^{-2}\right]}\Bigg\} \qquad (6)
\end{aligned}
$$

where N_1 and N_2 are odd.

B. Type 2-2 Frequency Sampling Filters

The Type 1-1 frequency sampling filters described in Equations (5) and (6) interpolate a frequency response through a set of frequency samples for $\boldsymbol{\omega} \in R_{\Omega}$ starting at $\omega_1 = \omega_2 = 0$. Another type of frequency sampling filter can be designed by interpolating a frequency response through a set of (N_1N_2) frequency samples starting at $\omega_1 = \pi/N_1$ and $\omega_2 = \pi/N_2$ instead of $\omega_1 = \omega_2 = 0$. This type of frequency sampling filter is called a Type 2-2 frequency sampling filter.

To develop Type 2-2 frequency sampling filters, the system function, $H(z_1,z_2)$, and the set, $H_k(\mathbf{k})$ for $\mathbf{k} \in R_K$, of frequency samples are related such that

$$H_k(\mathbf{k}) = |H_k(\mathbf{k})| e^{j\theta(\mathbf{k})}$$
$$= H(z_1, z_2)\Big|_{z_1 = e^{j(2\pi/N_1)(k_1+1/2)},\, z_2 = e^{j(2\pi/N_2)(k_2+1/2)}}$$

This relationship modifies the DFT into the form

$$H_k(\mathbf{k}) = \sum_{\mathbf{n}\in R_N} h(\mathbf{n}) e^{-j(2\pi/N_1)(k_1+1/2)n_1} e^{-j(2\pi/N_2)(k_2+1/2)n_2} \qquad (7)$$

where the modified IDFT is

$$h(\mathbf{n}) = \frac{1}{N_1 N_2} \sum_{\mathbf{k}\in R_K} H_k(\mathbf{k}) e^{j(2\pi/N_1)(k_1+1/2)n_1} e^{j(2\pi/N_2)(k_2+1/2)n_2}. \qquad (8)$$

Using a development similar to the one used for deriving the system function for Type 1-1 frequency sampling filters, Equation (8) can be substituted into the filter's z transform so that the system function, $H(z_1,z_2)$, of a Type 2-2 frequency sampling filter can be written as

$$H(z_1,z_2) = \frac{1+z_1^{-N_1}}{N_1}\frac{1+z_2^{-N_2}}{N_2}\times$$
$$\sum_{k_1=0}^{N_1-1}\sum_{k_2=0}^{N_2-1} \frac{H_k(k_1,k_2)}{\left(1-e^{j(2\pi/N_1)(k_1+1/2)}z_1^{-1}\right)\left(1-e^{j(2\pi/N_2)(k_2+1/2)}z_2^{-1}\right)} \qquad (9)$$

The frequency sampling filter in Equation (9) can be expressed in a computationally more efficient form by constraining the filter to have a real impulse response and linear phase. A Type 2-2 frequency sampling filter with a real impulse response has frequency samples of the form

$$|H_k(k_1,k_2)| = |H_k(N_1-1-k_1, N_2-1-k_2)|$$

$$\theta(k_1,k_2) = -\theta(N_1-1-k_1, N_2-1-k_2).$$

Using a development similar to the one used for deriving the system function for linear phase Type 1-1 frequency sampling filters, the system function of a linear phase Type 2-2 frequency sampling filter can be written as

$$H(z_1,z_2) = \frac{1+z_1^{-N_1}}{N_1}\frac{1+z_2^{-N_2}}{N_2}\left\{\frac{H_k(M_1,M_2)}{\left(1+z_1^{-1}\right)\left(1+z_2^{-1}\right)}\right.$$

$$+\sum_{k_1=0}^{M_1-1}\frac{(-1)^{k_1}2\left|H_k(k_1,M_2)\right|\cos\left[\frac{\pi}{N_1}(k_1+1/2)-\frac{\pi N_2}{2}\right]\left(1+z_1^{-1}\right)}{\left(1+z_2^{-1}\right)\left(1-2\cos\left[\frac{2\pi}{N_1}(k_1+1/2)\right]z_1^{-1}+z_1^{-2}\right)}$$

$$+\sum_{k_1=0}^{N_1-1}\sum_{k_2=0}^{M_2-1}(-1)^{k_1+k_2}2\left|H_k(k_1,k_2)\right| \times$$

$$\left[\frac{\cos\left[\frac{\pi}{N_1}\left(k_1+\frac{1}{2}\right)-\frac{\pi}{N_2}\left(k_2+\frac{1}{2}\right)\right]\left(z_1^{-1}+z_2^{-1}\right)}{\left(1-2\cos\left[\frac{2\pi}{N_1}\left(k_1+\frac{1}{2}\right)\right]z_1^{-1}+z_1^{-2}\right)\left(1-2\cos\left[\frac{2\pi}{N_2}\left(k_2+\frac{1}{2}\right)\right]z_2^{-1}+z_2^{-2}\right)}\right.$$

$$\left.\left.-\frac{\cos\left[\frac{\pi}{N_1}\left(k_1+\frac{1}{2}\right)+\frac{\pi}{N_2}\left(k_2+\frac{1}{2}\right)\right]\left(1+z_1^{-1}z_2^{-1}\right)}{\left(1-2\cos\left[\frac{2\pi}{N_1}\left(k_1+\frac{1}{2}\right)\right]z_1^{-1}+z_1^{-2}\right)\left(1-2\cos\left[\frac{2\pi}{N_2}\left(k_2+\frac{1}{2}\right)\right]z_2^{-1}+z_2^{-2}\right)}\right]\right\} \quad (10)$$

where N_1 and N_2 are odd.

The frequency sampling filter in Equation (10) can be expressed in a computationally more efficient form if we also constrain the filter to have fourfold symmetry. A Type 2-2 frequency sampling filter with fourfold symmetry has a DFT of the form

$$\begin{aligned}\left|H_k(k_1,k_2)\right| &= \left|H_k(k_1,N_2-1-k_2)\right| \\ &= \left|H_k(N_1-1-k_1,k_2)\right| \\ &= \left|H_k(N_1-1-k_1,N_2-1-k_2)\right|.\end{aligned}$$

Substituting these constraints into Equation (10) yields,

$$H(z_1,z_2)=\frac{1+z_1^{-N_1}}{N_1}\frac{1+z_2^{-N_2}}{N_2}\left\{\frac{H_k(M_1,M_2)}{\left(1+z_1^{-1}\right)\left(1+z_2^{-1}\right)}\right.$$

$$+\sum_{k_1=0}^{M_1-1}\frac{(-1)^{k_1}2\left|H_k(k_1,M_2)\right|\cos\left[\frac{\pi}{N_1}\left(k_1+\frac{1}{2}\right)-\frac{\pi N_2}{2}\right]\left(1+z_1^{-1}\right)}{\left(1-2\cos\left[\frac{2\pi}{N_1}\left(k_1+\frac{1}{2}\right)\right]z_1^{-1}+z_1^{-2}\right)\left(1+z_2^{-1}\right)}$$

$$+\sum_{k_2=0}^{M_2-1} \frac{(-1)^{k_2} 2\left|H_k(M_1,k_2)\right| \cos\left[\frac{\pi}{N_2}\left(k_2+\frac{1}{2}\right)-\frac{\pi N_1}{2}\right]\left(1+z_2^{-1}\right)}{\left(1+z_1^{-1}\right)\left(1-2\cos\left[\frac{2\pi}{N_2}\left(k_2+\frac{1}{2}\right)\right]z_2^{-1}+z_2^{-2}\right)}$$

$$+\sum_{k_1=0}^{M_1}\sum_{k_2=0}^{M_2}(-1)^{k_1+k_2} 4\left|H_k(k_1,k_2)\right| \times$$

$$\frac{\sin\left[\frac{\pi}{N_1}\left(k_1+\frac{1}{2}\right)\right]\sin\left[\frac{\pi}{N_2}\left(k_2+\frac{1}{2}\right)\right]\left(1+z_1^{-1}\right)\left(1+z_2^{-1}\right)}{\left(1-2\cos\left[\frac{2\pi}{N_1}\left(k_1+\frac{1}{2}\right)\right]z_1^{-1}+z_1^{-2}\right)\left(1-2\cos\left[\frac{2\pi}{N_2}\left(k_2+\frac{1}{2}\right)\right]z_2^{-1}+z_2^{-2}\right)} \qquad (11)$$

where N_1 and N_2 are odd.

C. Type 1-2 Frequency Sampling Filters

Another type of frequency sampling filter can be designed by interpolating a frequency response through a set of (N_1N_2) frequency samples starting at $\omega_1 = 0$ and $\omega_2 = \pi/N_2$. This type of frequency sampling filter is called a Type 1-2 frequency sampling filter.

To develop Type 1-2 frequency sampling filters, the system function, $H(z_1,z_2)$, and the set , $H_k(\mathbf{k})$ for $\mathbf{k} \in R_K$, of frequency samples are related such that

$$\begin{aligned} H_k(\mathbf{k}) &= \left|H_k(\mathbf{k})\right| e^{j\theta(\mathbf{k})} \\ &= H\left(z_1,z_2\right)\Big|_{z_1=e^{j(2\pi/N_1)k_1},\, z_2=e^{j(2\pi/N_2)(k_2+1/2)}} \end{aligned}$$

This relationship modifies the DFT into the form

$$H_k(\mathbf{k}) = \sum_{\mathbf{n}\in R_N} h(\mathbf{n}) e^{-j(2\pi/N_1)k_1n_1} e^{-j(2\pi/N_2)(k_2+1/2)n_2}$$

where the modified IDFT is

$$h(\mathbf{n}) = \frac{1}{N_1N_2}\sum_{\mathbf{k}\in R_K} H_k(\mathbf{k}) e^{j(2\pi/N_1)k_1n_1} e^{j(2\pi/N_2)(k_2+1/2)n_2}. \qquad (12)$$

Using a development similar to the one used for deriving the system function for Type 1-1 frequency sampling filters, Equation (12) can be substituted into the the filter's z transform so that the system function, $H(z_1,z_2)$, of a Type 1-2 frequency sampling filter can be written as

$$H(z_1,z_2)=\frac{1-z_1^{-N_1}}{N_1}\frac{1+z_2^{-N_2}}{N_2}\times \sum_{k_1=0}^{N_1-1}\sum_{k_2=0}^{N_2-1}\frac{H_k(k_1,k_2)}{\left(1-e^{j(2\pi/N_1)k_1}z_1^{-1}\right)\left(1-e^{j(2\pi/N_2)(k_2+1/2)}z_2^{-1}\right)} \tag{13}$$

The frequency sampling filter in Equation (13) can be expressed in a computationally more efficient form by constraining the filter to have a real impulse response and linear phase. A Type 1-2 frequency sampling filter with a real impulse response has frequency samples of the form

$$|H_k(k_1,k_2)|=|H_k(N_1-k_1,N_2-1-k_2)|$$

$$\theta(k_1,k_2)=-\theta(N_1-k_1,N_2-1-k_2).$$

Using a development similar to the one used for deriving the system function for linear phase Type 1-1 frequency sampling filters, the system function of a linear phase Type 1-2 frequency sampling filter can be written as

$$H(z_1,z_2)=\frac{1-z_1^{-N_1}}{N_1}\frac{1+z_2^{-N_2}}{N_2}\left\{\frac{H_k(0,M_2)}{\left(1-z_1^{-1}\right)\left(1+z_2^{-1}\right)}\right.$$

$$+\sum_{k_2=0}^{M_2-1}\frac{(-1)^{k_2}2|H_k(0,k_2)|\sin\left[\frac{\pi}{N_2}\left(k_2+\frac{1}{2}\right)\right]\left(1+z_2^{-1}\right)}{\left(1-z_1^{-1}\right)\left(1-2\cos\left[\frac{2\pi}{N_2}\left(k_2+\frac{1}{2}\right)\right]z_2^{-1}+z_2^{-2}\right)}$$

$$+\sum_{k_1=1}^{M_1}\sum_{k_2=0}^{N_2-1}(-1)^{k_1+k_2}2|\mathrm{H}_k(k_1,k_2)|\times$$

$$\left[\frac{\sin\left[\frac{\pi}{N_1}k_1+\frac{\pi}{N_2}\left(k_2+\frac{1}{2}\right)\right]\left(1-z_1^{-1}z_2^{-1}\right)}{\left[1-2\cos\left(\frac{2\pi}{N_1}k_1\right)z_1^{-1}+z_1^{-2}\right]\left(1-2\cos\left[\frac{2\pi}{N_2}\left(k_2+\frac{1}{2}\right)\right]z_2^{-1}+z_2^{-2}\right)}\right.$$

$$\left.\left.+\frac{\sin\left[-\frac{\pi}{N_1}k_1+\frac{\pi}{N_2}\left(k_2+\frac{1}{2}\right)\right]\left(z_2^{-1}-z_1^{-1}\right)}{\left[1-2\cos\left(\frac{2\pi}{N_1}k_1\right)z_1^{-1}+z_1^{-2}\right]\left(1-2\cos\left[\frac{2\pi}{N_2}\left(k_2+\frac{1}{2}\right)\right]z_2^{-1}+z_2^{-2}\right)}\right]\right\} \quad (14)$$

where N_1 and N_2 are odd.

The frequency sampling filter in Equation (14) can be expressed in a computationally more efficient form if we also constrain the filter to have fourfold symmetry. A Type 1-2 frequency sampling filter with fourfold symmetry has a DFT of the form

$$\begin{aligned}|H_k(k_1,k_2)| &= |H_k(k_1,N_2-1-k_2)| \\ &= |H_k(N_1-k_1,k_2)| \\ &= |H_k(N_1-k_1,N_2-1-k_2)|.\end{aligned}$$

Substituting these constraints into Equation (14) yields,

$$H(z_1,z_2)=\frac{1-z_1^{-N_1}}{N_1}\frac{1+z_2^{-N_2}}{N_2}\left\{\frac{H_k(0,M_2)}{\left(1-z_1^{-1}\right)\left(1+z_2^{-1}\right)}\right.$$

$$+\sum_{k_1=1}^{M_1}\frac{(-1)^{k_1+(N_2-1)/2}2|H_k(0,M_2)|\cos\left[\frac{\pi}{N_1}k_1\right]\left(1-z_1^{-1}\right)}{\left[1-2\cos\left(\frac{2\pi}{N_1}k_1\right)z_1^{-1}+z_1^{-2}\right]\left(1+z_2^{-1}\right)}$$

$$+\sum_{k_2=0}^{M_2-1}\frac{(-1)^{k_2}2|H_k(0,k_2)|\sin\left[\frac{\pi}{N_2}\left(k_2+\frac{1}{2}\right)\right]\left(1+z_2^{-1}\right)}{\left(1-z_1^{-1}\right)\left(1-2\cos\left[\frac{2\pi}{N_2}\left(k_2+\frac{1}{2}\right)\right]z_2^{-1}+z_2^{-2}\right)}$$

$$+\sum_{k_1=1}^{M_1}\sum_{k_2=0}^{M_2-1}(-1)^{k_1+k_2}4\left|H_k(k_1,k_2)\right|\times$$

$$\left.\frac{\cos\left(\frac{\pi}{N_1}k_1\right)\sin\left[\frac{\pi}{N_2}\left(k_2+\frac{1}{2}\right)\right]\left(1-z_1^{-1}\right)\left(1+z_2^{-1}\right)}{\left[1-2\cos\left(\frac{2\pi}{N_1}k_1\right)z_1^{-1}+z_1^{-2}\right]\left(1-2\cos\left[\frac{2\pi}{N_2}\left(k_2+\frac{1}{2}\right)\right]z_2^{-1}+z_2^{-2}\right)}\right\} \quad (15)$$

where N_1 and N_2 are odd.

D. Type 2-1 Frequency Sampling Filters

Another type of frequency sampling filter can be designed by interpolating a frequency response through a set of (N_1N_2) frequency samples starting at $\omega_1 = \pi/N_1$ and $\omega_2 = 0$. This type of frequency sampling filter is called a Type 2-1 frequency sampling filter.

To develop Type 2-1 frequency sampling filters, the system function, $H(z_1,z_2)$, and the set, $H_k(\mathbf{k})$ for $\mathbf{k} \in R_K$, of frequency samples are related such that

$$\begin{aligned} H_k(\mathbf{k}) &= \left|H_k(\mathbf{k})\right|e^{j\theta(\mathbf{k})} \\ &= H(z_1,z_2)\Big|_{z_1=e^{j(2\pi/N_1)(k_1+1/2)},\, z_2=e^{j(2\pi/N_2)k_2}} . \end{aligned}$$

This relationship modifies the DFT into the form

$$H_k(\mathbf{k}) = \sum_{\mathbf{n}\in R_N} h(\mathbf{n})e^{-j(2\pi/N_1)(k_1+1/2)n_1}e^{-j(2\pi/N_2)k_2n_2}$$

where the modified IDFT is

$$h(\mathbf{n}) = \frac{1}{N_1N_2}\sum_{\mathbf{k}\in R_K} H_k(\mathbf{k})e^{j(2\pi/N_1)(k_1+1/2)n_1}e^{j(2\pi/N_2)k_2n_2} . \quad (16)$$

Using a development similar to the one used for deriving the system function for Type 1-2 frequency sampling filters, Equation (16) can be substituted into the the filter's z transform so that the system function, $H(z_1,z_2)$, of a Type 2-1 frequency sampling filter can be written as

$$H(z_1,z_2)=\frac{1+z_1^{-N_1}}{N_1}\frac{1-z_2^{-N_2}}{N_2}\times$$
$$\sum_{k_1=0}^{N_1-1}\sum_{k_2=0}^{N_2-1}\frac{H_k(k_1,k_2)}{\left(1-e^{j(2\pi/N_1)(k_1+1/2)}z_1^{-1}\right)\left(1-e^{j(2\pi/N_2)k_2}z_2^{-1}\right)} \tag{17}$$

The frequency sampling filter in Equation (17) can be expressed in a computationally more efficient form by constraining the filter to have a real impulse response and linear phase. A Type 2-1 frequency sampling filter with a real impulse response has frequency samples of the form

$$|H_k(k_1,k_2)|=|H_k(N_1-1-k_1,N_2-k_2)|$$
$$\theta(k_1,k_2)=-\theta(N_1-1-k_1,N_2-k_2).$$

Using a development similar to the one used for deriving the system function for linear phase Type 1-2 frequency sampling filters, the system function of a linear phase Type 2-1 frequency sampling filter can be written as

$$H(z_1,z_2)=\frac{1+z_1^{-N_1}}{N_1}\frac{1-z_2^{-N_2}}{N_2}\left\{\frac{H_k(M_1,0)}{\left(1+z_1^{-1}\right)\left(1-z_2^{-1}\right)}\right.$$
$$+\sum_{k_1=0}^{M_1-1}\frac{(-1)^{k_1}2|H_k(k_1,0)|\sin\left[\frac{\pi}{N_1}\left(k_1+\frac{1}{2}\right)\right]\left(1+z_1^{-1}\right)}{\left(1-2\cos\left[\frac{2\pi}{N_1}\left(k_1+\frac{1}{2}\right)\right]z_1^{-1}+z_1^{-2}\right)\left(1-z_2^{-1}\right)}$$
$$+\sum_{k_1=1}^{N_1-1}\sum_{k_2=0}^{M_2}(-1)^{k_1+k_2}2|H_k(k_1,k_2)|\times$$
$$\left[\frac{\sin\left[\frac{\pi}{N_1}\left(k_1+\frac{1}{2}\right)+\frac{\pi}{N_2}k_2\right]\left(1-z_1^{-1}z_2^{-1}\right)}{\left(1-2\cos\left[\frac{2\pi}{N_1}\left(k_1+\frac{1}{2}\right)\right]z_1^{-1}+z_1^{-2}\right)\left[1-2\cos\left(\frac{2\pi}{N_2}k_2\right)z_2^{-1}+z_2^{-2}\right]}\right.$$
$$\left.\left.+\frac{\sin\left[\frac{\pi}{N_1}\left(k_1+\frac{1}{2}\right)-\frac{\pi}{N_2}k_2\right]\left(z_1^{-1}-z_2^{-1}\right)}{\left(1-2\cos\left[\frac{2\pi}{N_1}\left(k_1+\frac{1}{2}\right)\right]z_1^{-1}+z_1^{-2}\right)\left[1-2\cos\left(\frac{2\pi}{N_2}k_2\right)z_2^{-1}+z_2^{-2}\right]}\right]\right\} \tag{18}$$

where N_1 and N_2 are odd.

The frequency sampling filter in Equation (18) can be expressed in a computationally more efficient form if we also constrain the filter to have fourfold symmetry. A Type 2-1 frequency sampling filter with fourfold symmetry has a DFT of the form

$$\begin{aligned}|H_k(k_1,k_2)| &= |H_k(k_1,N_2-k_2)| \\ &= |H_k(N_1-1-k_1,k_2)| \\ &= |H_k(N_1-1-k_1,N_2-k_2)|.\end{aligned}$$

Substituting these constraints into Equation (18) yields,

$$\begin{aligned}H(z_1,z_2) = \frac{1+z_1^{-N_1}}{N_1}\frac{1-z_2^{-N_2}}{N_2}\Bigg\{ & \frac{H_k(M_1,0)}{\left(1+z_1^{-1}\right)\left(1-z_2^{-1}\right)} \\ & + \sum_{k_1=0}^{M_1-1} \frac{(-1)^{k_1}2|H_k(k_1,0)|\sin\left[\frac{\pi}{N_1}\left(k_1+\frac{1}{2}\right)\right]\left(1+z_1^{-1}\right)}{\left(1-2\cos\left[\frac{2\pi}{N_1}\left(k_1+\frac{1}{2}\right)\right]z_1^{-1}+z_1^{-2}\right)\left(1-z_2^{-1}\right)} \\ & + \sum_{k_2=1}^{M_2} \frac{(-1)^{k_2+(N_1-1)/2}2|H_k(M_1,0)|\cos\left[\frac{\pi}{N_2}k_2\right]\left(1-z_2^{-1}\right)}{\left(1+z_1^{-1}\right)\left[1-2\cos\left(\frac{2\pi}{N_2}k_2\right)z_2^{-1}+z_2^{-2}\right]} \\ & + \sum_{k_1=1}^{M_1-1}\sum_{k_2=0}^{M_2}(-1)^{k_1+k_2}4|H_k(k_1,k_2)| \times \\ & \frac{\sin\left[\frac{\pi}{N_1}\left(k_1+\frac{1}{2}\right)\right]\cos\left(\frac{\pi}{N_2}k_2\right)\left(1+z_1^{-1}\right)\left(1-z_2^{-1}\right)}{\left(1-2\cos\left[\frac{2\pi}{N_1}\left(k_1+\frac{1}{2}\right)\right]z_1^{-1}+z_1^{-2}\right)\left[1-2\cos\left(\frac{2\pi}{N_2}k_2\right)z_2^{-1}+z_2^{-2}\right]}\Bigg\}\end{aligned} \qquad (19)$$

where N_1 and N_2 are odd.

E. Computational Advantage of Frequency Sampling Filters

Each of the frequency sampling filter system functions described by Equations (3), (5), (6), (9)-(11), (13)-(15) and (17)-(19) describe an architecture that has a nonrecursive comb filter created by the terms, $\left(1 \pm z_1^{-N_1}\right)$ and $\left(1 \pm z_2^{-N_2}\right)$, in series with a number of parallel resonators[1]. When most of a frequency sampling filter's frequency samples, the $H_k(\mathbf{k})$'s, are exactly zero, most of the filter's resonators are not implemented. Therefore, in the case of a narrowband filter where most of the frequency samples are in the stopband and are set equal to zero, only a small number of the filter's resonators need to be implemented. In such instances, frequency sampling implementations usually require fewer arithmetic operations than the direct convolution implementations.

Each of the resonators in the frequency sampling filters described by Equations (3), (5), (6), (9)-(11), (13)-(15) and (17)-(19) can be represented by a separable system function[2]. If separable system function, $H(z_1,z_2)$, can be written as $H(z_1,z_2) = H_1(z_1)H_2(z_2)$, then it can be shown that that system is stable iff each of the 1-D system functions, $H_1(z_1)$ and $H_2(z_2)$, are stable[2]. Because each of the resonator system functions are separable, it is straight forward to observe that for each of the frequency sampling filters the zeroes created by the terms, $\left(1 \pm z_1^{-N_1}\right)$ and $\left(1 \pm z_2^{-N_2}\right)$, exactly cancel the resonators' poles that lie on the 4-D unit sphere. When implementing these filters with finite word length systems, exact pole zero cancellation generally does not occur, and uncanceled poles on the unit sphere will cause the filter to be unstable. To prevent this instability, each z_i^{-1} in each of the frequency sampling filter system functions is replaced by rz_i^{-1}; $r < 1$; $i = 1, 2$. This moves all the poles inside the 4-D unit sphere. To keep the frequency response of these modified filters as close as possible to the frequency response of the original filters, r should be chosen near to the value 1.

Substituting rz_i^{-1} for z_i^{-1} where $r < 1$ guarantees the stability of a frequency sampling filter at the cost of increasing its computational requirements. If implemented with z_i^{-1} replaced by rz_i^{-1}, a linear phase frequency sampling filter requires at most 8 multiplies per resonator, and a linear phase frequency sampling filter with fourfold symmetry requires at most 5 multiplies per resonator If only K of the frequency samples are non-zero, then a linear phase fre-

1. The term *resonator* is used in this paper to denote a system which has its poles on or near the 4-D unit sphere or has its poles on or near the 2-D unit circle.
2. A system function is said to be separable iff it can be expressed as the product of one dimensional (1-D) system functions.

quency sampling structure requires approximately $8K$ multiplies per output sample, and a linear phase frequency sampling filter with fourfold symmetry requires approximately $5K$ multiplies per output sample. If a linear phase FIR filter with region of support R_N is implemented using direct convolution, it requires approximately $N_1N_2/2$ multiplies per output sample; and if a linear phase FIR filter with fourfold symmetry and region of support R_N is implemented using direct convolution, it requires approximately $N_1N_2/4$ multiplies per output sample. Because multiplications typically require more time to compute and are more complex to implement than adds, a frequency sampling structure becomes computationally more efficient than a direct convolution implementation when it uses fewer multiplies. Therefore, if we wish to implement a linear phase filter with a region of support R_N, a frequency sampling filter can implement the filter more efficiently (in the sense of fewer multiplies) than a direct convolution implementation when $8K < N_1N_2/2$ or $K < N_1N_2/16$. If we wish to implement a linear phase filter with fourfold symmetry and region of support R_N, a frequency sampling filter can implement the filter more efficiently than a direct convolution implementation when $5K < N_1N_2/4$ or $K < N_1N_2/20$.

III. The Design of 2-D Frequency Sampling Filters

It was shown in section II that a frequency sampling filter approximates a desired frequency response by interpolating a frequency response through a set of frequency samples. In this section, a design technique that controls interpolation errors between samples and approximates a desired frequency response is developed for frequency sampling filters that have real impulse responses, linear phase and fourfold symmetry. This design technique approximates a desired frequency response by minimizing a weighted mean square error over the passbands and stopbands subject to constraints on the filter's amplitude response. To maintain the computational efficiency of the filter, this error criterion is minimized while constraining the frequency samples in the stopband to be identically zero. This design technique is developed as a constrained optimization problem which is solved by the method of Lagrange multipliers. The Lagrange multipliers optimization method results in a set of linear equations which are solved to determine the filter's coefficients. The design method applies to each of the four types of 2-D frequency sampling filters that have real impulse responses, linear phase and fourfold symmetry.

The frequency response of a frequency sampling filter which has an impulse response, $h(\mathbf{n})$ for $\mathbf{n} \in R_N$, is

$$H_\omega(\boldsymbol{\omega}) = H(e^{j\omega_1}, e^{j\omega_2}) = \sum_{\mathbf{n}\in R_N} h(\mathbf{n}) e^{-j\omega_1 n_1} e^{-j\omega_2 n_2}$$

If the filter has a real impulse response, fourfold symmetry and linear phase then

$$\begin{aligned} h(n_1, n_2) &= h(n_1, N_2 - 1 - n_2) \\ &= h(N_1 - 1 - n_1, n_2) \\ &= h(N_1 - 1 - n_1, N_2 - 1 - n_2), \end{aligned}$$

and it can then be shown that $H_\omega(\boldsymbol{\omega})$ can be written as

$$H_\omega(\omega_1, \omega_2) = e^{-j[\omega_1 (N_1 - 1)/2 + \omega_2 (N_2 - 1)/2]} H_r(\omega_1, \omega_2)$$

where $H_r(\omega_1,\omega_2)$ is a real function given by

$$H_r(\boldsymbol{\omega}) = \sum_{n_1=0}^{M_1} \sum_{n_2=0}^{M_2} 2^{\mathrm{sgn}(n_1)} 2^{\mathrm{sgn}(n_2)} h\big(M_1 - n_1, M_2 - n_2\big) \cos(\omega_1 n_1) \cos(\omega_2 n_2)$$

$M_1 = (N_1\text{-}1)/2$, $M_2 = (N_2\text{-}1)/2$, N_1 is odd, N_2 is odd and sgn(•) is the signum function. $H_r(\boldsymbol{\omega})$ is the amplitude of $H_\omega(\boldsymbol{\omega})$ and is often referred to as the zero phase frequency response or the amplitude response of the filter. If we let

$$\mathbf{x} = \big[h(M_1,M_2) \quad 2h(M_1-1,M_2) \quad \cdots \quad 4h(1,0) \quad 4h(0,0)\big]^T$$

where the superscript T denotes transpose, and let

$$\mathbf{s}(\omega_1,\omega_2) = \big[s_r(\omega_1,\omega_2)\big] = \big[\cos\big(a_1(r)\omega_1\big)\cos\big(a_2(r)\omega_2\big)\big]$$

where $r = 0, 1, \ldots, (M_1+1)(M_2+1) - 1$, $s_r(\boldsymbol{\omega})$ represents the rth element in the column vector, $\mathbf{s}(\boldsymbol{\omega})$,

$$a_1(r) = r \bmod (M_1 + 1)$$

$$a_2(r) = \mathrm{int}\left(\frac{r}{M_1+1}\right)$$

and int(x) truncates x, then

$$H_r(\boldsymbol{\omega}) = \mathbf{x}^T \mathbf{s}(\boldsymbol{\omega}) = \mathbf{s}^T(\boldsymbol{\omega})\mathbf{x}. \tag{20}$$

In Equations (1), (8), (12) and (16), the impulse response, $h(\mathbf{n})$, of each of the frequency sampling filters is expressed as a function of its DFT. By constraining the filters to have linear phase, the IDFT's can be written as

$$h(\mathbf{n}) = \begin{cases} \displaystyle\sum_{\mathbf{k}\in R_{11}} \frac{2^{\operatorname{sgn}(k_1+k_2)}(-1)^{(k_1+k_2)}}{N_1 N_2} |H_k(k_1,k_2)| \times \\ \qquad \cos\left[\frac{2\pi}{N_1} k_1\left(n_1+\frac{1}{2}\right) + \frac{2\pi}{N_2} k_2\left(n_2+\frac{1}{2}\right)\right] & \text{Type 1-1} \\ \displaystyle\sum_{\mathbf{k}\in R_{12}} \frac{2^{\operatorname{sgn}(k_1+M_2-k_2)}(-1)^{(k_1+k_2)}}{N_1 N_2} |H_k(k_1,k_2)| \times \\ \qquad \sin\left[\frac{2\pi}{N_1} k_1\left(n_1+\frac{1}{2}\right) + \frac{2\pi}{N_2}\left(k_2+\frac{1}{2}\right)\left(n_2+\frac{1}{2}\right)\right] & \text{Type 1-2} \\ \displaystyle\sum_{\mathbf{k}\in R_{21}} \frac{2^{\operatorname{sgn}(M_1-k_1+k_2)}(-1)^{(k_1+k_2)}}{N_1 N_2} |H_k(k_1,k_2)| \times \\ \qquad \sin\left[\frac{2\pi}{N_1}\left(k_1+\frac{1}{2}\right)\left(n_1+\frac{1}{2}\right) + \frac{2\pi}{N_2} k_2\left(n_2+\frac{1}{2}\right)\right] & \text{Type 2-1} \\ \displaystyle\sum_{\mathbf{k}\in R_{22}} \frac{2^{\operatorname{sgn}(M_1-k_1+M_2-k_2)}(-1)^{(k_1+k_2+1)}}{N_1 N_2} |H_k(k_1,k_2)| \times \\ \qquad \cos\left[\frac{2\pi}{N_1}\left(k_1+\frac{1}{2}\right)\left(n_1+\frac{1}{2}\right) + \frac{2\pi}{N_2}\left(k_2+\frac{1}{2}\right)\left(n_2+\frac{1}{2}\right)\right] & \text{Type 2-2} \end{cases}$$

where

$$R_{11} = \{(k_1,k_2): 0 \le k_1 \le N_1 - 1, 1 \le k_2 \le M_2 \cup 0 \le k_1 \le M_1, k_2 = 0\}$$

$$R_{12} = \{(k_1,k_2): 1 \le k_1 \le M_1, 0 \le k_2 \le N_2 - 1 \cup k_1 = 0, 0 \le k_2 \le M_2\}$$

$$R_{21} = \{(k_1,k_2): 0 \le k_1 \le N_1 - 1, 1 \le k_2 \le M_2 \cup 0 \le k_1 \le M_1, k_2 = 0\}$$

$$R_{22} = \{(k_1,k_2): 0 \le k_1 \le N_1 - 1, 0 \le k_2 \le M_2 - 1 \cup 0 \le k_1 \le M_1, k_2 = M_2\}$$

If the filters have linear phase and fourfold symmetry, then the IDFT transforms can be written as

$$
h(\mathbf{n})=\begin{cases}
\displaystyle\sum_{\mathbf{k}\in R_M}\frac{2^{\operatorname{sgn}(k_1)}2^{\operatorname{sgn}(k_2)}(-1)^{(k_1+k_2)}}{N_1N_2}\left|H_k(k_1,k_2)\right|\times \\
\qquad\cos\left[\frac{2\pi}{N_1}k_1\left(n_1+\frac{1}{2}\right)\right]\cos\left[\frac{2\pi}{N_2}k_2\left(n_2+\frac{1}{2}\right)\right] & \text{Type 1-1} \\
\displaystyle\sum_{\mathbf{k}\in R_M}\frac{2^{\operatorname{sgn}(k_1)}2^{\operatorname{sgn}(M_2-k_2)}(-1)^{(k_1+k_2)}}{N_1N_2}\left|H_k(k_1,k_2)\right|\times \\
\qquad\cos\left[\frac{2\pi}{N_1}k_1\left(n_1+\frac{1}{2}\right)\right]\sin\left[\frac{2\pi}{N_2}\left(k_2+\frac{1}{2}\right)\left(n_2+\frac{1}{2}\right)\right] & \text{Type 1-2} \\
\displaystyle\sum_{\mathbf{k}\in R_M}\frac{2^{\operatorname{sgn}(M_1-k_1)}2^{\operatorname{sgn}(k_2)}(-1)^{(k_1+k_2)}}{N_1N_2}\left|H_k(k_1,k_2)\right|\times \\
\qquad\sin\left[\frac{2\pi}{N_1}\left(k_1+\frac{1}{2}\right)\left(n_1+\frac{1}{2}\right)\right]\cos\left[\frac{2\pi}{N_2}k_2\left(n_2+\frac{1}{2}\right)\right] & \text{Type 2-1} \\
\displaystyle\sum_{\mathbf{k}\in R_M}\frac{2^{\operatorname{sgn}(M_1-k_1)}2^{\operatorname{sgn}(M_2-k_2)}(-1)^{(k_1+k_2)}}{N_1N_2}\left|H_k(k_1,k_2)\right|\times \\
\qquad\sin\left[\frac{2\pi}{N_1}\left(k_1+\frac{1}{2}\right)\left(n_1+\frac{1}{2}\right)\right]\sin\left[\frac{2\pi}{N_2}\left(k_2+\frac{1}{2}\right)\left(n_2+\frac{1}{2}\right)\right] & \text{Type 2-2}
\end{cases}
$$

where

$$R_M=\left\{(k_1,k_2):0\le k_1\le M_1,0\le k_2\le M_2\right\}.$$

If we define

$$\mathbf{X}=\left[H_k(0,0)\quad H_k(0,1)\quad\cdots\quad H_k(M_1,M_2-1)\quad H_k(M_1,M_2)\right]^T$$

and

$$
\mathbf{f}(\mathbf{n}) = \begin{cases}
\left[\dfrac{2^{\operatorname{sgn}(k_1)} 2^{\operatorname{sgn}(k_2)} (-1)^{(k_1+k_2)}}{N_1 N_2} \times \cos\left[\dfrac{2\pi}{N_1} k_1 \left(n_1 + \tfrac{1}{2}\right)\right] \cos\left[\dfrac{2\pi}{N_2} k_2 \left(n_2 + \tfrac{1}{2}\right)\right] \right] & \text{Type 1-1} \\
\left[\dfrac{2^{\operatorname{sgn}(k_1)} 2^{\operatorname{sgn}(M_2-k_2)} (-1)^{(k_1+k_2)}}{N_1 N_2} \times \cos\left[\dfrac{2\pi}{N_1} k_1 \left(n_1 + \tfrac{1}{2}\right)\right] \sin\left[\dfrac{2\pi}{N_2} \left(k_2 + \tfrac{1}{2}\right)\left(n_2 + \tfrac{1}{2}\right)\right] \right] & \text{Type 1-2} \\
\left[\dfrac{2^{\operatorname{sgn}(M_1-k_1)} 2^{\operatorname{sgn}(k_2)} (-1)^{(k_1+k_2)}}{N_1 N_2} \times \sin\left[\dfrac{2\pi}{N_1} \left(k_1 + \tfrac{1}{2}\right)\left(n_1 + \tfrac{1}{2}\right)\right] \cos\left[\dfrac{2\pi}{N_2} k_2 \left(n_2 + \tfrac{1}{2}\right)\right] \right] & \text{Type 2-1} \\
\left[\dfrac{2^{\operatorname{sgn}(M_1-k_1)} 2^{\operatorname{sgn}(M_2-k_2)} (-1)^{(k_1+k_2)}}{N_1 N_2} \times \sin\left[\dfrac{2\pi}{N_1} \left(k_1 + \tfrac{1}{2}\right)\left(n_1 + \tfrac{1}{2}\right)\right] \sin\left[\dfrac{2\pi}{N_2} \left(k_2 + \tfrac{1}{2}\right)\left(n_2 + \tfrac{1}{2}\right)\right] \right] & \text{Type 2-2}
\end{cases}
$$

for $r = 0, 1, \ (M_1+1)(M_2+1) - 1$; $k_1 = \operatorname{int}[r/(M_2+1)]$; and $k_2 = r \bmod (M_2+1)$, then

$$h(\mathbf{n}) = \mathbf{X}^T \mathbf{f}(\mathbf{n}) = \mathbf{f}^T(\mathbf{n})\mathbf{X}$$

where the appropriate expression for $\mathbf{f}(n)$ is used depending upon the type of frequency sampling filter being designed. If we also define

$$\mathbf{F} = \left[\mathbf{f}(M_1, M_2) \quad 2\mathbf{f}(M_1 - 1, M_2) \quad \cdots \quad 4\mathbf{f}(1,0) \quad 4\mathbf{f}(0,0)\right]^T$$

then

$$\mathbf{x} = \mathbf{F}\mathbf{X} \tag{21}$$

Substituting (21) into (20),

$$H_r(\boldsymbol{\omega}) = \mathbf{X}^T \mathbf{F}^T \mathbf{s}(\boldsymbol{\omega}) = \mathbf{s}^T(\boldsymbol{\omega})\mathbf{F}\mathbf{X}\,. \tag{22}$$

An approximation of a linear phase filter with a desired passband amplitude, $H_d(\boldsymbol{\omega})$, can be obtained by minimizing the mean square error between the desired amplitude response, $H_d(\boldsymbol{\omega})$, and the FIR filter's amplitude response, $H_r(\boldsymbol{\omega})$. If we let

$$\Omega_{pb} = \{(\omega_1, \omega_2) : \omega_1 \in \text{passband}, \omega_2 \in \text{passband}\},$$

and let J_{pb} represent the mean square error over the region of passband frequencies, then

$$J_{pb} = \frac{1}{m(\Omega_{pb})} \iint_{\boldsymbol{\omega} \in \Omega_{pb}} |H_r(\boldsymbol{\omega}) - H_d(\boldsymbol{\omega})|^2 d\boldsymbol{\omega} \tag{23}$$

where m(•) is the 2-D Lebesgue measure. Substituting Equation (22) into Equation (23), J_{pb} can be written as

$$J_{pb}(\mathbf{X}) = \frac{1}{m(\Omega_{pb})} \left[\mathbf{X}^T \mathbf{F}^T \iint_{\boldsymbol{\omega} \in \Omega_{pb}} \mathbf{s}(\boldsymbol{\omega}) \mathbf{s}^T(\boldsymbol{\omega}) d\boldsymbol{\omega} \, \mathbf{F} \mathbf{X} \right.$$

$$\left. -2\mathbf{X}^T \mathbf{F}^T \iint_{\boldsymbol{\omega} \in \Omega_{pb}} H_d(\boldsymbol{\omega}) \mathbf{s}(\boldsymbol{\omega}) d\boldsymbol{\omega} + \iint_{\boldsymbol{\omega} \in \Omega_{pb}} H_d^2(\boldsymbol{\omega}) d\boldsymbol{\omega} \right] \tag{24}$$

If we let

$$\mathbf{W}(\boldsymbol{\omega}) = \mathbf{s}(\boldsymbol{\omega}) \mathbf{s}^T(\boldsymbol{\omega})$$

then

$$\mathbf{W}(\boldsymbol{\omega}) = [W_{rc}(\boldsymbol{\omega})] = [\cos(a_1(r)\omega_1)\cos(a_2(r)\omega_2)\cos(a_1(c)\omega_1)\cos(a_2(c)\omega_2)]$$

where $W_{rc}(\boldsymbol{\omega})$ represent the element in the rth row and the cth column of the matrix, $\mathbf{W}(\boldsymbol{\omega})$, and $r, c = 0, 1, \ldots, (M_1+1)(M_2+1) - 1$. Defining $\mathbf{Q}(\boldsymbol{\omega})$ as the matrix,

$$\mathbf{Q}(\boldsymbol{\omega}) = \int\int \mathbf{W}(\boldsymbol{\omega}) \, d\boldsymbol{\omega},$$

the element in the rth row and the cth column of the matrix, $\mathbf{Q}(\boldsymbol{\omega})$, can be written as

$$Q_{rc}(\omega_1, \omega_2) = Q_{rc}^1(\omega_1) Q_{rc}^2(\omega_2)$$

where

$$Q_{rc}^{i}(\omega_i) = \begin{cases} \omega_i & a_i(r) = a_i(c) = 0 \\ \dfrac{\omega_i}{2} + \dfrac{\sin 2a_i(r)\omega_i}{4a_i(r)} & a_i(r) = a_i(c) \neq 0 \\ \dfrac{\sin[a_i(r)+a_i(c)]\omega_i}{2[a_i(r)+a_i(c)]} + \dfrac{\sin[a_i(r)-a_i(c)]\omega_i}{2[a_i(r)-a_i(c)]} & a_i(r) \neq a_i(c) \end{cases}$$

and $r, c = 0, 1, \ldots, (M_1+1)(M_2+1) - 1$. If we define

$$\mathbf{Q}_p = \iint_{\boldsymbol{\omega}\in\Omega_{pb}} \mathbf{W}(\boldsymbol{\omega})\, d\boldsymbol{\omega}$$

where $\mathbf{Q}_p$ is calculated by evaluating $\mathbf{Q}_1(\omega_1) = \left[Q_{rc}^{1}(\omega_1)\right]$ and $\mathbf{Q}_2(\omega_2) = \left[Q_{rc}^{2}(\omega_2)\right]$ at the appropriate limits and performing an element by element product of the resulting matrices, and also define the terms,

$$\mathbf{R}(\boldsymbol{\omega}) = \iint H_d(\boldsymbol{\omega})\mathbf{s}(\boldsymbol{\omega})d\boldsymbol{\omega} \qquad \mathbf{R}_p = \iint_{\boldsymbol{\omega}\in\Omega_{pb}} H_d(\boldsymbol{\omega})\mathbf{s}(\boldsymbol{\omega})d\boldsymbol{\omega}$$

$$\gamma(\boldsymbol{\omega}) = \iint H_d^2(\boldsymbol{\omega})d\boldsymbol{\omega} \qquad \gamma_p = \iint_{\boldsymbol{\omega}\in\Omega_{pb}} H_d^2(\boldsymbol{\omega})d\boldsymbol{\omega}$$

then Equation (24) can be written as

$$J_{pb}(\mathbf{X}) = \frac{1}{m(\Omega_{pb})}\left[\mathbf{X}^T\mathbf{F}^T\mathbf{Q}_p\mathbf{F}\mathbf{X} - 2\mathbf{X}^T\mathbf{F}^T\mathbf{R}_p + \gamma_p\right] \tag{25}$$

For filters which approximate constant values in their passbands, the expression for $\mathbf{R}(\boldsymbol{\omega})$ can be simplified. If $H_d(\boldsymbol{\omega})$ is equal to a constant in the passband, then without loss of generality, we can let $H_d(\boldsymbol{\omega}) = 1$, and Equation (25) becomes

$$J_{pb}(\mathbf{X}) = \frac{1}{m(\Omega_{pb})}\left[\mathbf{X}^T\mathbf{F}^T\mathbf{Q}_p\mathbf{F}\mathbf{X} - 2\mathbf{X}^T\mathbf{F}^T\mathbf{R}_p\right] + 1 \tag{26}$$

Because the elements in the vector, $\mathbf{s}(\boldsymbol{\omega})$, are separable, the matrix, $\mathbf{R}(\boldsymbol{\omega})$, can be written as

$$\mathbf{R}(\boldsymbol{\omega}) = \iint \mathbf{s}(\boldsymbol{\omega})d\boldsymbol{\omega} = \left[R_{rc}(\omega_1,\omega_2)\right] = \left[R_{rc}^{1}(\omega_1)R_{rc}^{2}(\omega_2)\right]$$

where

$$R_{rc}^{i}(\omega_i)=\begin{cases}\omega_i & a_i(r)=0\\ \sin[a_i(r)\omega_i]/a_i(r) & a_i(r)\neq 0\end{cases}$$

This expression for $\mathbf{R}(\boldsymbol{\omega})$ eliminates the need for integrating $H_d(\boldsymbol{\omega})\mathbf{s}(\boldsymbol{\omega})$ when calculating $\mathbf{R}_p$. The matrix, $\mathbf{R}_p$, in Equation (26) can be calculated by evaluating $\mathbf{R}_1(\omega_1)=\left[R_{rc}^{1}(\omega_1)\right]$ and $\mathbf{R}_2(\omega_2)=\left[R_{rc}^{2}(\omega_2)\right]$ at the appropriate limits and performing an element by element product of the resulting matrices.

An approximation of a desired amplitude response can also be obtained by imposing constraints on the amplitude response and its derivatives at particular frequencies. For Example,

$$\begin{array}{ccc}
H_r(\boldsymbol{\omega})\big|_{\boldsymbol{\omega}=\boldsymbol{\omega}_0}=K_{000} & \cdots & H_r(\boldsymbol{\omega})\big|_{\boldsymbol{\omega}=\boldsymbol{\omega}_l}=K_{00l}\\
\vdots & & \vdots\\
\left.\dfrac{\partial^{(m+n)}H_r(\boldsymbol{\omega})}{\partial\omega_1^m\partial\omega_2^n}\right|_{\boldsymbol{\omega}=\boldsymbol{\omega}_0}=K_{mn0} & \cdots & \left.\dfrac{\partial^{(m+n)}H_r(\boldsymbol{\omega})}{\partial\omega_1^m\partial\omega_2^n}\right|_{\boldsymbol{\omega}=\boldsymbol{\omega}_l}=K_{mnl}\\
\vdots & & \vdots\\
\left.\dfrac{\partial^{2p}H_r(\boldsymbol{\omega})}{\partial\omega_1^p\partial\omega_2^p}\right|_{\boldsymbol{\omega}=\boldsymbol{\omega}_0}=K_{pp0} & \cdots & \left.\dfrac{\partial^{2p}H_r(\boldsymbol{\omega})}{\partial\omega_1^p\partial\omega_2^p}\right|_{\boldsymbol{\omega}=\boldsymbol{\omega}_l}=K_{ppl}
\end{array}$$

where the K's are constants. Using Equation (22), these passband constraints can be written as

$$\frac{\partial^{(m+n)}H_r(\boldsymbol{\omega})}{\partial\omega_1^m\partial\omega_2^n}=\frac{\partial^{(m+n)}\mathbf{s}^T(\boldsymbol{\omega})}{\partial\omega_1^m\partial\omega_2^n}\mathbf{FX}.$$

If we let

$$\mathbf{C}^T=\left[\mathbf{s}(\boldsymbol{\omega}_0)\quad \mathbf{s}(\boldsymbol{\omega}_1)\quad\cdots\quad \left.\frac{\partial^2\mathbf{s}(\boldsymbol{\omega})}{\partial\omega_1\partial\omega_2}\right|_{\boldsymbol{\omega}=\boldsymbol{\omega}_0}\quad\cdots\quad \left.\frac{\partial^{(2p)}\mathbf{s}(\boldsymbol{\omega})}{\partial\omega_1^p\partial\omega_2^p}\right|_{\boldsymbol{\omega}=\boldsymbol{\omega}_l}\right]$$

and

$$\mathbf{K}^T=\left[K_{000}\quad K_{001}\quad\cdots\quad K_{110}\quad\cdots\quad K_{ppl}\right],$$

the passband constraints can be written in the matrix form,

$$\mathbf{CFX}=\mathbf{K}. \tag{27}$$

If we let J_{sb} represent the mean square error over the stopband frequencies, then

$$J_{sb} = \frac{1}{m(\Omega_{sb})} \iint_{\boldsymbol{\omega} \in \Omega_{sb}} |H_r(\boldsymbol{\omega}) - H_d(\boldsymbol{\omega})|^2 \, d\boldsymbol{\omega} \tag{28}$$

where

$$\Omega_{sb} = \{(\omega_1, \omega_2) : \omega_1 \in \text{stopband}, \omega_2 \in \text{stopband}\}.$$

Because $H_d(\boldsymbol{\omega})$ equals zero in the stopband, Equation (28) can be written as

$$J_{sb} = \frac{1}{m(\Omega_{sb})} \iint_{\boldsymbol{\omega} \in \Omega_{sb}} H_r^2(\boldsymbol{\omega}) d\boldsymbol{\omega} \tag{29}$$

Substituting Equation (22) into Equation (29),

$$J_{sb}(\mathbf{X}) = \frac{1}{m(\Omega_{sb})} \mathbf{X}^T \mathbf{F}^T \iint_{\boldsymbol{\omega} \in \Omega_{sb}} \mathbf{s}(\boldsymbol{\omega}) \mathbf{s}^T(\boldsymbol{\omega}) \, d\boldsymbol{\omega} \quad \mathbf{FX} \tag{30}$$

Recall that earlier, we let $\mathbf{W}(\boldsymbol{\omega}) = \mathbf{s}(\boldsymbol{\omega})\mathbf{s}^T(\boldsymbol{\omega})$ and

$$\mathbf{Q}(\boldsymbol{\omega}) = \int\int \mathbf{W}(\boldsymbol{\omega}) \, d\boldsymbol{\omega}.$$

Thus, if we define

$$\mathbf{Q}_s = \iint_{\boldsymbol{\omega} \in \Omega_{sb}} \mathbf{W}(\boldsymbol{\omega}) \, d\boldsymbol{\omega}$$

where $\mathbf{Q}_s$ is calculated by evaluating $\mathbf{Q}_1(\omega_1)$ and $\mathbf{Q}_2(\omega_2)$ at the appropriate limits and performing an element by element product of the resulting matrices, Equation (30) can be written as

$$J_{sb}(\mathbf{X}) = \frac{1}{m(\Omega_{sb})} \mathbf{X}^T \mathbf{F}^T \mathbf{Q}_s \, \mathbf{FX}. \tag{31}$$

To generate an efficient realization of a linear phase FIR filter with a frequency sampling structure, the frequency samples in the stopband are constrained equal to zero. Recall that the vector, $\mathbf{X}$, contains the frequency samples. In Equations (25), (27) and (31), constraining frequency samples to zero is equivalent to removing elements in the vector, $\mathbf{X}$, that contain the corresponding frequency samples and also removing the corresponding columns from the matrix, $\mathbf{F}$. If we define these reduced terms as $\mathbf{X}_c$ and $\mathbf{F}_c$, then the design problem can now be stated as follows. Minimize the error function

$$J(\mathbf{X}_c) = \alpha J_{pb}(\mathbf{X}_c) + (1-\alpha) J_{sb}(\mathbf{X}_c)$$
$$= \frac{1-\alpha}{m(\Omega_{sb})} \mathbf{X}_c^T \mathbf{F}_c^T \mathbf{Q}_s \mathbf{F}_c \mathbf{X}_c$$
$$+ \frac{\alpha}{m(\Omega_{pb})} \left[\mathbf{X}_c^T \mathbf{F}_c^T \mathbf{Q}_p \mathbf{F}_c \mathbf{X}_c - 2 \mathbf{X}_c^T \mathbf{F}_c^T \mathbf{R}_p + \gamma_p \right] \quad (32)$$

where $0 \leq \alpha \leq 1$ subject to the constraint, $\mathbf{CF}_c\mathbf{X}_c = \mathbf{K}$. The scalar term, α, in Equation (32) allows the designer to weight the relative importance between the mean square error over the passbands and the mean square error over the stopbands.

If the number of non-redundant constraints in Equation (27) is less than the number of elements in the vector $\mathbf{X}_c$, then the remaining degrees of freedom are used to minimize $J(\mathbf{X}_c)$. If the number of non-redundant constraints is equal to the number of elements in the vector $\mathbf{X}_c$, then the design becomes a linear algebra problem and the frequency samples are unique for the constraints. And if the number of non-redundant constraints is greater than the number of elements in the vector $\mathbf{X}_c$, then no solution will exist which satisfies all the constraints.

Assuming that the number of non-redundant constraints in Equation (27) is less than the number of elements in the vector $\mathbf{X}_c$, the design problem can be solved by defining a Lagrange multiplier vector as

$$\boldsymbol{\lambda}^T = \left[\lambda_{000} \quad \lambda_{001} \quad \cdots \quad \lambda_{110} \quad \cdots \quad \lambda_{ppl} \right]$$

and minimizing the augmented cost function,

$$J_a(\mathbf{X}_c, \boldsymbol{\lambda}) = \alpha J_{pb}(\mathbf{X}_c) + (1-\alpha) J_{sb}(\mathbf{X}_c) + \boldsymbol{\lambda}^T (\mathbf{CF}_c \mathbf{X}_c - \mathbf{K}). \quad (33)$$

Substituting the appropriate expressions for J_{pb} and J_{sb} into Equation (33),

$$J_a(\mathbf{X}_c, \boldsymbol{\lambda}) = \frac{\alpha}{m(\Omega_{pb})} \left[\mathbf{X}_c^T \mathbf{F}_c^T \mathbf{Q}_p \mathbf{F}_c \mathbf{X}_c - 2 \mathbf{X}_c^T \mathbf{F}_c^T \mathbf{R}_p + \gamma_p \right]$$
$$+ \frac{1-\alpha}{m(\Omega_{sb})} \mathbf{X}_c^T \mathbf{F}_c^T \mathbf{Q}_s \mathbf{F}_c \mathbf{X}_c + \boldsymbol{\lambda}^T (\mathbf{CF}_c \mathbf{X}_c - \mathbf{K})$$

The necessary conditions for an optimal solution are

$$\frac{\partial J_a(\mathbf{X}_c, \boldsymbol{\lambda})}{\partial \mathbf{X}_c} = \mathbf{0} \quad (34)$$

and

$$\frac{\partial J_a(\mathbf{X}_c, \boldsymbol{\lambda})}{\partial \boldsymbol{\lambda}} = \mathbf{0}. \tag{35}$$

Because $\mathbf{F}_c^T \mathbf{Q}_p \mathbf{F}_c$ and $\mathbf{F}_c^T \mathbf{Q}_s \mathbf{F}_c$ are symmetric matrices, Equation (34) becomes

$$\frac{2\alpha}{m(\Omega_{pb})}\left[\mathbf{F}_c^T \mathbf{Q}_p \mathbf{F}_c \mathbf{X}_c - \mathbf{F}_c^T \mathbf{R}_p\right] + \frac{2(1-\alpha)}{m(\Omega_{sb})} \mathbf{F}_c^T \mathbf{Q}_s \mathbf{F}_c \mathbf{X}_c + \mathbf{F}_c^T \mathbf{C}^T \boldsymbol{\lambda} = \mathbf{0}. \tag{36}$$

Equation (35) implies

$$\frac{\partial J_a(\mathbf{X}_c, \boldsymbol{\lambda})}{\partial \boldsymbol{\lambda}} = \mathbf{C}\mathbf{F}_c \mathbf{X}_c - \mathbf{K} = \mathbf{0}. \tag{37}$$

Equations (36) and (37) can be written in the matrix form,

$$\left[\begin{array}{c|c} \frac{2\alpha}{m(\Omega_{pb})}\mathbf{F}_c^T \mathbf{Q}_p \mathbf{F}_c + \frac{2(1-\alpha)}{m(\Omega_{sb})}\mathbf{F}_c^T \mathbf{Q}_s \mathbf{F}_c & \mathbf{F}_c^T \mathbf{C}^T \\ \hline \mathbf{C}\mathbf{F}_c & \mathbf{0} \end{array}\right] \left[\begin{array}{c} \mathbf{X}_c \\ \hline \boldsymbol{\lambda} \end{array}\right] = \left[\begin{array}{c} \frac{2\alpha}{m(\Omega_{pb})}\mathbf{F}_c^T \mathbf{R}_p \\ \hline \mathbf{K} \end{array}\right] \tag{38}$$

A solution to Equation (38) will exist when $\mathbf{C}$ has full rank which occurs when $\mathbf{C}$ contains no redundant or trivial constraints.

IV. Example

A. Example of a Type 1-1 Frequency Sampling Filter

In this example, a Type 1-1 linear phase fourfold symmetric frequency sampling filter that approximates the amplitude response,

$$H_d(\boldsymbol{\omega}) = \begin{cases} 1 & \boldsymbol{\omega} \in \Omega_{pb} \\ 0 & \boldsymbol{\omega} \in \Omega_{sb} \end{cases},$$

where

$$\Omega_{pb} = \{(\omega_1, \omega_2) : 0 \le \omega_1 \le 0.1\pi, 0 \le \omega_2 \le 0.1\pi\}$$

$$\Omega_{sb} = \{(\omega_1, \omega_2) : 0.13\pi \le \omega_1 \le \pi \cup 0.13\pi \le \omega_2 \le \pi\}$$

is designed. The filter will be designed to have an impulse response that has a region of support R_N where

$$R_N = \{(n_1, n_2) : 0 \le n_1 \le 100, 0 \le n_2 \le 100\}.$$

To approximate the passband response, the mean square error over the passband will be minimized subject to the constraints

$$H_r(\boldsymbol{\omega})\Big|_{\boldsymbol{\omega}=(0,0)} = 1$$

$$\left.\frac{\partial^{(m+n)} H_r(\boldsymbol{\omega})}{\partial\omega_1^m \partial\omega_2^n}\right|_{\boldsymbol{\omega}=(0,0)} = 0$$

where $n, m = 0, 1, 2, 3, 4$ excluding $m = n = 0$. To approximate the stopband response, the scalar term, α, which weights the relative importance between the mean square error over the passbands and the mean square error over the stopbands, will be set to 0.1. Substituting $\alpha = 0.1$ into Equation (38) gives

$$\left[\begin{array}{c|c} \frac{0.2}{m(\Omega_{pb})}\mathbf{F}_c^T\mathbf{Q}_p\mathbf{F}_c + \frac{1.8}{m(\Omega_{sb})}\mathbf{F}_c^T\mathbf{Q}_s\mathbf{F}_c & \mathbf{F}_c^T\mathbf{C}^T \\ \hline \mathbf{C}\mathbf{F}_c & \mathbf{0} \end{array}\right]\left[\begin{array}{c} \mathbf{X}_c \\ \hline \boldsymbol{\lambda} \end{array}\right] = \left[\begin{array}{c} \frac{0.2}{m(\Omega_{pb})}\mathbf{F}_c^T\mathbf{R}_p \\ \hline \mathbf{K} \end{array}\right] \quad (39)$$

where $m\left(\Omega_{pb}\right)=(0.1\pi)^2$ and $m\left(\Omega_{sb}\right)=\pi^2-(0.13\pi)^2$.

To solve Equation (39), the matrix, $\mathbf{C}$, must have full rank, and the nonzero frequency samples must be determined. Because the odd ordered partial derivatives of a linear phase filter evaluated at $\boldsymbol{\omega} = (0,0)$ are zero, constraining them is redundant, and they must be omitted. Thus, for the matrix, $\mathbf{C}$, to have full rank, only the constraints

$$H_r(\boldsymbol{\omega})\Big|_{\boldsymbol{\omega}=(0,0)} = 1$$

$$\left.\frac{\partial^{(m+n)} H_r(\boldsymbol{\omega})}{\partial\omega_1^m \partial\omega_2^n}\right|_{\boldsymbol{\omega}=(0,0)} = 0$$

where $n, m = 0, 2, 4$ excluding $m = n = 0$ are considered. The nonzero frequency samples include all of the frequency samples not in the stopbands. In this example, the stopband frequency samples are

$$H(\mathbf{k}) \text{ for } \mathbf{k} \in \left\{(k_1, k_2) : 7 \le k_1 \le 50 \bigcup 7 \le k_2 \le 50\right\}$$

which implies that the nonzero frequency samples are

$$H(\mathbf{k}) \text{ for } \mathbf{k} \in \left\{(k_1, k_2) : 0 \le k_1 \le 6, 0 \le k_2 \le 6\right\}.$$

Substituting the appropriate matrix values into Equation (39), the vector, $\mathbf{X}_c$, which contains the values of the nonzero frequency samples can be calculated. Table 1 shows the filter's nonzero frequency samples, and Figure 1 shows the filter's magnitude response. Figure 2 shows the filter's magnitude response in dB and Figure 3 shows the magnitude response of the filter's passband in detail. Using Equation(4), the filter's phase is

$$\arg\left[H_{\omega}(\omega_1,\omega_2)\right] = -50\omega_1 - 50\omega_2 .$$

V. Summary and Conclusions

In this chapter, four types of frequency sampling filters were developed. For each type of filter, system functions were developed for filters with arbitrary impulse responses and arbitrary frequency responses; filters with real impulse responses and linear phase; and filters with real impulse responses, linear phase and fourfold symmetry. Also, in this chapter, a technique for designing frequency sampling filters with real impulse responses, linear phase and fourfold symmetry was developed. This design technique approximates a desired frequency response by minimizing a weighted mean square error over the stopbands and passbands subject to constraints on the filter's amplitude response. This approximation technique results in a set of linear equations which are solved to determine the filter's coefficients. This design method applies to each of the four types of frequency sampling filters developed in this chapter.

Table 1. Nonzero Frequency Samples for the Example.

	$H(0,0)$	= 1.0000000	$H(2,3)$	= $H(3,2)$	= 0.9967544
$H(0,1)$	= $H(1,0)$	= 0.9999942	$H(2,4)$	= $H(4,2)$	= 0.9926905
$H(0,2)$	= $H(2,0)$	= 0.9998744	$H(2,5)$	= $H(5,2)$	= 0.9701703
$H(0,3)$	= $H(3,0)$	= 0.9940741	$H(2,6)$	= $H(6,2)$	= 0.5011298
$H(0,4)$	= $H(4,0)$	= 0.9953709		$H(3,3)$	= 1.0131842
$H(0,5)$	= $H(5,0)$	= 0.9704475	$H(3,4)$	= $H(4,3)$	= 0.9991182
$H(0,6)$	= $H(6,0)$	= 0.5266080	$H(3,5)$	= $H(5,3)$	= 0.9777624
	$H(1,1)$	= 1.0005610	$H(3,6)$	= $H(6,3)$	= 0.4530710
$H(1,2)$	= $H(2,1)$	= 1.0064317		$H(4,4)$	= 0.9915667
$H(1,3)$	= $H(3,1)$	= 1.0057280	$H(4,5)$	= $H(5,4)$	= 0.9719154
$H(1,4)$	= $H(4,1)$	= 1.0035502	$H(4,6)$	= $H(6,4)$	= 0.4941636
$H(1,5)$	= $H(5,1)$	= 0.9791734		$H(5,5)$	= 0.9312007
$H(1,6)$	= $H(6,1)$	= 0.5152585	$H(5,6)$	= $H(6,5)$	= 0.3930366
	$H(2,2)$	= 0.9951068		$H(6,6)$	= 0.0884227

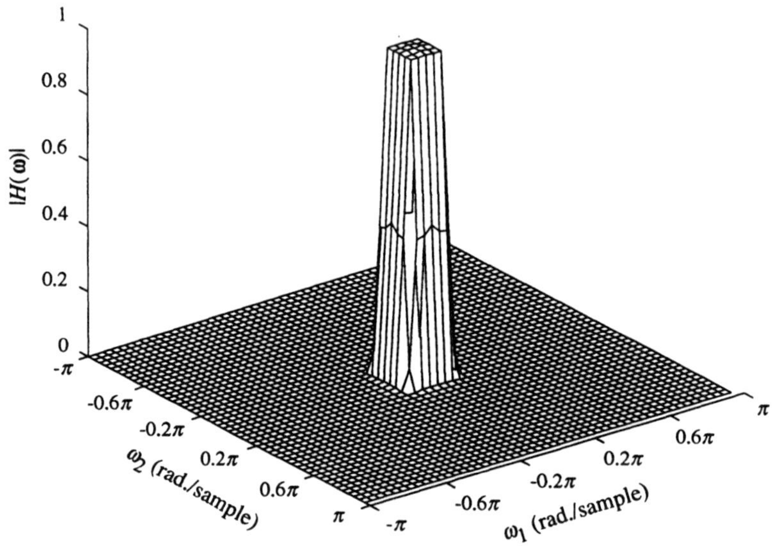

Figure 1. Magnitude Response of the Type 1-1 Frequency Sampling Filter in the Example.

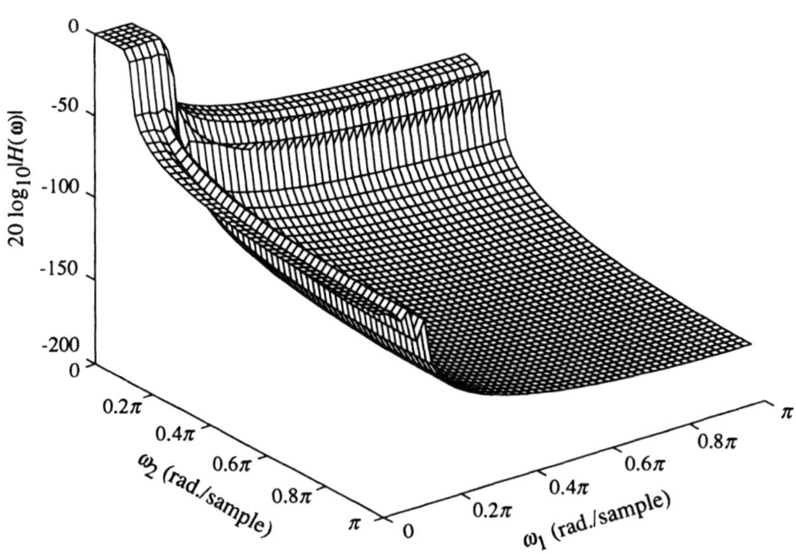

Figure 2. Magnitude Response of the Type 1-1 Frequency Sampling Filter in the Example.

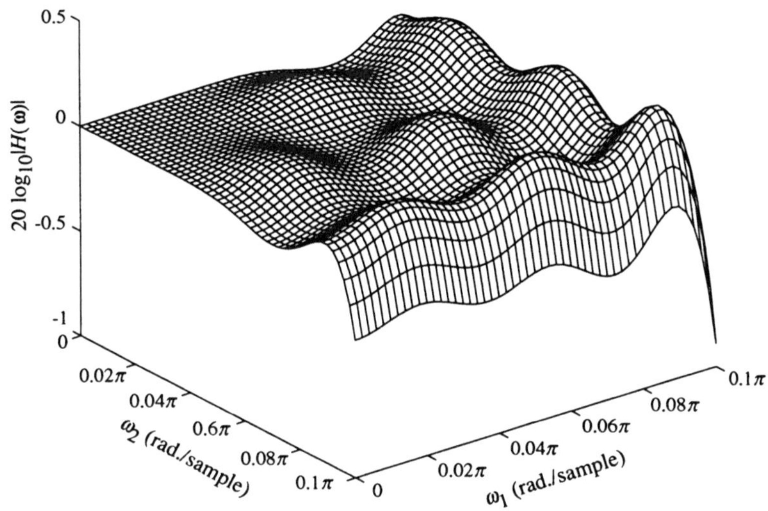

Figure 3. Magnitude Response of the Passband for the Type 1-1 Frequency Sampling Filter in the Example.

VI. References

1. L. R. Rabiner and B. Gold, *Theory and Application of Digital Signal Processing*, Prentice Hall, New Jersey (1975).

2. D. E. Dudgeon and R. M. Mersereau, *Multidimensional Digital Signal Processing*, Prentice Hall, New Jersey (1984).

3. J. V. Hu and L. R. Rabiner, "Design Techniques for Two-Dimensional Filters," *IEEE Transactions on Audio and Electroacoustics*, Vol. AU-20, No. 4, pp. 249-257 (1972).

4. J. S. Lim, *Two-Dimensional Signal and Image Processing*, Prentice Hall, New Jersey (1990).

Unified Bias Analysis of Subspace-Based DOA Estimation Algorithms

Fu Li
Yang Lu
Department of Electrical Engineering
Portland State University
Portland, OR 97207-0751

I. INTRODUCTION

Motivated by the increasing demands in applications such as radar and sonar detection, geophysical exploration, telecommunications, and biomedical science, senor array signal processing has been a very active research field for several decades. One of the principal task in array processing is to estimate directions of signals impinging simultaneously on an array of senors. Many direction-of-arrival (DOA) estimation algorithms have been proposed which can be classified as pre-subspace algorithms such as Maximum Entropy Method and Minimum Variance Method, and subspace-based algorithms. Among them, subspace-based DOA estimation algorithms are the most popular ones due to their relatively high resolution capability and low computational complexity.

There exist different subspace algorithms which estimate DOAs by exploiting underlying signal models in different fashions. Therefore the performance analysis of these algorithms for the purpose of justification and comparison is important. Early studies on performance were based on simulations. Pioneered by Kaveh and Barabell [1], analytical evaluation of mean-squared DOA estimation error for subspace algorithms has attracted much excellent research (see [2] for a complete references). However, the bias behavior of subspace-based estimators remains largely unexplored, even though bias is of great importance in many applications such as predicting resolution threshold. This is because of theoretical difficulty and

CONTROL AND DYNAMIC SYSTEMS, VOL. 77

mathematical complexity involved in bias analysis. Kaveh and Barabell [1] addressed bias of estimated spectrum instead of estimated parameters. Wang and Wakefield [3] analyzed bias of estimated non-asymptotically, but their predicted bias does not match empirical results well even in the case where asymptotic measurements are used; Xu and Buckley [4] derived the bias and variance expressions for MUSIC and Min-Norm estimates using the asymptotic distribution of the estimated subspace vectors. Li and Lu [5] analyzed the bias for ESPRIT, TAM, and Matrix-Pencil based on a finite amount of array measurements.

In this paper, efforts have been directed at developing a unified bias analysis of DOA estimation for several subspace-based algorithms in terms of physical parameters such as signal-to-noise ratio as opposed to previous individual analysis expressed in parameters which depend on the specific data trials such as eigenvectors. Thus this analysis can provide us with insights into different algorithms to observe the direct relationship between the performance of algorithms and signal measurement conditions.

NOMENCLATURE:

1. Matrix Operations:

   $[\cdot]^T$, $[\cdot]^H$, $[\cdot]^\dagger$ – transpose, conjugate transpose, left pseudo-inverse of matrix $[\cdot]$

   $[\cdot]^\uparrow$ or $[\cdot]^\downarrow$ – matrix $[\cdot]$ with the first or last row deleted

   $\Re[\cdot]$ or $\Im[\cdot]$ – real or imaginary part of matrix $[\cdot]$

   $Tr[\cdot]$ – trace of matrix $[\cdot]$

2. Principal Symbols:

   K, L, M – number of sources, sensors, and snapshots

   θ_k – the direction of the k-th arrival signal

   Θ – location region of interest

   σ_n^2 – noise power

   $\mathbf{a}(\theta_k)$, $\mathbf{A}(\theta)$ – array response vector or matrix

$\mathbf{e}_k^H$ – $[0 \cdots 1 \cdots 0]$, 1 in the k-th position

I or **W** – identity or weighting matrix

S or **N** – signal or noise matrix

$\mathbf{R}_s$ – Source covariance matrix

$\boldsymbol{\Omega}_s$ or $\boldsymbol{\Omega}_o$ – signal- or noise- subspace projection matrix. $\boldsymbol{\Omega}_s = \mathbf{A}(\mathbf{A}^H\mathbf{A})^{-1}\mathbf{A}^H = \mathbf{I} - \boldsymbol{\Omega}_o$

3. Statistical Symbols:

$E\{\cdot\}$ – mathematical expectation of $\{\cdot\}$

$\stackrel{i}{=}$ – equal up to the order of $||\mathbf{N}||^i$

$[\cdot]_1$ or $[\cdot]_2$ – 1-th or 2-nd order approximation of noise-corrupted matrix $[\cdot]$

II. REVIEW OF DOA ESTIMATION ALGORITHMS

Consider K signals (usually random processes) simultaneously incident on an uniform line array of L equally spaced sensors. The objective is to estimate directions of arrival signals from the measurements taken simultaneously from all the senors at M different time instants. The following assumptions are made about signal, noise and array structure.

- The K signal wavefronts are narrow-band planar wavefronts with the same center wavelength λ_c and are uncorrelated to each other. The number of signals is predetermined which satisfies $K < L$.
- The observation noise n_i at each senor is circular, stationary, additive complex white Gaussian process with zero mean and variance σ_n^2, uncorrelated to the signals.
- The array configuration is known and the response has been calibrated. For simplicity, we only consider uniform line array.

Under these assumptions, the signals arrived at the i-th senor at time t is

$$
\begin{aligned}
y_i(t) &= \sum_{k=1}^{K} e^{j\frac{2\pi d}{\lambda_c}(i-1)\sin\theta_k} s_k(t) \\
&= \left(e^{j\frac{2\pi d}{\lambda_c}(i-1)\sin\theta_1} \cdots e^{j\frac{2\pi d}{\lambda_c}(i-1)\sin\theta_K}\right) \begin{pmatrix} s_1(t) \\ \vdots \\ s_K(t) \end{pmatrix}, \qquad (1)
\end{aligned}
$$

where $s_k(t)$ is the k-th signal arrived at the angle θ_k. The signals arrived at all sensors at time t are

$$
\mathbf{y}(t) = \begin{pmatrix} 1 & \cdots & 1 \\ e^{j\frac{2\pi d}{\lambda_c}\sin\theta_1} & \cdots & e^{j\frac{2\pi d}{\lambda_c}\sin\theta_K} \\ \vdots & \ddots & \vdots \\ e^{j\frac{2\pi d}{\lambda_c}(L-1)\sin\theta_1} & \cdots & e^{j\frac{2\pi d}{\lambda_c}(L-1)\sin\theta_K} \end{pmatrix} \begin{pmatrix} s_1(t) \\ \vdots \\ s_K(t) \end{pmatrix} \qquad (2)
$$

or in vector form

$$
\mathbf{y}(t) \stackrel{\text{def}}{=} [\mathbf{a}(\theta_1)\cdots\mathbf{a}(\theta_K)]\mathbf{s}(t) \stackrel{\text{def}}{=} \mathbf{A}(\theta)\mathbf{s}(t) \qquad (3)
$$

where $\mathbf{A}(\theta)$ (and $\mathbf{a}(\theta)$) represents the array characteristics. Take M snapshots of signals at each senor to form the data matrix

$$
\mathbf{Y} = \begin{pmatrix} y_1(1) & \cdots & y_1(M) \\ \vdots & \ddots & \vdots \\ y_L(M) & \cdots & y_L(M) \end{pmatrix} = \mathbf{A}(\theta)\mathbf{S} \qquad (4)
$$

where

$$
\mathbf{S} \stackrel{\text{def}}{=} [\mathbf{s}(1)\cdots\mathbf{s}(M))] = \begin{pmatrix} s_1(1) & \cdots & s_1(M)) \\ \vdots & \ddots & \vdots \\ s_K(1) & \cdots & s_K(M) \end{pmatrix}. \qquad (5)
$$

The direction-finding problem can be accomplished by estimating $(\theta_1\cdots\theta_K)$ from array output $\mathbf{Y}$.

A. SUBSPACE DECOMPOSITION

It is noticed from Eq.(4) that both the matrices $\mathbf{A}(\theta)$ $(L\times K)$ and $\mathbf{S}$ $(K\times M)$ have a rank of K, data matrix $\mathbf{Y}$ $(L\times M)$ is a rank-deficient matrix, thus the subspace decomposition can be identically performed [6] either directly on $\mathbf{Y}$ by singular value decomposition (SVD) or on the sample covariance matrix $\tilde{\mathbf{R}} = \frac{1}{M}\tilde{\mathbf{Y}}\tilde{\mathbf{Y}}^H$ by an eigenvalue decomposition. For simplicity, we will employ SVD in our analysis. The subspace decomposition using SVD on the direct data matrix $\mathbf{Y}$ is as follows:

$$\mathbf{Y} = \mathbf{U}\mathbf{\Sigma}\mathbf{V}^H = \begin{pmatrix} \mathbf{U}_s & \mathbf{U}_o \end{pmatrix} \begin{pmatrix} \mathbf{\Sigma}_s & 0 \\ 0 & 0 \end{pmatrix} \begin{pmatrix} \mathbf{V}_s^H \\ \mathbf{V}_o^H \end{pmatrix} \tag{6}$$

where $\mathbf{\Sigma}$ are singular value matrix with decreasing singular values on the diagonal as

$$\sigma_1 \geq \cdots \geq \sigma_K \geq 0 \tag{7}$$

and $\mathbf{U}_s$ $(L \times K)$ are singular vectors associated with the K non-zero singular values while $\mathbf{U}_o$ $(L \times (L-K))$ are singular vectors associated with $L-K$ zero singular values. There are two fundamental properties of the subspace upon which subspace-based DOA algorithms are:

- The vectors in $\mathbf{U}_s$ span the signal subspace which is the column space of the array manifold $\mathbf{A}(\theta)$.
- The vectors in $\mathbf{U}_o$ span the orthogonal subspace which is the orthogonal complement of signal subspace spanned by the array manifold $\mathbf{A}(\theta)$, i. e.

$$\mathbf{a}(\theta_k)^H \mathbf{U}_o = 0, \quad k = 1, \cdots, K. \tag{8}$$

B. MUSIC

MUSIC (MUitiple SIgnal Classification algorithm) was proposed by Schmidt [7]. It utilizes Eq.(8) to perform an one-dimensional search for K zeros over the null-spectrum

$$P_{MU}(\theta) = \mathbf{a}(\theta)^H \mathbf{U}_o \mathbf{U}_o^H \mathbf{a}(\theta) \tag{9}$$

when noise presents, null-spectrum $P_{MU}(\theta)$ reaches K minima around the true DOAs.

In this and future equations, the symbol θ without a subscript is a scalar variable which represents a possible direction of arrival, while the subscripted symbol θ_k, $(k = 1, \cdots, K)$ is referred to the actual directions of arrival in the noise-free data.

C. MIN-NORM

Min-Norm (Minimum-Norm algorithm) was introduced by Kumaresan and Tufts [8] to identify a single vector $\mathbf{d}$ in the span of orthogonal subspace with unity first element and minimum Euclidian norm. This vector is constructed by linear combination of all $L - K$ vectors in $\mathbf{U}_o$ as

$$\mathbf{d} = \mathbf{U}_o \frac{\mathbf{c}}{||\mathbf{c}||^2} \tag{10}$$

where $\mathbf{c}^H$ is the first row of $\mathbf{U}_o$. For uniform line array, DOA estimation is performed through polynomial rooting

$$D(z) = \mathbf{a}^T(z)\mathbf{d} = \prod_{i=1}^{L-1}(1 - r_i^* z). \tag{11}$$

The K roots on the unit circle with $r_i = e^{j\frac{2\pi d}{\lambda_c}\sin\theta_i}$ $(i = 1, \cdots, K)$ contain the DOA information while the rest $L - K - 1$ are regarded as extraneous roots. In the noisy case, K roots with the largest amplitudes are chosen as the signal-roots and the rest are referred as noise roots.

Min-Norm is also applicable to arbitrary array geometry as proposed by Li *et al* [9] by searching for the K zeros of the null-spectrum over θ

$$P_{MN}(\theta) = |\mathbf{a}^H(\theta)\mathbf{d}|^2. \tag{12}$$

D. ROOT-MUSIC

Root-MUSIC [10] only applies to uniform line array. It forms and roots the null-spectrum polynomial

$$P_{RM}(z) = \mathbf{a}(z^{-1})^T \mathbf{U}_o \mathbf{U}_o^H \mathbf{a}(z) = A \prod_{i=1}^{L-1}(1 - r_i z^{-1})(1 - r_i^* z). \tag{13}$$

Root-MUSIC, unlike Min-Norm, always chooses the K roots with largest amplitudes *inside* the unit-circle.

E. ESPRIT

ESPRIT (Estimation of Signal Parameters via Rotational Invariant Techniques) [11] utilizes two identical sub-arrays which are physically displaced from each other by a known displacement Δ to obtain a shift-invariant structure. When a uniform line array is used, we can choose the first $L-1$ sensors as first sub-array and last $L-1$ sensors as second sub-array for the maximum sub-array apertures (then $\Delta = d$). The shift-invariance can be expressed using array manifold

$$\mathbf{A}^{\downarrow}\mathbf{D} = \mathbf{A}^{\uparrow} \tag{14}$$

where $\mathbf{A}^{\downarrow}$ and $\mathbf{A}^{\uparrow}$ are the first $L-1$ rows and last $L-1$ rows of $\mathbf{A}$. The diagonal matrix $\mathbf{D}$ has the diagonal element $\lambda_k = e^{j\frac{2\pi d}{\lambda_c}\sin\theta_k}$. In practice, we obtain $\mathbf{D}$ as the eigenvalue matrix of $\mathbf{F}_{es}$

$$\mathbf{U}_s^{\downarrow}\mathbf{F}_{es} = \mathbf{U}_s^{\uparrow} \tag{15}$$

where $\mathbf{U}_s^{\downarrow}$ and $\mathbf{U}_s^{\uparrow}$ are the first $L-1$ rows and last $L-1$ rows of $\mathbf{U}_s$, respectively. $\mathbf{F}_{es}$ can be obtained as

$$\mathbf{F}_{es} = \mathbf{U}_s^{\downarrow\dagger}\mathbf{U}_s^{\uparrow} \tag{16}$$

$\mathbf{D}$ is related to $\mathbf{F}_{es}$ through an eigenvalue decomposition

$$\mathbf{D} = \mathbf{L}^{-1}\mathbf{F}_{es}\mathbf{L}. \tag{17}$$

F. STATE-SPACE REALIZATION

Under the assumption of using uniform line array, the state space realization approach (SSR, i.e. TAM) [12] and Matrix-Pencil Method [13] have the same DOA estimates [2].

III. SUBSPACE PERTURBATION

In [6], Li extensively studied the first-order subspace perturbations upon which a unified formula for mean-square error of subspace algorithms is derived. Unfortunately, zero bias results from the first-order subspace perturbations analysis, which indicates higher order perturbations should be included for an accurate bias prediction. In this section, the analysis for second-order subspace perturbations is developed, which provides a common foundation for bias analysis.

A. PERTURBATION DUE TO NOISE CORRUPTIONS

Various data perturbations are always presented in practice which result in the perturbation of estimated subspaces. The perturbed data matrix can be written as

$$\tilde{\mathbf{Y}} = \mathbf{Y} + \mathbf{N}$$

where $\mathbf{N}$ is the observation noise which is assumed to be *circular Gaussian* with zero mean. The subspace decomposition of perturbed data by SVD is

$$\tilde{\mathbf{Y}} = \tilde{\mathbf{U}}\tilde{\mathbf{\Sigma}}\tilde{\mathbf{V}}^H = \begin{pmatrix} \tilde{\mathbf{U}}_s & \tilde{\mathbf{U}}_o \end{pmatrix} \begin{pmatrix} \tilde{\mathbf{\Sigma}}_s & 0 \\ 0 & \tilde{\mathbf{\Sigma}}_o \end{pmatrix} \begin{pmatrix} \tilde{\mathbf{V}}_s^H \\ \tilde{\mathbf{V}}_o^H \end{pmatrix}. \tag{18}$$

We can now write

$$\tilde{\mathbf{U}}_o = \mathbf{U}_o + \Delta\mathbf{U}_o \quad \text{and} \quad \tilde{\mathbf{U}}_s = \mathbf{U}_s + \Delta\mathbf{U}_s \tag{19}$$

where $\Delta\mathbf{U}_o$ and $\Delta\mathbf{U}_s$ are the perturbations in the estimated orthogonal- and signal- subspaces. The following lemma was proved in [2].

Lemma: The perturbed orthogonal subspace is spanned by $\mathbf{U}_o + \mathbf{U}_s\mathbf{Q}$ and the perturbed signal-subspace is spanned by $\mathbf{U}_s + \mathbf{U}_o\mathbf{P}$, where $\mathbf{P}$ and $\mathbf{Q}$ are matrices whose norms are of the order of $||\mathbf{N}||$. The matrix norm can be any submultiplicative norm such as the Euclidean 2-norm or the Frobenius norm.

Notice that

$$\begin{aligned} \tilde{\mathbf{U}}_o^H\tilde{\mathbf{U}}_o &= (\mathbf{U}_o + \mathbf{U}_s\mathbf{Q})^H(\mathbf{U}_o + \mathbf{U}_s\mathbf{Q}) = \mathbf{I} + \mathbf{Q}^H\mathbf{Q} \\ \tilde{\mathbf{U}}_s^H\tilde{\mathbf{U}}_s &= (\mathbf{U}_s + \mathbf{U}_o\mathbf{P})^H(\mathbf{U}_s + \mathbf{U}_o\mathbf{P}) = \mathbf{I} + \mathbf{P}^H\mathbf{P} \end{aligned} \tag{20}$$

which imply that $\mathbf{U}_o + \mathbf{U}_s\mathbf{Q}$ and $\mathbf{U}_s + \mathbf{U}_o\mathbf{P}$ are not orthonormal. The orthonormal bases of perturbed subspaces are thus given as

$$\tilde{\mathbf{U}}_o = (\mathbf{U}_o + \mathbf{U}_s\mathbf{Q})(\mathbf{I} + \mathbf{Q}^H\mathbf{Q})^{-\frac{1}{2}} \tag{21}$$

$$\tilde{\mathbf{U}}_s = (\mathbf{U}_s + \mathbf{U}_o\mathbf{P})(\mathbf{I} + \mathbf{P}^H\mathbf{P})^{-\frac{1}{2}}. \tag{22}$$

The relationship between $\mathbf{P}$ and $\mathbf{Q}$ can be derived using the orthogonality between the perturbed orthogonal- and signal- subspaces

$$\begin{aligned}(\mathbf{I} + \mathbf{Q}^H\mathbf{Q})^{-\frac{1}{2}}(\mathbf{U}_o + \mathbf{U}_s\mathbf{Q})^H(\mathbf{U}_s + \mathbf{U}_o\mathbf{P})(\mathbf{I} + \mathbf{P}^H\mathbf{P})^{-\frac{1}{2}} &= 0 \\ (\mathbf{U}_o + \mathbf{U}_s\mathbf{Q})^H(\mathbf{U}_s + \mathbf{U}_o\mathbf{P}) &= 0 \\ \mathbf{P} + \mathbf{Q}^H &= 0. \end{aligned} \tag{23}$$

Similarly, it can be proved that

$$\tilde{\mathbf{V}}_o = (\mathbf{V}_o + \mathbf{V}_s\bar{\mathbf{Q}})(\mathbf{I} + \bar{\mathbf{Q}}^H\bar{\mathbf{Q}})^{-\frac{1}{2}} \tag{24}$$

$$\tilde{\mathbf{V}}_s = (\mathbf{V}_s + \mathbf{V}_o\bar{\mathbf{P}})(\mathbf{I} + \bar{\mathbf{P}}^H\bar{\mathbf{P}})^{-\frac{1}{2}} \tag{25}$$

where $\bar{\mathbf{Q}}$ and $\bar{\mathbf{P}}$ are matrices whose norms are also in the order of $||\mathbf{N}||$.

In the bias analysis, we are interested in the first-order and the second-order approximations of subspace perturbations. Define $[\cdot]_1$ and $[\cdot]_2$ as the first- and second- order approximations of $[\cdot]$ in terms of data perturbation $\mathbf{N}$, respectively, and use the notation $\stackrel{i}{=}$ to mean "equal up to the order of $||\mathbf{N}||^i$". The first-order approximation of orthogonal perturbed subspace is obtained by first-order expansion of Eq.(21) which is illustrated as follows: Considering

$$(\mathbf{I} + \mathbf{Q}^H\mathbf{Q})^{-\frac{1}{2}} = \mathbf{I} - \frac{1}{2}\mathbf{Q}^H\mathbf{Q} + \cdots \tag{26}$$

keep Eq.(26) up to zeroth-order and substitute it into Eq.(21) so as to expand the perturbation expansion to first-order,

$$\tilde{\mathbf{U}}_{o1} = \mathbf{U}_o + \mathbf{U}_s\mathbf{Q}_1. \tag{27}$$

Compare Eq.(27) with Eq.(19) to obtain a first-order orthogonal subspace perturbation as

$$\Delta\mathbf{U}_{o1} = \mathbf{U}_s\mathbf{Q}_1. \tag{28}$$

Similarly, the first-order signal subspace perturbation is expressed as

$$\Delta\mathbf{U}_{s1} = \mathbf{U}_o\mathbf{P}_1. \tag{29}$$

The second-order approximation of perturbed orthogonal subspace is obtained by keeping Eq.(26) up to the second-order as

$$(\mathbf{I}+\mathbf{Q}^H\mathbf{Q})^{-\frac{1}{2}} \stackrel{2}{=} \mathbf{I}-\frac{1}{2}\mathbf{Q}_1^H\mathbf{Q}_1. \tag{30}$$

Substitute Eq.(30) into Eq.(21) then keep the equation up to the second-order

$$\tilde{\mathbf{U}}_{o2} \stackrel{2}{=} (\mathbf{U}_o+\mathbf{U}_s\mathbf{Q}_2)(\mathbf{I}-\frac{1}{2}\mathbf{Q}_1^H\mathbf{Q}_1). \tag{31}$$

Deleting the fourth-order term in Eq.(31) yields

$$\tilde{\mathbf{U}}_{o2} = \mathbf{U}_o+\mathbf{U}_s\mathbf{Q}_2-\frac{1}{2}\mathbf{U}_o\mathbf{Q}_1^H\mathbf{Q}_1. \tag{32}$$

From Eqs.(19) and (32), the second-order orthogonal subspace perturbation is given as

$$\Delta\mathbf{U}_{o2} = \mathbf{U}_s\mathbf{Q}_2-\frac{1}{2}\mathbf{U}_o\mathbf{Q}_1^H\mathbf{Q}_1. \tag{33}$$

Similarly,

$$\Delta\mathbf{U}_{s2} = \mathbf{U}_o\mathbf{P}_2-\frac{1}{2}\mathbf{U}_s\mathbf{P}_1^H\mathbf{P}_1. \tag{34}$$

B. FIRST-ORDER SUBSPACE PERTURBATION

We now briefly review the first-order subspace perturbations developed by Li and Vaccaro in [2] for future reference in bias analysis. Pre-multiply Eq.(18) with $\tilde{\mathbf{U}}_o^H$

$$\tilde{\mathbf{U}}_o^H\tilde{\mathbf{Y}} = \tilde{\mathbf{\Sigma}}_o\tilde{\mathbf{V}}_o^H. \tag{35}$$

Using Eqs.(19) and (28) and the fact that $\tilde{\mathbf{\Sigma}}_o = \Delta\mathbf{\Sigma}_o$ (since the noise-free value of $\mathbf{\Sigma}_o$ is $\mathbf{\Sigma}_o = 0$), Eq.(35) can be written as

$$(\mathbf{U}_o+\mathbf{U}_s\mathbf{Q}_1)^H(\mathbf{Y}+\mathbf{N}) = \Delta\mathbf{\Sigma}_o(\mathbf{V}_o+\mathbf{V}_s\bar{\mathbf{Q}}_1)^H. \tag{36}$$

Using the fact that $\mathbf{U}_o^H\mathbf{Y} = 0$, we can obtain

$$\mathbf{U}_o^H\mathbf{N}+\mathbf{Q}_1^H\mathbf{U}_s^H\mathbf{Y}+\mathbf{Q}_1^H\mathbf{U}_s^H\mathbf{N} = \Delta\mathbf{\Sigma}_o\mathbf{V}_o^H+\Delta\mathbf{\Sigma}_o\bar{\mathbf{Q}}_1^H\mathbf{V}_s^H. \tag{37}$$

Next, post-multiply Eq.(37) with $\mathbf{V}_s$, we have

$$\mathbf{U}_o^H\mathbf{N}\mathbf{V}_s+\mathbf{Q}_1^H\mathbf{\Sigma}_s+\mathbf{Q}_1^H\mathbf{U}_s^H\mathbf{N}\mathbf{V}_s = \Delta\mathbf{\Sigma}_o\bar{\mathbf{Q}}_1^H. \tag{38}$$

The first-order approximation of $\mathbf{Q}$ is then obtained from

$$\mathbf{U}_o^H\mathbf{N}\mathbf{V}_s + \mathbf{Q}_1^H\mathbf{\Sigma}_s \stackrel{1}{=} 0 \tag{39}$$

which gives

$$\mathbf{Q}_1 = -\mathbf{\Sigma}_s^{-1}\mathbf{V}_s^H\mathbf{N}^H\mathbf{U}_o. \tag{40}$$

Using Eq.(23), $\mathbf{P}_1$ can be obtained as

$$\mathbf{P}_1 = \mathbf{U}_o^H\mathbf{N}\mathbf{V}_s\mathbf{\Sigma}_s^{-1}. \tag{41}$$

The first-order subspace perturbations are obtained as

$$\Delta\mathbf{U}_{o1} = \mathbf{U}_s\mathbf{Q}_1 = -\mathbf{U}_s\mathbf{\Sigma}_s^{-1}\mathbf{V}_s^H\mathbf{N}^H\mathbf{U}_o \tag{42}$$

$$\Delta\mathbf{U}_{s1} = \mathbf{U}_0\mathbf{P}_1 = \mathbf{U}_o\mathbf{U}_o^H\mathbf{N}\mathbf{V}_s\mathbf{\Sigma}_s^{-1}. \tag{43}$$

C. SECOND-ORDER SUBSPACE PERTURBATION

We will now derive the explicit expressions for subspace perturbations which are valid up to the second-order with respect to data perturbation $\mathbf{N}$. We begin at Eq.(35). Using Eqs.(21), (24), and the fact that $\tilde{\mathbf{\Sigma}}_o = \Delta\mathbf{\Sigma}_o$ (since the noise-free value of $\mathbf{\Sigma}_o$ is $\mathbf{\Sigma}_o = 0$), Eq.(35) can be written as

$$(\mathbf{I} + \mathbf{Q}^H\mathbf{Q})^{-\frac{1}{2}}(\mathbf{U}_o + \mathbf{U}_s\mathbf{Q})^H(\mathbf{Y} + \mathbf{N}) = \Delta\mathbf{\Sigma}_o(\mathbf{I} + \bar{\mathbf{Q}}^H\bar{\mathbf{Q}})^{-\frac{1}{2}}(\mathbf{V}_o + \mathbf{V}_s\bar{\mathbf{Q}})^H. \tag{44}$$

Using the fact that $\mathbf{U}_o^H\mathbf{Y} = 0$, we get

$$\begin{aligned} & (\mathbf{I} + \mathbf{Q}^H\mathbf{Q})^{-\frac{1}{2}}(\mathbf{U}_o^H\mathbf{N} + \mathbf{Q}^H\mathbf{U}_s^H\mathbf{Y} + \mathbf{Q}^H\mathbf{U}_s^H\mathbf{N}) \\ = \; & \Delta\mathbf{\Sigma}_o(\mathbf{I} + \bar{\mathbf{Q}}^H\bar{\mathbf{Q}})^{-\frac{1}{2}}(\mathbf{V}_o + \mathbf{V}_s\bar{\mathbf{Q}})^H. \end{aligned} \tag{45}$$

$\mathbf{Q}_2$ can be obtained by expanding Eq.(45) into a second-order equation

$$\mathbf{U}_o^H\mathbf{N} + \mathbf{Q}_2^H\mathbf{U}_s^H\mathbf{Y} + \mathbf{Q}_1^H\mathbf{U}_s^H\mathbf{N} \stackrel{2}{=} \Delta\mathbf{\Sigma}_{o2}\mathbf{V}_o^H + \Delta\mathbf{\Sigma}_{o1}\bar{\mathbf{Q}}_1^H\mathbf{V}_s^H. \tag{46}$$

Again, post-multiply Eq.(46) with $\mathbf{V}_s$,

$$\mathbf{U}_o^H\mathbf{N}\mathbf{V}_s + \mathbf{Q}_2^H\mathbf{\Sigma}_s + \mathbf{Q}_1^H\mathbf{U}_s^H\mathbf{N}\mathbf{V}_s \stackrel{2}{=} \Delta\mathbf{\Sigma}_{o1}\bar{\mathbf{Q}}_1^H \tag{47}$$

where $\Delta\mathbf{\Sigma}_{o1}$ can be obtained by post-multiplying $\mathbf{V}_o$ to both sides of Eq.(46) and retaining it up to the first-order as

$$\Delta\mathbf{\Sigma}_{o1} = \mathbf{U}_o^H \mathbf{N} \mathbf{V}_o \tag{48}$$

and $\bar{\mathbf{Q}}_1$ can be obtained by pre-multiplying $\mathbf{U}_s^H$ and post-multiplying $\tilde{\mathbf{V}}_o$ to $\tilde{\mathbf{Y}}$, then going through similar derivation. We now have

$$\bar{\mathbf{Q}}_1 = -\mathbf{\Sigma}_s^{-1}\mathbf{U}_s^H \mathbf{N} \mathbf{V}_o. \tag{49}$$

Substitute Eqs.(40), (48), and (49) into Eq.(47), we obtain

$$\begin{aligned} \mathbf{Q}_2 &= -\mathbf{\Sigma}_s^{-1}\mathbf{V}_s^H\mathbf{N}^H\mathbf{U}_o + \mathbf{\Sigma}_s^{-1}\mathbf{V}_s^H\mathbf{N}^H\mathbf{U}_s\mathbf{\Sigma}_s^{-1}\mathbf{V}_s^H\mathbf{N}^H\mathbf{U}_o \\ &\quad - \mathbf{\Sigma}_s^{-2}\mathbf{U}_s^H\mathbf{N}\mathbf{V}_o\mathbf{V}_o^H\mathbf{N}^H\mathbf{U}_o. \end{aligned} \tag{50}$$

Again $\mathbf{P}_2$ can be obtained from Eq.(23). An important statistical quantity is $E(\mathbf{Q}_2)$. We notice that if the observation noise is *circular Gaussian* with zero-mean, then the expectation of first term in Eq.(50) is equal to zero because of zero-mean. The expectation of second term is equal to zero because of circularity [15]. The expectation of the third term is also zero because

$$E(\mathbf{\Sigma}_s^{-2}\mathbf{U}_s^H\mathbf{N}\mathbf{V}_o\mathbf{V}_o^H\mathbf{N}^H\mathbf{U}_o) = Tr(\mathbf{V}_o\mathbf{V}_o^H)\mathbf{\Sigma}_s^{-2}\mathbf{U}_s^H\mathbf{U}_o\sigma_n^2 \tag{51}$$

where $\mathbf{U}_s^H\mathbf{U}_o = 0$ and Tr stands for matrix trace. A detailed proof is in Appendix A of [5]. Therefore we conclude

$$E(\mathbf{Q}_2) \;=\; E(-\mathbf{P}_2^H) = 0. \tag{52}$$

Now we can study the statistics properties of subspace perturbations by substituting Eqs.(52), (40), (41), and

$$E(\mathbf{N}\mathbf{U}\mathbf{V}^H\mathbf{N}^H) = \sigma_n^2 Tr(\mathbf{U}\mathbf{V}^H)\mathbf{I} \tag{53}$$

from Appendix A of [5] into Eqs.(33) and (34), respectively,

$$\begin{aligned} E(\Delta\mathbf{U}_{o2}) &= -\frac{1}{2}\mathbf{U}_o\mathbf{U}_o^H E(\mathbf{N}\mathbf{V}_s\mathbf{\Sigma}_s^{-2}\mathbf{V}_s^H\mathbf{N}^H)\mathbf{U}_o \\ &= -\frac{1}{2}\mathbf{U}_o Tr(\mathbf{V}_s\mathbf{\Sigma}_s^{-2}\mathbf{V}_s^H)\sigma_n^2 \end{aligned} \tag{54}$$

$$
\begin{aligned}
E(\Delta \mathbf{U}_{s2}) &= -\frac{1}{2}\mathbf{U}_s\mathbf{\Sigma}_s^{-1}\mathbf{V}_s^H E(\mathbf{N}^H\mathbf{U}_o\mathbf{U}_o^H\mathbf{N})\mathbf{V}_s\mathbf{\Sigma}_s^{-1} \\
&= -\frac{1}{2}\mathbf{U}_s\mathbf{\Sigma}_s^{-2}Tr(\mathbf{U}_o\mathbf{U}_o^H)\sigma_n^2.
\end{aligned} \tag{55}
$$

IV. ANALYSIS OF DOA ESTIMATION BIAS

Performance analysis of DOA estimation bias is important yet challenging due to the theoretical difficulty and mathematical complexity involved. Previous work on bias analysis has one or more the following limitations: (1) The asymptotic assumption that unlimited amount of data is available may not be realistic in many array processing applications. (2) The inclusion of singular values and singular vectors in bias expressions which are obtained from nonlinear transformation, i.e. SVD, of data prevents us from observing the relationship between the estimated DOAs and physical parameters, such as source separation, signal coherence, numbers of senors and snapshots on estimation bias.

In this section, bias performance analysis for extrema-searching approach, polynomial rooting approach and matrix-shifting approach are developed based on the common model underlying each approach. We will finally express DOA estimation bias in terms of fundamental parameters such as array manifold, source covariance and number of snapshots.

A. BIAS ANALYSIS FOR EXTREMA-SEARCH ALGORITHMS

Extrema searching algorithms obtain DOAs through searching for minima in null spectrum. A common model for the null spectrum function associated with MUSIC and Min-Norm searching algorithms is [6]

$$
P(\theta, \mathbf{U}_o) = \mathbf{a}^H(\theta)\mathbf{U}_o\mathbf{W}\mathbf{U}_o^H\mathbf{a}(\theta) \tag{56}
$$

where the weighting matrix $\mathbf{W}$ is specified as $\mathbf{I}$ for MUSIC and $\mathbf{W} = \bar{\mathbf{c}}\bar{\mathbf{c}}^H$ for Min-Norm in which $\bar{\mathbf{c}} = \mathbf{c}/||\mathbf{c}||^2$ where $\mathbf{c}^H$ is the first row of $\mathbf{U}_o$. From the orthogonality $\mathbf{a}^H(\theta_k)\mathbf{U}_o = 0$, the noise-free null-spectrum satisfies

$$
P(\theta_k, \mathbf{U}_o) = 0, \quad k = 1, \cdots, K, \tag{57}
$$

where θ_k is the k-th direction of arrival. In practice, with noise perturbation, null spectrum is no longer zero which results in error in DOA estimation. DOA estimates for MUSIC and Min-Norm searching algorithms are obtained by

$$\tilde{\theta}_k = \arg\min_{\theta_k \in \Theta} P(\tilde{\theta}_k, \tilde{\mathbf{U}}_o), \tag{58}$$

where

$$\tilde{\theta}_k = \theta_k + \Delta\theta_k.$$

$\Delta\theta_k$ is the estimation error of k-th DOA. From Eq.(58), it can be seen that $P(\tilde{\theta}_k, \tilde{\mathbf{U}}_o)$ satisfies

$$\frac{\partial P(\tilde{\theta}_k, \tilde{\mathbf{U}}_o)}{\partial \theta} = 0. \tag{59}$$

A second-order expression for $E(\Delta\theta_k)$ which is accurate for SNR down to threshold region can be attained through two steps.

Step 1. Approximate $\Delta\theta_k$ by expanding $\frac{\partial P(\tilde{\theta}_k, \tilde{\mathbf{U}}_o)}{\partial \theta}$ to the second-order using Taylor series as

$$\frac{\partial P(\tilde{\theta}_k, \tilde{\mathbf{U}}_o)}{\partial \theta} = \frac{\partial P(\theta_k, \tilde{\mathbf{U}}_o)}{\partial \theta} + \frac{\partial^2 P(\theta_k, \tilde{\mathbf{U}}_o)}{\partial \theta^2}\Delta\theta_k + \frac{\partial^3 P(\theta_k, \tilde{\mathbf{U}}_o)}{\partial \theta^3}\Delta^2\theta_k/2. \tag{60}$$

Using Eq.(59), Eq.(60) is reduced to

$$\frac{\partial P(\theta_k, \tilde{\mathbf{U}}_o)}{\partial \theta} + \frac{\partial^2 P(\theta_k, \tilde{\mathbf{U}}_o)}{\partial \theta^2}\Delta\theta_k + \frac{\partial^3 P(\theta_k, \tilde{\mathbf{U}}_o)}{\partial \theta^3}\Delta^2\theta_k/2 = 0. \tag{61}$$

Let

$$\begin{aligned} \frac{\partial P(\theta_k, \mathbf{U}_o)}{\partial \theta} &\stackrel{\text{def}}{=} N & \frac{\partial P(\theta_k, \tilde{\mathbf{U}}_o)}{\partial \theta} &\stackrel{\text{def}}{=} N + \Delta N \\ \frac{\partial^2 P(\theta_k, \mathbf{U}_o)}{\partial \theta^2} &\stackrel{\text{def}}{=} D & \frac{\partial^2 P(\theta_k, \tilde{\mathbf{U}}_o)}{\partial \theta^2} &\stackrel{\text{def}}{=} D + \Delta D \\ \frac{\partial^3 P(\theta_k, \mathbf{U}_o)}{\partial \theta^3} &\stackrel{\text{def}}{=} B & \frac{\partial^3 P(\theta_k, \tilde{\mathbf{U}}_o)}{\partial \theta^3} &\stackrel{\text{def}}{=} B + \Delta B. \end{aligned} \tag{62}$$

Then Eq.(61) can be written as

$$(N + \Delta N) + (D + \Delta D)\Delta\theta_k + (B + \Delta B)\Delta^2\theta_k/2 = 0. \tag{63}$$

Keep the terms up to the second-order

$$N + \Delta N_2 + D\Delta\theta_{k2} + \Delta D_1 \Delta\theta_{k1} + B\Delta^2\theta_{k1}/2 \stackrel{2}{=} 0. \tag{64}$$

where $\Delta\theta_{k1} = -\frac{\Delta N_1}{D}$ [6]. The second-order approximation of $\Delta\theta_k$ is therefore expressed as

$$\Delta\theta_{k2} = -\frac{N + \Delta N_2}{D} + \frac{\Delta N_1 \Delta D_1}{D^2} - \frac{B}{2D}\Delta^2\theta_{k1}. \tag{65}$$

Step 2. Take expectation on both sides to obtain general bias expression which is applicable to any searching algorithm

$$E(\Delta\theta_{k2}) = -E(\frac{N + \Delta N_2}{D}) + E(\frac{\Delta N_1 \Delta D_1}{D^2}) - \frac{B}{2D}E(\Delta^2\theta_{k1}). \tag{66}$$

1. BIAS FOR MUSIC SEARCHING ALGORITHM

For MUSIC, the specific terms in Eq.(66) are given as

$$\begin{aligned}
N &= 0 && (67)\\
D &= 2\|\mathbf{a}^{(1)}(\theta_k)^H \boldsymbol{\Omega}_o\|^2 && (68)\\
B &= 6\Re\{\mathbf{a}^{(1)}(\theta_k)^H \boldsymbol{\Omega}_o \mathbf{a}^{(2)}(\theta_k)\} && (69)\\
E(\Delta N_2) &= 2(L-K)\sigma_n^2 \Re\{\mathbf{a}^{(1)}(\theta_k)^H (\mathbf{A}^\dagger)^H \hat{\mathbf{R}}_s^{-1} \mathbf{e}_k\} && (70)\\
E(\Delta N_1 \Delta D_1) &= 2\hat{\mathbf{R}}_s^{-1}(k,k)\sigma_n^2 \Re\{\mathbf{a}^{(1)}(\theta_k)^H \boldsymbol{\Omega}_o \mathbf{a}^{(2)}(\theta_k)\} \\
&\quad + 4\sigma_n^2 \|\mathbf{a}^{(1)}(\theta_k)^H \boldsymbol{\Omega}_o\|^2 \Re\{\mathbf{e}_k \hat{\mathbf{R}}_s^{-1} \mathbf{A}^\dagger \mathbf{a}^{(1)}(\theta_k)\}. && (71)
\end{aligned}$$

Details of the derivation are referred to Appendix I.A. Substitute Eqs.(67)-(71) into Eq.(66) and use the relationship $\hat{\mathbf{R}}_s^{-1} = \frac{1}{M}\mathbf{R}_s^{-1}$ (see [16]) which gives

$$\begin{aligned}
E(\Delta\theta_{k2}) &= -\frac{\sigma_n^2(L-K-1)}{M\|\mathbf{a}^{(1)}(\theta_k)^H \boldsymbol{\Omega}_o\|^2} \Re\{\mathbf{a}^{(1)}(\theta_k)^H (\mathbf{A}^\dagger)^H \mathbf{R}_s^{-1} \mathbf{e}_k\} \\
&\quad - \frac{\Re\{\mathbf{a}^{(1)}(\theta_k)^H \boldsymbol{\Omega}_o \mathbf{a}^{(2)}(\theta_k)\}}{2\|\mathbf{a}^{(1)}(\theta_k)^H \boldsymbol{\Omega}_o\|^2} E(\Delta\theta_{k1}^2). && (72)
\end{aligned}$$

2. BIAS FOR MIN-NORM SEARCHING ALGORITHM

For Min-Norm, the specific terms in Eq.(66) are given as

$$\begin{aligned}
N &= 0 && (73)\\
D &= \frac{2}{\|\mathbf{e}_1^H \boldsymbol{\Omega}_o\|^4} |\mathbf{a}^{(1)}(\theta_k)^H \boldsymbol{\Omega}_o \mathbf{e}_1|^2 && (74)\\
B &= \frac{6}{\|\mathbf{e}_1^H \boldsymbol{\Omega}_o\|^4} \Re\{\mathbf{a}^{(1)}(\theta_k)^H \boldsymbol{\Omega}_o \mathbf{e}_1 \mathbf{e}_1^H \boldsymbol{\Omega}_o \mathbf{a}^{(2)}(\theta_k)\} && (75)
\end{aligned}$$

$$
\begin{aligned}
E(\Delta N_2) &= \frac{2(L-K)\sigma_n^2}{\|\mathbf{e}_1^H \boldsymbol{\Omega}_o\|^4} \Re\{\mathbf{a}^{(1)}(\theta_k)^H \boldsymbol{\Omega}_o \mathbf{e}_1 \mathbf{e}_1^H (\mathbf{A}^{\dagger})^H \hat{\mathbf{R}}_s^{-1} \mathbf{e}_k\} \\
&+ \frac{2\sigma_n^2}{\|\mathbf{e}_1^H \boldsymbol{\Omega}_o\|^2} \Re\{\mathbf{a}^{(1)}(\theta_k)^H (\mathbf{A}^{\dagger})^H \hat{\mathbf{R}}_s^{-1} \mathbf{e}_k\} \qquad (76) \\
E(\Delta N_1 \Delta D_1) &= \frac{4\sigma_n^2 |\mathbf{a}^{(1)}(\theta_k)^H \boldsymbol{\Omega}_o \mathbf{e}_1|^2}{\|\mathbf{e}_1^H \boldsymbol{\Omega}_o\|^8} \Re\{\mathbf{a}^{(1)}(\theta_k)^H \boldsymbol{\Omega}_o \mathbf{e}_1 \mathbf{e}_1^H (\mathbf{A}^{\dagger})^H \hat{\mathbf{R}}_s^{-1} \mathbf{e}_k\} \\
&+ \frac{4\sigma_n^2 |\mathbf{a}^{(1)}(\theta_k)^H \boldsymbol{\Omega}_o \mathbf{e}_1|^2}{\|\mathbf{e}_1^H \boldsymbol{\Omega}_o\|^6} \Re\{\mathbf{a}^{(1)}(\theta_k)^H (\mathbf{A}^{\dagger})^H \hat{\mathbf{R}}_s^{-1} \mathbf{e}_k\} \\
&+ \frac{2\sigma_n^2 \hat{\mathbf{R}}_s^{-1}(k,k)}{\|\mathbf{e}_1^H \boldsymbol{\Omega}_o\|^6} \Re\{\mathbf{a}^{(1)}(\theta_k)^H \boldsymbol{\Omega}_o \mathbf{e}_1 \mathbf{e}_1^H \boldsymbol{\Omega}_o \mathbf{a}^{(2)}(\theta_k)\}. \qquad (77)
\end{aligned}
$$

Detailed derivation is referred to Appendix I.B. Substitute Eqs.(73)-(77) into Eq.(66) and use the relationship $\hat{\mathbf{R}}_s^{-1} = \frac{1}{M}\mathbf{R}_s^{-1}$ which gives

$$
\begin{aligned}
E(\Delta\theta_{k2}) &= -\frac{\sigma_n^2(L-K-1)}{M|\mathbf{a}^{(1)}(\theta_k)^H \boldsymbol{\Omega}_o \mathbf{e}_1|^2} \Re\{\mathbf{a}^{(1)}(\theta_k)^H \boldsymbol{\Omega}_o \mathbf{e}_1 \mathbf{e}_1^H (\mathbf{A}^{\dagger})^H \mathbf{R}_s^{-1} \mathbf{e}_k\} \\
&- \frac{\Re\{\mathbf{a}^{(1)}(\theta_k)^H \boldsymbol{\Omega}_o \mathbf{e}_1 \mathbf{e}_1^H \boldsymbol{\Omega}_o \mathbf{a}^{(2)}(\theta_k)\}}{2M|\mathbf{a}^{(1)}(\theta_k)^H \boldsymbol{\Omega}_o \mathbf{e}_1|^2} E(\Delta\theta_{k1}^2). \qquad (78)
\end{aligned}
$$

B. BIAS ANALYSIS OF POLYNOMIAL ROOTING ALGORITHMS

DOAs can be determined from the roots of spectral polynomial. A common spectral polynomial for Min-Norm and Root-MUSIC is found in [6]

$$
P(z) = \mathbf{a}(z^{-1})^T \mathbf{U}_o \mathbf{W} \mathbf{U}_o^H \mathbf{a}(z) = A \prod_{i=1}^{L-1} (1 - r_i z^{-1})(1 - r_i^* z) \qquad (79)
$$

where the weighting matrix $\mathbf{W}$ is the same as in extrema-searching algorithm and A is a scaling factor. The relationship between the roots of the spectral polynomial and the directions of arrival is

$$
r_i = e^{j\frac{2\pi d}{\lambda_c}\sin\theta_i}. \qquad (80)
$$

The difference between searching algorithms and rooting algorithms is that searching algorithms search for minima of a function in one dimension while rooting algorithms search for global minima over multi-dimensional space to obtain roots simultaneously, thus when two DOAs are closely spaced, rooting method may be able to give true DOAs while searching algorithm may have only one extremum resulting in the failure of resolving two DOAs near threshold.

When the signal is perturbed, the signal roots are off unit circle which can be obtained by setting the first derivative of spectral polynomial to zero i.e.

$$\frac{\partial P(\tilde{r}_k, \tilde{\mathbf{U}}_o)}{\partial z} = 0 \quad k = 1, \cdots, K, \tag{81}$$

where $\tilde{r}_k = r_k + \Delta r_k$, and Δr_k is the perturbation of the true root which induces error in DOA estimation.

The derivation for the second-order $E(\Delta\theta_k)$ can be divided into three steps.

Step 1. Expand $\frac{\partial P(\tilde{r}_k, \tilde{\mathbf{U}}_o)}{\partial z}$ into the second-order using Taylor series to derive the first-order and the second-order expressions of Δr_k,

$$\frac{\partial P(\tilde{r}_k, \tilde{\mathbf{U}}_o)}{\partial z} = \frac{\partial P(r_k, \tilde{\mathbf{U}}_o)}{\partial z} + \frac{\partial^2 P(r_k, \tilde{\mathbf{U}}_o)}{\partial z^2}\Delta r_k + \frac{\partial^3 P(r_k, \tilde{\mathbf{U}}_o)}{\partial z^3}\Delta^2 r_k/2. \tag{82}$$

Using Eq.(81), Eq.(82) is reduced to

$$\frac{\partial P(r_k, \tilde{\mathbf{U}}_o)}{\partial z} + \frac{\partial^2 P(r_k, \tilde{\mathbf{U}}_o)}{\partial z^2}\Delta r_k + \frac{\partial^3 P(r_k, \tilde{\mathbf{U}}_o)}{\partial z^3}\Delta^2 r_k/2 = 0. \tag{83}$$

Let

$$\begin{aligned}
\frac{\partial P(r_k, \mathbf{U}_o)}{\partial z} &\stackrel{\text{def}}{=} N & \frac{\partial P(r_k, \tilde{\mathbf{U}}_o)}{\partial z} &\stackrel{\text{def}}{=} N + \Delta N \\
\frac{\partial^2 P(r_k, \mathbf{U}_o)}{\partial z^2} &\stackrel{\text{def}}{=} D & \frac{\partial^2 P(r_k, \tilde{\mathbf{U}}_o)}{\partial z^2} &\stackrel{\text{def}}{=} D + \Delta D \\
\frac{\partial^3 P(r_k, \mathbf{U}_o)}{\partial z^3} &\stackrel{\text{def}}{=} B & \frac{\partial^3 P(r_k, \tilde{\mathbf{U}}_o)}{\partial z^3} &\stackrel{\text{def}}{=} B + \Delta B.
\end{aligned} \tag{84}$$

Then Eq.(83) can be written as

$$(N + \Delta N) + (D + \Delta D)\Delta r_k + (B + \Delta B)\Delta^2 r_k/2 = 0. \tag{85}$$

Keep the equation up to the second-order as

$$N + \Delta N_2 + D\Delta r_{k2} + \Delta D_1 \Delta r_{k1} + B\Delta^2 r_{k1}/2 \stackrel{2}{=} 0 \tag{86}$$

where Δr_{k1} can be obtained by keeping Eq.(85) up to the first-order as

$$\Delta r_{k1} = -\frac{(N + \Delta N_1)}{D}. \tag{87}$$

Substitute Eq.(87) into Eq.(86) to obtain Δr_{k2}

$$\Delta r_{k2} = -\frac{(N+\Delta N_2)}{D} + \frac{(\Delta D_1 N + \Delta D_1 \Delta N_1)}{D^2} - \frac{B(N+\Delta N_1)^2}{2D^3}. \quad (88)$$

Step 2. A second-order approximation of $\Delta\theta_k$ can be derived using the angle-root relation in Appendix C of [5] as

$$\Delta\theta_{k2} = C_k \Im(\frac{\Delta r_{k2}}{r_k} - \frac{\Delta^2 r_{k1}}{2r_k^2}) + \frac{\tan\theta_k}{2}\Delta^2\theta_{k1}. \quad (89)$$

Substitute Eqs.(87) and (88) into Eq.(89)

$$\begin{aligned} \Delta\theta_{k2} &= \frac{C_k}{r_k}\Im[-\frac{(N+\Delta N_2)}{D} + \frac{(N\Delta D_1 + \Delta D_1 \Delta N_1)}{D^2} - \frac{B(N+\Delta N_1)^2}{2D^3}] \\ &\quad -\frac{C_k}{2r_k^2}\Im[\frac{(N+\Delta N_1)^2}{D^2}] + \frac{\tan\theta_k}{2}\Delta^2\theta_{k1}. \end{aligned}$$

Step 3. Take expectation on both sides of equation to obtain general bias expression for any polynomial rooting algorithms

$$\begin{aligned} & E(\Delta\theta_{k2}) \\ = \; & \frac{C_k}{r_k}\Im\{E[-\frac{(N+\Delta N_2)}{D} + \frac{(N\Delta D_1 + \Delta D_1\Delta N_1)}{D^2} \\ & -\frac{B(N+\Delta N_1)^2}{2D^3}]\} - \frac{C_k}{2r_k^2}\Im\{E[\frac{(N+\Delta N_1)^2}{D^2}]\} + \frac{\tan\theta_k}{2}E(\Delta^2\theta_{k1}). \end{aligned} \quad (90)$$

1. BIAS FOR ROOT MUSIC

For Root MUSIC, the specific terms in Eq.(90) are given as

$$N = 0 \quad (91)$$

$$D = -2r_k^{-2}\|\mathbf{a}^{(1)}(r_k^{-1})^T\mathbf{\Omega}_o\|^2 \quad (92)$$

$$B = 6r_k^{-3}\|\mathbf{a}^{(1)}(r_k^{-1})^T\mathbf{\Omega}_o\|^2 + 6jr_k^{-3}\Im\{r_k^{-1}\mathbf{a}^{(2)}(r_k^{-1})^T\mathbf{\Omega}_o\mathbf{a}^{(1)}(r_k)\} \quad (93)$$

$$E(\Delta N_2) = -2j(L-K)\sigma_n^2 r_k^{-1}\Im\{r_k^{-1}\mathbf{a}^{(1)}(r_k^{-1})^T(\mathbf{A}^\dagger)^H\hat{\mathbf{R}}_s^{-1}\mathbf{e}_k\} \quad (94)$$

$$E(\Delta^2 N_1) = -2\sigma_n^2 r_k^{-2}\|\mathbf{a}^{(1)}(r_k^{-1})^T\mathbf{\Omega}_o\|^2\hat{\mathbf{R}}_s^{-1}(k,k) \quad (95)$$

$$\begin{aligned} E(\Delta N_1 \Delta D_1) &= 4j\sigma_n^2 r_k^{-3}\|\mathbf{a}^{(1)}(r_k^{-1})^T\mathbf{\Omega}_o\|^2\Im\{r_k^{-1}\mathbf{a}^{(1)}(r_k^{-1})^T(\mathbf{A}^\dagger)^H\hat{\mathbf{R}}_s^{-1}\mathbf{e}_k\} \\ &\quad +2j\sigma_n^2 r_k^{-3}\hat{\mathbf{R}}_s^{-1}(k,k)\Im\{r_k^{-1}\mathbf{a}^{(2)}(r_k^{-1})^T\mathbf{\Omega}_o\mathbf{a}^{(1)}(r_k)\} \\ &\quad +2\sigma_n^2 r_k^{-3}\hat{\mathbf{R}}_s^{-1}(k,k)\|\mathbf{a}^{(1)}(r_k^{-1})^T\mathbf{\Omega}_o\|^2. \end{aligned} \quad (96)$$

The derivation of Eqs.(91)-(96) is in Appendix II.A. Substitute Eqs.(91)-(96)into Eq.(90) and use the fact $\hat{\mathbf{R}}_s^{-1} = \frac{1}{M}\mathbf{R}_s^{-1}$ to gives an explicit bias expression

$$\begin{aligned} E(\Delta\theta_{k2}) &= -\frac{C_k(L-K-1)\sigma_n^2}{M\|\mathbf{a}^{(1)}(r_k^{-1})^T\boldsymbol{\Omega}_o\|^2}\Im\{r_k^{-1}\mathbf{a}^{(1)}(r_k^{-1})^T(\mathbf{A}^\dagger)^H\mathbf{R}_s^{-1}\mathbf{e}_k\} \\ &\quad -\frac{C_k\mathbf{R}_s^{-1}(k,k)\sigma_n^2}{4M\|\mathbf{a}^{(1)}(r_k^{-1})^T\boldsymbol{\Omega}_o\|^4}\Im\{r_k^{-1}\mathbf{a}^{(2)}(r_k^{-1})^T\boldsymbol{\Omega}_o\mathbf{a}^{(1)}(r_k)\} \\ &\quad +\frac{\tan\theta_k}{2}E(\Delta\theta_{k1}^2). \end{aligned}$$

2. BIAS FOR ROOT MIN-NORM

For Root Min-Norm, the specific terms are given as

$$N = 0 \tag{97}$$

$$D = \frac{-2r_k^{-2}}{\|\mathbf{e}_1^H\boldsymbol{\Omega}_o\|^4}|\mathbf{a}^{(1)}(r_k^{-1})^T\boldsymbol{\Omega}_o\mathbf{e}_1|^2 \tag{98}$$

$$\begin{aligned} B &= \frac{1}{\|\mathbf{e}_1^H\boldsymbol{\Omega}_o\|^4}\{6r_k^{-3}|\mathbf{a}^{(1)}(r_k^{-1})^T\boldsymbol{\Omega}_o\mathbf{e}_1|^2 \\ &\quad +6jr_k^{-3}\Im[r_k^{-1}\mathbf{a}^{(2)}(r_k^{-1})^T\boldsymbol{\Omega}_o\mathbf{e}_1\mathbf{e}_1^H\boldsymbol{\Omega}_o\mathbf{a}^{(1)}(r_k)]\} \end{aligned} \tag{99}$$

$$\begin{aligned} & E(\Delta N_2) \\ &= -\frac{2j(L-K)r_k^{-1}\sigma_n^2}{\|\mathbf{e}_1^H\boldsymbol{\Omega}_o\|^4}\Im\{r_k^{-1}\mathbf{a}^{(1)}(r_k^{-1})^T\boldsymbol{\Omega}_o\mathbf{e}_1\mathbf{e}_1^H(\mathbf{A}^\dagger)^H\hat{\mathbf{R}}_s^{-1}\mathbf{e}_k \\ &\quad +\|\mathbf{e}_1^H\boldsymbol{\Omega}_o\|^2 r_k^{-1}\mathbf{a}^{(1)}(r_k^{-1})^T(\mathbf{A}^\dagger)^H\hat{\mathbf{R}}_s^{-1}\mathbf{e}_k\} \end{aligned} \tag{100}$$

$$E(\Delta^2 N_1) = -\frac{2\sigma_n^2 r_k^{-2}}{\|\mathbf{e}_1^H\boldsymbol{\Omega}_o\|^6}|\mathbf{a}^{(1)}(r_k^{-1})^T\boldsymbol{\Omega}_o\mathbf{e}_1|^2\hat{\mathbf{R}}_s^{-1}(k,k) \tag{101}$$

$$\begin{aligned} & E(\Delta N_1\Delta D_1) \\ &= \frac{\sigma_n^2}{\|\mathbf{e}_1^H\boldsymbol{\Omega}_o\|^8}\{2r_k^{-3}\hat{\mathbf{R}}_s^{-1}(k,k)\|\mathbf{e}_1^H\boldsymbol{\Omega}_o\|^2|\mathbf{a}^{(1)}(r_k^{-1})^T\boldsymbol{\Omega}_o\mathbf{e}_1|^2 \\ &\quad +4jr_k^{-3}\|\mathbf{e}_1^H\boldsymbol{\Omega}_o\|^2|\mathbf{a}^{(1)}(r_k^{-1})^T\boldsymbol{\Omega}_o\mathbf{e}_1|^2\Im\{r_k^{-1}\mathbf{a}^{(1)}(r_k^{-1})^T(\mathbf{A}^\dagger)^H\hat{\mathbf{R}}_s^{-1}\mathbf{e}_k\} \\ &\quad +4jr_k^{-3}|\mathbf{a}^{(1)}(r_k^{-1})^T\boldsymbol{\Omega}_o\mathbf{e}_1|^2\Im\{r_k^{-1}\mathbf{a}^{(1)}(r_k^{-1})^T\boldsymbol{\Omega}_o\mathbf{e}_1\mathbf{e}_1^H(\mathbf{A}^\dagger)^H\hat{\mathbf{R}}_s^{-1}\mathbf{e}_k\} \\ &\quad +2jr_k^{-3}\|\mathbf{e}_1^H\boldsymbol{\Omega}_o\|^2\Im\{r_k^{-1}\hat{\mathbf{R}}_s^{-1}(k,k)\mathbf{a}^{(2)}(r_k^{-1})^T\boldsymbol{\Omega}_o\mathbf{e}_1\mathbf{e}_1^H\boldsymbol{\Omega}_o\mathbf{a}^{(1)}(r_k)\}\}. \end{aligned} \tag{102}$$

The derivation of Eqs.(97)-(102) is in Appendix II.B. Substitute Eqs.(97)-(102) into Eq.(90) and use the fact $\hat{\mathbf{R}}_s^{-1} = \frac{1}{M}\mathbf{R}_s^{-1}$ to obtain an concise bias expression

$$\begin{aligned} E(\Delta\theta_{k2}) &= -\frac{C_k\sigma_n^2(L-K-1)}{M|\mathbf{a}^{(1)}(r_k^{-1})^T\boldsymbol{\Omega}_o\mathbf{e}_1|^2}\Im\{r_k^{-1}\mathbf{a}^{(1)}(r_k^{-1})^T\boldsymbol{\Omega}_o\mathbf{e}_1\mathbf{e}_1^H(\mathbf{A}^\dagger)^H\mathbf{R}_s^{-1}\mathbf{e}_k\} \\ &-\frac{C_k\sigma_n^2\mathbf{R}_s^{-1}(k,k)\|\mathbf{e}_1^H\boldsymbol{\Omega}_o\|^2}{4M|\mathbf{a}^{(1)}(r_k^{-1})^T\boldsymbol{\Omega}_o\mathbf{e}_1|^2}\Im\{r_k^{-1}\mathbf{a}^{(2)}(r_k^{-1})^T\boldsymbol{\Omega}_o\mathbf{e}_1\mathbf{e}_1^H\boldsymbol{\Omega}_o\mathbf{a}^{(1)}(r_k)\} \\ &+\frac{\tan\theta_k}{2}E(\Delta\theta_{k1}^2). \end{aligned}$$

C. EQUATING EXTREMA SEARCHING AND POLYNOMIAL ROOTING

When a uniform line array is used, polynomial rooting algorithms have the same performance as extrema searching algorithms if the two DOAs are well separated. Using Eq.(80), it can be shown

$$\begin{aligned} \mathbf{a}^{(1)}(\theta_k) &= j\frac{2\pi d\cos\theta_k}{\lambda_c}r_k\mathbf{a}^{(1)}(r_k) \\ &= \frac{jr_k\mathbf{a}^{(1)}(r_k)}{C_k} \end{aligned} \tag{103}$$

$$\begin{aligned} \mathbf{a}^{(2)}(\theta_k) &= -j\frac{2\pi d\sin\theta_k}{\lambda_c}r_k\mathbf{a}^{(1)}(r_k) - (\frac{2\pi d\cos\theta_k}{\lambda_c})^2 r_k\mathbf{a}^{(1)}(r_k) \\ &-(\frac{2\pi d\cos\theta_k}{\lambda_c})^2 r_k^2\mathbf{a}^{(2)}(r_k) \\ &= -\frac{j\tan\theta_k r_k\mathbf{a}^{(1)}(r_k)}{C_k} - \frac{r_k\mathbf{a}^{(1)}(r_k)}{C_k^2} - \frac{r_k^2\mathbf{a}^{(2)}(r_k)}{C_k^2}. \end{aligned} \tag{104}$$

Substituting Eqs.(103) and (104) into Eq.(72) and refer $E(\Delta\theta_k^2)$ to Table I.

$$\begin{aligned} &E(\Delta\theta_{k2}) \\ =& -\frac{C_k^4\mathbf{R}_s^{-1}(k,k)\sigma_n^2}{4M\|\mathbf{a}^{(1)}(r_k^{-1})^T\boldsymbol{\Omega}_o\|^4}\Re\{-\frac{\tan\theta_k}{C_k^2}\mathbf{a}^{(1)}(r_k^{-1})^T\boldsymbol{\Omega}_o\mathbf{a}^{(1)}(r_k) \\ &+\frac{j}{C_k^3}\mathbf{a}^{(1)}(r_k^{-1})^T\boldsymbol{\Omega}_o\mathbf{a}^{(1)}(r_k) + \frac{jr_k}{C_k^3}\mathbf{a}^{(1)}(r_k^{-1})^T\boldsymbol{\Omega}_o\mathbf{a}^{(2)}(r_k)\} \\ &+\frac{C_k\sigma_n^2(L-P-1)}{M\|\mathbf{a}^{(1)}(r_k^{-1})^T\boldsymbol{\Omega}_o\|^2}\Re\{jr_k^{-1}\mathbf{a}^{(1)}(r_k^{-1})^T(\mathbf{A}^\dagger)^H\mathbf{R}_s^{-1}\mathbf{e}_k\} \\ =& -\frac{C_k\mathbf{R}_s^{-1}(k,k)\sigma_n^2}{4M\|\mathbf{a}^{(1)}(r_k^{-1})^T\boldsymbol{\Omega}_o\|^4}\Re\{jr_k\mathbf{a}^{(1)}(r_k^{-1})^T\boldsymbol{\Omega}_o\mathbf{a}^{(2)}(r_k)\} \end{aligned}$$

$$+\frac{C_k\sigma_n^2(L-P-1)}{M\|\mathbf{a}^{(1)}(r_k^{-1})^T\mathbf{\Omega}_o\|^2}\Re\{jr_k^{-1}\mathbf{a}^{(1)}(r_k^{-1})^T(\mathbf{A}^\dagger)^H\mathbf{R}_s^{-1}\mathbf{e}_k\}.$$
$$+\frac{\tan\theta_k C_k^2\mathbf{R}_s^{-1}(k,k)\sigma_n^2}{4M\|\mathbf{a}^{(1)}(r_k^{-1})^T\mathbf{\Omega}_o\|^2}$$

Clearly, the bias for MUSIC and Root MUSIC are identical. Similarly, substitute Eqs.(103) and (104) into Eq.(78) and use Table I to equate Min-Norm search algorithm with Min-Norm.

Table I: MSE FOR MUSIC, MIN-NORM, ROOT MUSIC, AND ROOT MIN-NORM

$E(\Delta\theta_k^2)$	MUSIC	Min-Norm
Searching	$\frac{\sigma_n^2\mathbf{R}_s^{-1}(k,k)}{2M\|\mathbf{a}^{(1)}(\theta_k)^H\mathbf{\Omega}_o\|^2}$	$\frac{\sigma_n^2\mathbf{R}_s^{-1}(k,k)\|\mathbf{e}_1^H\mathbf{\Omega}_o\|^2}{2M\|\mathbf{a}^{(1)}(\theta_k)^H\mathbf{\Omega}_o\mathbf{e}_1\|^2}$
Rooting	$\frac{C_k^2\sigma_n^2\mathbf{R}_s^{-1}(k,k)}{2M\|\mathbf{a}^{(1)}(r_k^{-1})^T\mathbf{\Omega}_o\|^2}$	$\frac{C_k^2\sigma_n^2\mathbf{R}_s^{-1}(k,k)\|\mathbf{e}_1^H\mathbf{\Omega}_o\|^2}{2M\|\mathbf{a}^{(1)}(r_k^{-1})^T\mathbf{\Omega}_o\mathbf{e}_1\|^2}$

D. BIAS FOR MATRIX-SHIFTING ALGORITHMS

For ESPRIT, SSR, and Matrix-Pencil algorithms, we use the results that we derived in [5]. The expression of DOA estimation bias is

$$\begin{aligned} E(\Delta\theta_{k2}) &= \frac{C_k\sigma_n^2}{M}\Im\{\mathbf{e}_k^H(\mathbf{A}^{\downarrow H}\mathbf{A}^{\downarrow})^{-1}\mathbf{R}_s^{-1}\mathbf{e}_k[-(L-K-1) \\ &\quad -\lambda_k^{-1}Tr(\mathbf{P}_{\mathbf{A}^{\downarrow}}\mathbf{I}^{\uparrow}\mathbf{I}^{\downarrow H})]\} + \frac{\tan\theta_k}{2}E(\Delta\theta_{k1}{}^2) \end{aligned} \tag{105}$$

where $\mathbf{P}_{\mathbf{A}^{\downarrow}}$ is the projection matrix of $\mathbf{A}^{\downarrow}$.

E. NUMERICAL SIMULATIONS

We now compare the bias performance of DOA estimators predicted by our analysis with the Monte Carlo simulations.

The general configuration of the experiments is: a uniform line array of eight sensors (with $d = \lambda_c/2$) with two sources at 0.2 and 0.35 radians (angles are measured with respect to the normal of the array). The signals

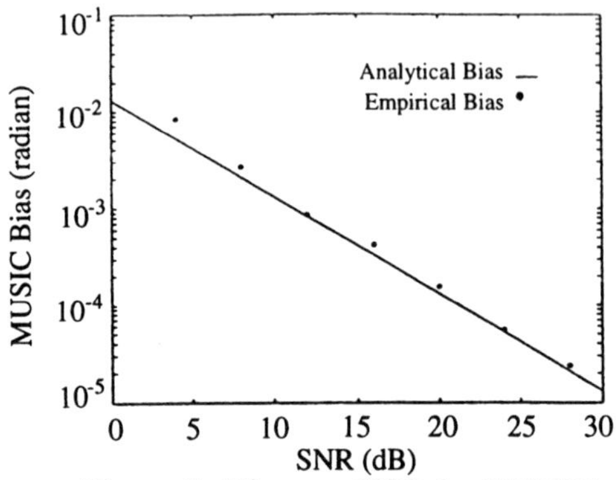

Figure 1: Bias vs. SNR for MUSIC

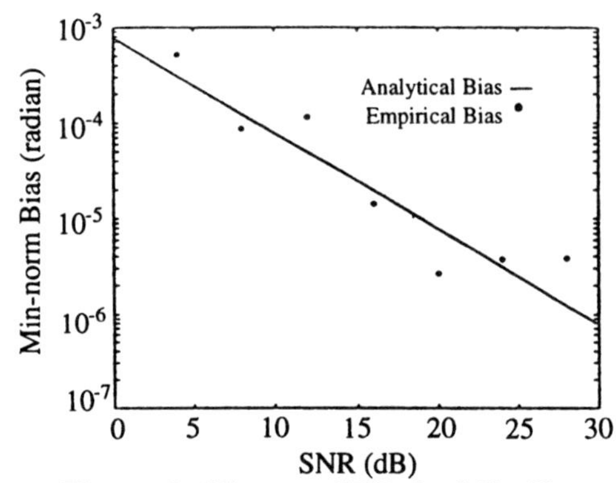

Figure 2: Bias vs. SNR for Min-Norm

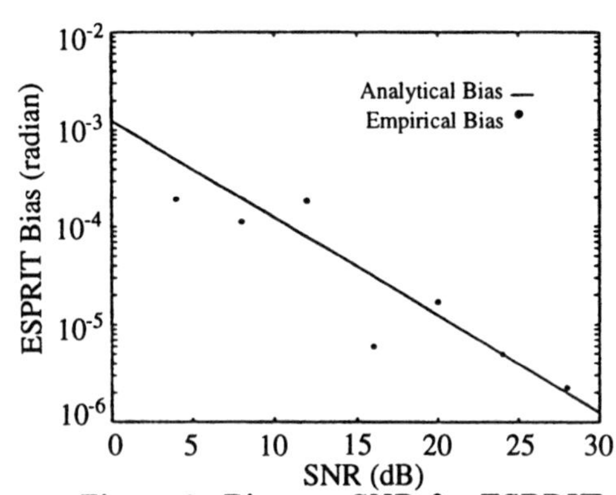

Figure 3: Bias vs. SNR for ESPRIT

are $s_k(t) = e^{j(\omega_c t+\phi_{kn})}$, where ϕ_{kn} are independent random phase angles uniformly distributed in the interval $(-\pi, \pi)$. Twenty snapshots of array data were simulated for 10 thousand trials.

Figures 1, 2, and 3 show bias of DOA estimation versus signal-to-noise-ratio (SNR) for MUSIC, Min-Norm, and ESPRIT algorithms, respectively. The SNR is defined as $SNR = 10log\frac{\sigma_s^2}{\sigma_n^2}$.

It is clearly shown from the simulation results (plotted in discrete symbols) that our analytical bias (plotted in solid lines) for DOA estimation algorithms are valid over a wide range of SNR.

V. UNIFICATION OF BIAS ANALYSES

We now unify the bias expressions developed in the previous sections for subspace-based DOA estimators including MUSIC, Min-Norm, and ESPRIT-type algorithms into a single, tractable formula

$$E(\Delta\theta_k) = \frac{\sigma_n^2}{M}\Im\{\mathbf{e}_k^H \mathbf{W}_B \mathbf{R}_s^{-1}\mathbf{e}_k\} + D_k E(\Delta\theta_k^2) \tag{106}$$

where weighting matrix $\mathbf{W}_B$ and constant D_k are specified for different algorithms as in Tables II and $E(\Delta\theta_k^2)$ is given in [16] as

$$E(\Delta{\theta_k}^2) = \frac{\mathbf{R}_s^{-1}(k,k)\sigma_n^2}{2M[\mathbf{a}^{(1)}(\theta_k)^H \mathbf{\Omega}_o \mathbf{W}_M \mathbf{\Omega}_o \mathbf{a}^{(1)}(\theta_k)]}. \tag{107}$$

where the weighting matrix $\mathbf{W}_M$ is specialized for different algorithms as in Table II. In [16], α^H is given as $\alpha^H = C_k \mathbf{e}_k^H(\mathbf{A}^{\uparrow\dagger}\mathbf{I}^{\uparrow} - \mathbf{A}^{\downarrow\dagger}\mathbf{I}^{\downarrow})$.

The strength of performance analysis is that one does not have to run hundreds or thousands simulation trials to discover a phenomenon. In this section, we will observe and compare the bias performance of MUSIC, Min-Norm, and ESPRIT as functions of physical conditions using the analytical formula Eq.(106), since we are interested in the performance of algorithms at low SNR, insufficient snapshots, and small number of available senors, which are the realistic limitations in array applications. The general configuration of the experiments is the same as described in Section IV except SNR is set to be 10 dB. We will predict the bias versus the variation of one physical parameter while fixing other parameters.

Table II: PARAMETERS FOR THE UNIFIED EXPRESSION

	$\mathbf{W}_B$
MUSIC	$-\frac{j(L-K-1)}{\lVert\mathbf{a}^{(1)}(\theta_k)^H\mathbf{\Omega}_o\rVert^2}\mathbf{A}^{(1)H}(\mathbf{A}^\dagger)^H$
Min-Norm	$-\frac{j(L-K-1)}{\lVert\mathbf{a}^{(1)}(\theta_k)^H\mathbf{\Omega}_o\mathbf{e}_1\rVert^2}\mathbf{A}^{(1)H}\mathbf{\Omega}_o\mathbf{e}_1\mathbf{e}_1^H(\mathbf{A}^\dagger)^H$
ESPRIT	$-C_k\{(L-K-1)+\lambda_k^{-1}Tr(\mathbf{P}_{\mathbf{A}^\downarrow}\mathbf{I}^\uparrow\mathbf{I}^{\downarrow H})\}(\mathbf{A}^{\downarrow H}\mathbf{A}^\downarrow)^{-1}$
	D_k
MUSIC	$-\frac{\Re\{\mathbf{a}^{(1)}(\theta_k)^H\mathbf{\Omega}_o\mathbf{a}^{(2)}(\theta_k)\}}{2\lVert\mathbf{a}^{(1)}(\theta_k)^H\mathbf{\Omega}_o\rVert^2}$
Min-Norm	$-\frac{\Re\{\mathbf{a}^{(1)}(\theta_k)^H\mathbf{\Omega}_o\mathbf{e}_1\mathbf{e}_1^H\mathbf{\Omega}_o\mathbf{a}^{(2)}(\theta_k)\}}{2\lVert\mathbf{a}^{(1)}(\theta_k)^H\mathbf{\Omega}_o\mathbf{e}_1\rVert^2}$
ESPRIT	$\frac{\tan\theta_k}{2}$
	$\mathbf{W}_M$
MUSIC	$\mathbf{I}$
Min-Norm	$\frac{\mathbf{\Omega}_o\mathbf{e}_1\mathbf{e}_1^H\mathbf{\Omega}_o}{\lVert\mathbf{e}_1^H\mathbf{\Omega}_o\rVert^2}$
ESPRIT	$\frac{\alpha\alpha^H}{\lVert\alpha\rVert^2}$

Figure 4, 5, and 6 show the bias for MUSIC, Min-Norm, and ESPRIT vs. SNR, the number of senors, and the number of snapshots, respectively. The observations are summarized as the following:

- At low SNR, MUSIC has large bias which is comparable to the root-mean-squared error (RMSE) while Min-Norm and ESPRIT have negligible bias compared to RMSE.
- When the number of senors is small, MUSIC has the largest bias which is again comparable to RMSE, and bias of MUSIC decreases fastest with increase of number of senors.
- The bias is not monotonically decreasing for MUSIC and Min-Norm with increase of number of senors.
- Compared to RMSE, bias of MUSIC is very large at small number of snapshots while bias of Min-Norm and ESPRIT are alway small

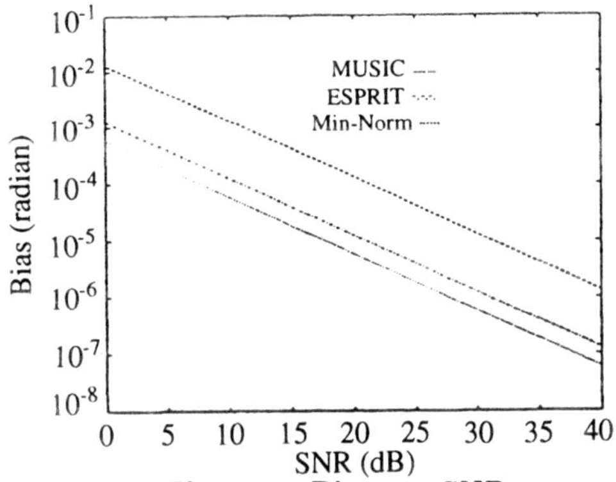

Figure 4: Bias vs. SNR

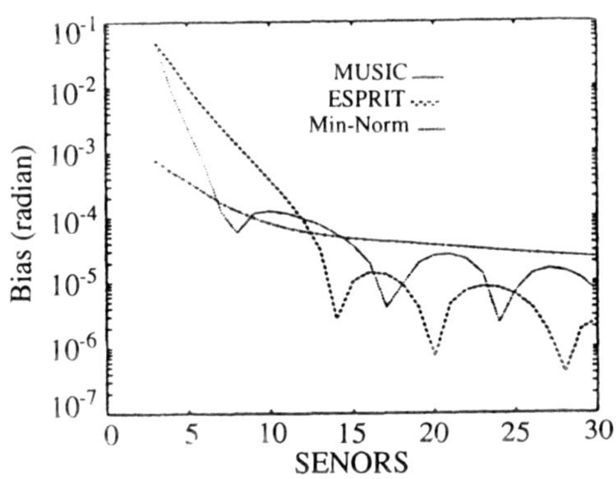

Figure 5: Bias vs. Number of Senors

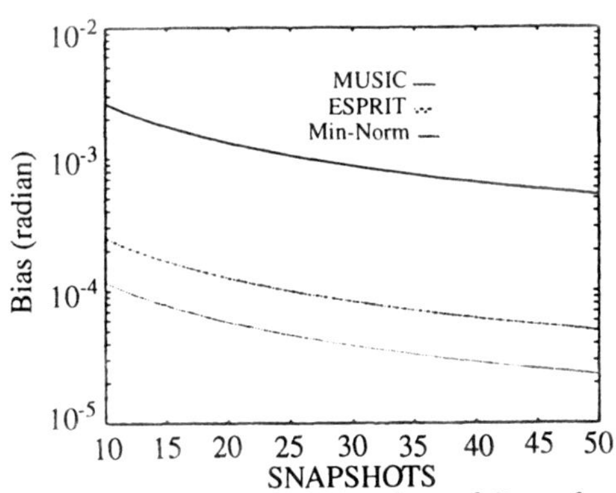

Figure 6: Bias vs. Number of Snapshots

enough to be neglected.

VI. CONCLUSION

In this paper, we have presented unified bias expression for subspace-based DOA estimation algorithms including MUSIC, Min-Norm, and Subspace Rotation. The expression is obtained through the exploration of the relationship between the estimated DOAs and the underlying estimated subspace. The major work includes:

- deriving the second-order subspace perturbations induced by noise.
- equating DOA estimation bias for extrema searching and polynomial rooting algorithms .
- unifying DOA estimation bias of subspace-based algorithms in terms of signal, noise and array characteristics (*physical parameters*).
- providing additional insights into the performance of DOA algorithms from the unified bias formula.

APPENDIX I. BIAS DERIVATION FOR EXTREMA SEARCHING ALGORITHMS

In this section, we will present the detailed derivation of Eqs.(67)-(71) for MUSIC and derivation of Eqs.(73)-(77) for Min-Norm. It is recognized that the method used here can be applied to other extrema searching algorithms with slight modification of perturbed weighting matrix. For later convenience, we first introduce some formulas.
Formula 1. Projection Matrix Perturbation

$$\begin{aligned}
\tilde{\boldsymbol{\Omega}}_o &\stackrel{\text{def}}{=} \boldsymbol{\Omega}_o + \Delta\boldsymbol{\Omega}_o \\
&= \tilde{\mathbf{U}}_o\tilde{\mathbf{U}}_o^H \\
&= (\mathbf{U}_o + \Delta\mathbf{U}_o)(\mathbf{U}_o + \Delta\mathbf{U}_o)^H \\
&= \mathbf{U}_o\mathbf{U}_o^H + \Delta\mathbf{U}_o\mathbf{U}_o^H + \mathbf{U}_o\Delta\mathbf{U}_o^H + \Delta\mathbf{U}_o\Delta\mathbf{U}_o^H.
\end{aligned}$$

Keep $\Delta\boldsymbol{\Omega}_o$ to the first-order to obtain

$$\Delta\boldsymbol{\Omega}_{o1} = \Delta\mathbf{U}_{o1}\mathbf{U}_o^H + \mathbf{U}_o\Delta\mathbf{U}_{o1}^H.$$

Keep $\Delta\boldsymbol{\Omega}_o$ to the second-order to obtain

$$\Delta\boldsymbol{\Omega}_{o2} = \Delta\mathbf{U}_{o2}\mathbf{U}_o^H + \mathbf{U}_o\Delta\mathbf{U}_{o2}^H + \Delta\mathbf{U}_{o1}\Delta\mathbf{U}_{o1}^H.$$

Formula 2. [16]

$$\mathbf{U}_s\boldsymbol{\Sigma}_s^{-2}\mathbf{U}_s^H = (\mathbf{A}^\dagger)^H\hat{\mathbf{R}}_s^{-1}\mathbf{A}^\dagger. \tag{108}$$

A. MUSIC: For MUSIC, the null spectrum

$$P(\theta_k, \tilde{\mathbf{U}}_o) = \mathbf{a}^H(\theta_k)\tilde{\boldsymbol{\Omega}}_o\mathbf{a}(\theta_k).$$

By definition Eq.(62),

$$\begin{aligned} N + \Delta N &\stackrel{\text{def}}{=} \frac{\partial P(\theta_k, \tilde{\mathbf{U}}_o)}{\partial\theta} \\ &= \mathbf{a}^H(\theta_k)\tilde{\boldsymbol{\Omega}}_o\mathbf{a}^{(1)}(\theta_k) + \mathbf{a}^{(1)}(\theta_k)^H\tilde{\boldsymbol{\Omega}}_o\mathbf{a}(\theta_k) \\ &= 2\Re\{\mathbf{a}^H(\theta_k)\tilde{\boldsymbol{\Omega}}_o\mathbf{a}^{(1)}(\theta_k)\}. \end{aligned} \tag{109}$$

The zeroth-order term is N, i.e., without noise perturbation,

$$N = 2\Re\{\mathbf{a}^H(\theta_k)\boldsymbol{\Omega}_o\mathbf{a}^{(1)}(\theta_k)\}.$$

Use Eq.(8) to obtain Eq.(67). ΔN_1 is obtained by approximating Eq.(109) to the first-order and throwing away the zeroth-order term which is N,

$$\Delta N_1 = 2\Re\{\mathbf{a}^H(\theta_k)\Delta\boldsymbol{\Omega}_{o1}\mathbf{a}^{(1)}(\theta_k)\}. \tag{110}$$

Substitute $\Delta\boldsymbol{\Omega}_{o1}$ into Eq.(110) and use Eq.(8) to obtain

$$\begin{aligned} \Delta N_1 &= 2\Re\{\mathbf{a}^H(\theta_k)[\Delta\mathbf{U}_{o1}\mathbf{U}_o^H + \mathbf{U}_o\Delta\mathbf{U}_{o1}^H]\mathbf{a}^{(1)}(\theta_k)\} \\ &= 2\Re\{\mathbf{a}^H(\theta_k)\Delta\mathbf{U}_{o1}\mathbf{U}_o^H\mathbf{a}^{(1)}(\theta_k)\} \end{aligned} \tag{111}$$

and ΔN_2 is the second-order term excluding the zeroth-order term,

$$\Delta N_2 = 2\Re\{\mathbf{a}^H(\theta_k)\Delta\boldsymbol{\Omega}_{o2}\mathbf{a}^{(1)}(\theta_k)\}. \tag{112}$$

Substitute $\Delta\boldsymbol{\Omega}_{o2}$ into Eq.(112) and use Eq.(8) to obtain

$$\begin{aligned}&\Delta N_2\\ =\;& 2\Re\{\mathbf{a}^H(\theta_k)[\Delta\mathbf{U}_{o2}\mathbf{U}_o^H+\mathbf{U}_o\Delta\mathbf{U}_{o2}^H+\Delta\mathbf{U}_{o1}\Delta\mathbf{U}_{o1}^H]\mathbf{a}^{(1)}(\theta_k)\}\\ =\;& 2\Re\{\mathbf{a}^H(\theta_k)\Delta\mathbf{U}_{o2}\mathbf{U}_o^H\mathbf{a}^{(1)}(\theta_k)+\mathbf{a}^H(\theta_k)\Delta\mathbf{U}_{o1}\Delta\mathbf{U}_{o1}^H\mathbf{a}^{(1)}(\theta_k)\}.\end{aligned} \tag{113}$$

By definition Eq.(62),

$$\begin{aligned}&D+\Delta D\\ \stackrel{\text{def}}{=}\;& \frac{\partial^2 P(\theta_k,\tilde{\mathbf{U}}_o)}{\partial\theta^2}\\ =\;& \mathbf{a}^H(\theta_k)\tilde{\boldsymbol{\Omega}}_o\mathbf{a}^{(2)}(\theta_k)+\mathbf{a}^{(2)}(\theta_k)^H\tilde{\boldsymbol{\Omega}}_o\mathbf{a}(\theta_k)+2\mathbf{a}^{(1)}(\theta_k)^H\tilde{\boldsymbol{\Omega}}_o\mathbf{a}^{(1)}(\theta_k)\\ =\;& 2\Re\{\mathbf{a}^H(\theta_k)\tilde{\boldsymbol{\Omega}}_o\mathbf{a}^{(2)}(\theta_k)\}+2\mathbf{a}^{(1)}(\theta_k)^H\tilde{\boldsymbol{\Omega}}_o\mathbf{a}^{(1)}(\theta_k).\end{aligned} \tag{114}$$

The zeroth-order term of Eq.(114) is D, i.e.,

$$D = 2\Re\{\mathbf{a}^H(\theta_k)\boldsymbol{\Omega}_o\mathbf{a}^{(2)}(\theta_k)\}+2\mathbf{a}^{(1)}(\theta_k)^H\boldsymbol{\Omega}_o\mathbf{a}^{(1)}(\theta_k). \tag{115}$$

Use Eq.(8) to obtain Eq.(68). ΔD_1 is equals to the residual of the first-order term of Eq.(114) as

$$\Delta D_1 = 2\Re\{\mathbf{a}^H(\theta_k)\Delta\boldsymbol{\Omega}_{o1}\mathbf{a}^{(2)}(\theta_k)\}+2\mathbf{a}^{(1)}(\theta_k)^H\Delta\boldsymbol{\Omega}_{o1}\mathbf{a}^{(1)}(\theta_k). \tag{116}$$

Substitute $\Delta\boldsymbol{\Omega}_{o1}$ into Eq.(116) and use Eq.(8) to obtain

$$\begin{aligned}&\Delta D_1\\ =\;& 2\Re\{\mathbf{a}^H(\theta_k)[\Delta\mathbf{U}_{o1}\mathbf{U}_o^H+\mathbf{U}_o\Delta\mathbf{U}_{o1}^H]\mathbf{a}^{(2)}(\theta_k)\}\\ &+2\mathbf{a}^{(1)}(\theta_k)^H[\Delta\mathbf{U}_{o1}\mathbf{U}_o^H+\mathbf{U}_o\Delta\mathbf{U}_{o1}^H]\mathbf{a}^{(1)}(\theta_k)\\ =\;& 2\Re\{\mathbf{a}^H(\theta_k)\Delta\mathbf{U}_{o1}\mathbf{U}_o^H\mathbf{a}^{(2)}(\theta_k)\}+4\Re\{\mathbf{a}^{(1)}(\theta_k)^H\Delta\mathbf{U}_{o1}\mathbf{U}_o^H\mathbf{a}^{(1)}(\theta_k)\}.\end{aligned} \tag{117}$$

By definition Eq.(62),

$$\begin{aligned}B+\Delta B \;\stackrel{\text{def}}{=}\;& \frac{\partial^3 P(\theta_k,\tilde{\mathbf{U}}_o)}{\partial\theta^3}\\ =\;& \mathbf{a}^H(\theta_k)\tilde{\boldsymbol{\Omega}}_o\mathbf{a}^{(3)}(\theta_k)+\mathbf{a}^{(3)}(\theta_k)^H\tilde{\boldsymbol{\Omega}}_o\mathbf{a}(\theta_k)\\ & 3\mathbf{a}^{(2)}(\theta_k)^H\tilde{\boldsymbol{\Omega}}_o\mathbf{a}^{(1)}(\theta_k)+3\mathbf{a}^{(1)}(\theta_k)^H\tilde{\boldsymbol{\Omega}}_o\mathbf{a}^{(2)}(\theta_k).\end{aligned} \tag{118}$$

B is the zeroth-order term of Eq.(118), i.e.,

$$\begin{aligned} B &= \mathbf{a}^H(\theta_k)\boldsymbol{\Omega}_o\mathbf{a}^{(3)}(\theta_k) + \mathbf{a}^{(3)}(\theta_k)^H\boldsymbol{\Omega}_o\mathbf{a}(\theta_k) \\ &\quad 3\mathbf{a}^{(2)}(\theta_k)^H\boldsymbol{\Omega}_o\mathbf{a}^{(1)}(\theta_k) + 3\mathbf{a}^{(1)}(\theta_k)^H\boldsymbol{\Omega}_o\mathbf{a}^{(2)}(\theta_k). \end{aligned}$$

Use Eq.(8) to obtain Eq.(69). Substitute Eq.(54) into Eq.(113) and take expectation on both sides of the equation. Notice $\mathbf{a}^H(\theta_k)\mathbf{U}_o = 0$ so that the expectation of the first term in Eq.(113) is zero, we have

$$E(\Delta N_2) = 2E\{\Re\{\mathbf{a}^H(\theta_k)\Delta\mathbf{U}_{o1}\Delta\mathbf{U}_{o1}^H\mathbf{a}^{(1)}(\theta_k)\}\}.$$

Substitute Eqs.(42) and (53) back into the equation

$$\begin{aligned} E(\Delta N_2) &= 2\Re\{\mathbf{a}^H(\theta_k)\mathbf{U}_s\boldsymbol{\Sigma}_s^{-1}\mathbf{V}_s^H E(\mathbf{N}^H\mathbf{U}_o\mathbf{U}_o^H\mathbf{N})\mathbf{V}_s\boldsymbol{\Sigma}_s^{-1}\mathbf{U}_s^H\mathbf{a}^{(1)}(\theta_k)\} \\ &= 2\Re\{\mathbf{a}^H(\theta_k)\mathbf{U}_s\boldsymbol{\Sigma}_s^{-2}\mathbf{U}_s^H\mathbf{a}^{(1)}(\theta_k)\}Tr(\mathbf{U}_o\mathbf{U}_o^H)\sigma_n^2 \\ &= 2(L-P)\sigma_n^2\Re\{\mathbf{a}^H(\theta_k)\mathbf{U}_s\boldsymbol{\Sigma}_s^{-2}\mathbf{U}_s^H\mathbf{a}^{(1)}(\theta_k)\}. \end{aligned} \tag{119}$$

Substitute Eq.(108) into Eq.(119) to obtain Eq.(70). Now we begin to derive Eq.(71),

$$\begin{aligned} & E(\Delta N_1 \Delta D_1) \\ = \; & 4E\{\Re\{\mathbf{a}^H(\theta_k)\Delta\mathbf{U}_{o1}\mathbf{U}_o^H\mathbf{a}^{(1)}(\theta_k)\}\Re\{\mathbf{a}^H(\theta_k)\Delta\mathbf{U}_{o1}\mathbf{U}_o^H\mathbf{a}^{(2)}(\theta_k)\}\} \\ & +8E\{\Re\{\mathbf{a}^H(\theta_k)\Delta\mathbf{U}_{o1}\mathbf{U}_o^H\mathbf{a}^{(1)}(\theta_k)\}\Re\{\mathbf{a}^{(1)}(\theta_k)^H\Delta\mathbf{U}_{o1}\mathbf{U}_o^H\mathbf{a}^{(1)}(\theta_k)\}\}. \end{aligned} \tag{120}$$

Since $\Delta\mathbf{U}_{o1}$ is linear to $\mathbf{N}$, $E(\Delta\mathbf{U}_{o1}\Delta\mathbf{U}_{o1}) = 0$ because of the circularity of the noise. Thus Eq.(120) can be reduced to

$$\begin{aligned} & E(\Delta N_1 \Delta D_1) \\ = \; & 2E\{\Re\{\mathbf{a}^H(\theta_k)\Delta\mathbf{U}_{o1}\mathbf{U}_o^H\mathbf{a}^{(1)}(\theta_k)\mathbf{a}^{(2)}(\theta_k)^H\mathbf{U}_o\Delta\mathbf{U}_{o1}^H\mathbf{a}(\theta_k)\}\} \\ & +4E\{\Re\{\mathbf{a}^H(\theta_k)\Delta\mathbf{U}_{o1}\mathbf{U}_o^H\mathbf{a}^{(1)}(\theta_k)\mathbf{a}^{(1)}(\theta_k)^H\mathbf{U}_o\Delta\mathbf{U}_{o1}^H\mathbf{a}^{(1)}(\theta_k)\}\}. \end{aligned}$$

Substitute Eqs.(42) and (53) back into the equation

$$E(\Delta N_1 \Delta D_1)$$

$$
\begin{aligned}
&= 2E\{\Re\{\mathbf{a}^H(\theta_k)\mathbf{U}_s\mathbf{\Sigma}_s^{-1}\mathbf{V}_s^H\mathbf{N}^H\mathbf{U}_o\mathbf{U}_o^H\mathbf{a}^{(1)}(\theta_k) \\
&\quad \mathbf{a}^{(2)}(\theta_k)^H\mathbf{U}_o\mathbf{U}_o^H\mathbf{N}\mathbf{V}_s\mathbf{\Sigma}_s^{-1}\mathbf{U}_s^H\mathbf{a}(\theta_k)\}\} \\
&\quad +4E\{\Re\{\mathbf{a}^H(\theta_k)\mathbf{U}_s\mathbf{\Sigma}_s^{-1}\mathbf{V}_s^H\mathbf{N}^H\mathbf{U}_o\mathbf{U}_o^H\mathbf{a}^{(1)}(\theta_k) \\
&\quad \mathbf{a}^{(1)}(\theta_k)^H\mathbf{U}_o\mathbf{U}_o^H\mathbf{N}\mathbf{V}_s\mathbf{\Sigma}_s^{-1}\mathbf{U}_s^H\mathbf{a}^{(1)}(\theta_k)\}\} \\
&= 2\sigma_n^2\Re\{\mathbf{a}^H(\theta_k)\mathbf{U}_s\mathbf{\Sigma}_s^{-1}\mathbf{V}_s^H\mathbf{V}_s\mathbf{\Sigma}_s^{-1}\mathbf{U}_s^H\mathbf{a}(\theta_k) \\
&\quad Tr(\mathbf{U}_o\mathbf{U}_o^H\mathbf{a}^{(1)}(\theta_k)\mathbf{a}^{(2)}(\theta_k)^H\mathbf{U}_o\mathbf{U}_o^H)\} \\
&\quad +4\sigma_n^2\Re\{\mathbf{a}^H(\theta_k)\mathbf{U}_s\mathbf{\Sigma}_s^{-1}\mathbf{V}_s^H\mathbf{V}_s\mathbf{\Sigma}_s^{-1}\mathbf{U}_s^H\mathbf{a}^{(1)}(\theta_k) \\
&\quad Tr(\mathbf{U}_o\mathbf{U}_o^H\mathbf{a}^{(1)}(\theta_k)\mathbf{a}^{(1)}(\theta_k)^H\mathbf{U}_o\mathbf{U}_o^H)\} \\
&= 2\sigma_n^2\Re\{\mathbf{a}^H(\theta_k)\mathbf{U}_s\mathbf{\Sigma}_s^{-2}\mathbf{U}_s^H\mathbf{a}(\theta_k)\mathbf{a}^{(2)}(\theta_k)^H\mathbf{\Omega}_o\mathbf{a}^{(1)}(\theta_k)\} \\
&\quad +4\sigma_n^2\Re\{\mathbf{a}^H(\theta_k)\mathbf{U}_s\mathbf{\Sigma}_s^{-2}\mathbf{U}_s^H\mathbf{a}^{(1)}(\theta_k)\mathbf{a}^{(1)}(\theta_k)^H\mathbf{\Omega}_o\mathbf{a}^{(1)}(\theta_k)\}. \quad (121)
\end{aligned}
$$

Substitute Eq.(108) into Eq.(121) to obtain Eq.(71).

B. MIN-NORM: For Min-Norm, the null spectrum

$$
\begin{aligned}
P(\theta_k,\tilde{\mathbf{U}}_o) &= \mathbf{a}^H(\theta_k)\tilde{\mathbf{U}}_o\bar{\mathbf{c}}\bar{\mathbf{c}}^H\tilde{\mathbf{U}}_o^H\mathbf{a}(\theta_k) \\
&= \frac{1}{\|\mathbf{e}_1^H\tilde{\mathbf{\Omega}}_o\|^4}\mathbf{a}^H(\theta_k)\tilde{\mathbf{\Omega}}_o\mathbf{e}_1\mathbf{e}_1^H\tilde{\mathbf{\Omega}}_o\mathbf{a}(\theta_k).
\end{aligned}
$$

By definition Eq.(62),

$$
\begin{aligned}
&N+\Delta N \\
&\stackrel{\text{def}}{=} \frac{\partial P(\theta_k,\tilde{\mathbf{U}}_o)}{\partial\theta} \\
&= \frac{1}{\|\mathbf{e}_1^H\tilde{\mathbf{\Omega}}_o\|^4}\{\mathbf{a}^{(1)}(\theta_k)^H\tilde{\mathbf{\Omega}}_o\mathbf{e}_1\mathbf{e}_1^H\tilde{\mathbf{\Omega}}_o\mathbf{a}(\theta_k)+\mathbf{a}^H(\theta_k)\tilde{\mathbf{\Omega}}_o\mathbf{e}_1\mathbf{e}_1^H\tilde{\mathbf{\Omega}}_o\mathbf{a}^{(1)}(\theta_k)\} \\
&= \frac{2}{\|\mathbf{e}_1^H\tilde{\mathbf{\Omega}}_o\|^4}\Re\{\mathbf{a}^{(1)}(\theta_k)^H\tilde{\mathbf{\Omega}}_o\mathbf{e}_1\mathbf{e}_1^H\tilde{\mathbf{\Omega}}_o\mathbf{a}(\theta_k)\}. \quad (122)
\end{aligned}
$$

The zeroth-order term is N, i.e.,

$$
N = \frac{2}{\|\mathbf{e}_1^H\mathbf{\Omega}_o\|^4}\Re\{\mathbf{a}^{(1)}(\theta_k)^H\mathbf{\Omega}_o\mathbf{e}_1\mathbf{e}_1^H\mathbf{\Omega}_o\mathbf{a}(\theta_k)\}. \quad (123)
$$

Use Eq.(8) to obtain Eq.(73). Keep Eq.(122) up to the first-order and throw away the zeroth-order term to obtain ΔN_1,

$$
\Delta N_1 = \frac{2}{\|\mathbf{e}_1^H\mathbf{\Omega}_o\|^4}\Re\{\mathbf{a}^{(1)}(\theta_k)^H[\mathbf{\Omega}_o\mathbf{e}_1\mathbf{e}_1^H\Delta\mathbf{\Omega}_{o1}+\Delta\mathbf{\Omega}_{o1}\mathbf{e}_1\mathbf{e}_1^H\mathbf{\Omega}_o]\mathbf{a}(\theta_k)\}. \quad (124)
$$

Substitute $\Delta\mathbf{\Omega}_{o1}$ into Eq.(124) and use Eq.(8) to obtain

$$\begin{aligned}\Delta N_1 &= \frac{2}{\|\mathbf{e}_1^H\mathbf{\Omega}_o\|^4}\Re\{\mathbf{a}^{(1)}(\theta_k)^H\mathbf{\Omega}_o\mathbf{e}_1\mathbf{e}_1^H\Delta\mathbf{\Omega}_{o1}\mathbf{a}(\theta_k)\}\\ &= \frac{2}{\|\mathbf{e}_1^H\mathbf{\Omega}_o\|^4}\Re\{\mathbf{a}^{(1)}(\theta_k)^H\mathbf{\Omega}_o\mathbf{e}_1\mathbf{e}_1^H[\Delta\mathbf{U}_{o1}\mathbf{U}_o^H+\mathbf{U}_o\Delta\mathbf{U}_{o1}^H]\mathbf{a}(\theta_k)\}\\ &= \frac{2}{\|\mathbf{e}_1^H\mathbf{\Omega}_o\|^4}\Re\{\mathbf{a}^{(1)}(\theta_k)^H\mathbf{\Omega}_o\mathbf{e}_1\mathbf{e}_1^H\mathbf{U}_o\Delta\mathbf{U}_{o1}^H\mathbf{a}(\theta_k)\}. \qquad (125)\end{aligned}$$

Keep Eq.(122) up to the second-order and exclude the zeroth-order term to obtain ΔN_2

$$\begin{aligned}\Delta N_2 &= \frac{2}{\|\mathbf{e}_1^H\mathbf{\Omega}_o\|^4}\Re\{\mathbf{a}^{(1)}(\theta_k)^H[\mathbf{\Omega}_o\mathbf{e}_1\mathbf{e}_1^H\Delta\mathbf{\Omega}_{o2}+\Delta\mathbf{\Omega}_{o2}\mathbf{e}_1\mathbf{e}_1^H\mathbf{\Omega}_o\\ &\quad +\Delta\mathbf{\Omega}_{o1}\mathbf{e}_1\mathbf{e}_1^H\Delta\mathbf{\Omega}_{o1}]\mathbf{a}(\theta_k)\}. \qquad (126)\end{aligned}$$

Substitute $\Delta\mathbf{\Omega}_{o1}$ and $\Delta\mathbf{\Omega}_{o2}$ into Eq.(126) and use Eq.(8) to obtain

$$\begin{aligned}&\Delta N_2\\ =\ & \frac{2}{\|\mathbf{e}_1^H\mathbf{\Omega}_o\|^4}\Re\{\mathbf{a}^{(1)}(\theta_k)^H[\mathbf{\Omega}_o\mathbf{e}_1\mathbf{e}_1^H\Delta\mathbf{\Omega}_{o2}+\Delta\mathbf{\Omega}_{o1}\mathbf{e}_1\mathbf{e}_1^H\Delta\mathbf{\Omega}_{o1}]\mathbf{a}(\theta_k)\}\\ =\ & \frac{2}{\|\mathbf{e}_1^H\mathbf{\Omega}_o\|^4}\Re\{\mathbf{a}^{(1)}(\theta_k)^H\mathbf{\Omega}_o\mathbf{e}_1\mathbf{e}_1^H[\Delta\mathbf{U}_{o2}\mathbf{U}_o^H+\mathbf{U}_o\Delta\mathbf{U}_{o2}^H+\Delta\mathbf{U}_{o1}\Delta\mathbf{U}_{o1}^H]\mathbf{a}(\theta_k)\\ &+\mathbf{a}^{(1)}(\theta_k)^H[\Delta\mathbf{U}_{o1}\mathbf{U}_o^H+\mathbf{U}_o\Delta\mathbf{U}_{o1}^H]\mathbf{e}_1\mathbf{e}_1^H[\Delta\mathbf{U}_{o1}\mathbf{U}_o^H+\mathbf{U}_o\Delta\mathbf{U}_{o1}^H]\mathbf{a}(\theta_k)\}\\ =\ & \frac{2}{\|\mathbf{e}_1^H\mathbf{\Omega}_o\|^4}\Re\{\mathbf{a}^{(1)}(\theta_k)^H\mathbf{\Omega}_o\mathbf{e}_1\mathbf{e}_1^H[\mathbf{U}_o\Delta\mathbf{U}_{o2}^H+\Delta\mathbf{U}_{o1}\Delta\mathbf{U}_{o1}^H]\mathbf{a}(\theta_k)\\ &+\mathbf{a}^{(1)}(\theta_k)^H[\Delta\mathbf{U}_{o1}\mathbf{U}_o^H+\mathbf{U}_o\Delta\mathbf{U}_{o1}^H]\mathbf{e}_1\mathbf{e}_1^H\mathbf{U}_o\Delta\mathbf{U}_{o1}^H\mathbf{a}(\theta_k)\}. \qquad (127)\end{aligned}$$

By definition Eq.(62),

$$\begin{aligned}D+\Delta D &\stackrel{\text{def}}{=} \frac{\partial^2 P(\theta_k,\tilde{\mathbf{U}}_o)}{\partial\theta^2}\\ &= \frac{1}{\|\mathbf{e}_1^H\tilde{\mathbf{\Omega}}_o\|^4}\{\mathbf{a}^{(2)}(\theta_k)^H\tilde{\mathbf{\Omega}}_o\mathbf{e}_1\mathbf{e}_1^H\tilde{\mathbf{\Omega}}_o\mathbf{a}(\theta_k)\\ &\quad +2\mathbf{a}^{(1)}(\theta_k)^H\tilde{\mathbf{\Omega}}_o\mathbf{e}_1\mathbf{e}_1^H\tilde{\mathbf{\Omega}}_o\mathbf{a}^{(1)}(\theta_k)+\mathbf{a}(\theta_k)^H\tilde{\mathbf{\Omega}}_o\mathbf{e}_1\mathbf{e}_1^H\tilde{\mathbf{\Omega}}_o\mathbf{a}^{(2)}(\theta_k)\}\\ &= \frac{1}{\|\mathbf{e}_1^H\tilde{\mathbf{\Omega}}_o\|^4}\{2\Re\{\mathbf{a}^{(2)}(\theta_k)^H\tilde{\mathbf{\Omega}}_o\mathbf{e}_1\mathbf{e}_1^H\tilde{\mathbf{\Omega}}_o\mathbf{a}(\theta_k)\}\\ &\quad +2\mathbf{a}^{(1)}(\theta_k)^H\tilde{\mathbf{\Omega}}_o\mathbf{e}_1\mathbf{e}_1^H\tilde{\mathbf{\Omega}}_o\mathbf{a}^{(1)}(\theta_k)\}.\end{aligned}$$

The zeroth-order term is D, i.e.,

$$D=\frac{1}{\|\mathbf{e}_1^H\mathbf{\Omega}_o\|^4}\{2\Re\{\mathbf{a}^{(2)}(\theta_k)^H\mathbf{\Omega}_o\mathbf{e}_1\mathbf{e}_1^H\mathbf{\Omega}_o\mathbf{a}(\theta_k)\}+2\mathbf{a}^{(1)}(\theta_k)^H\mathbf{\Omega}_o\mathbf{e}_1\mathbf{e}_1^H\mathbf{\Omega}_o\mathbf{a}^{(1)}(\theta_k)\}. \qquad (128)$$

Use Eq.(8) to obtain Eq.(74). Keep Eq.(128) up to the first-order and throw away the zeroth-order term to obtain ΔD_1

$$\begin{aligned} & \Delta D_1 \\ = & \frac{1}{\|\mathbf{e}_1^H \boldsymbol{\Omega}_o\|^4} \{2\Re\{\mathbf{a}^{(2)}(\theta_k)^H [\Delta\boldsymbol{\Omega}_{o1} \mathbf{e}_1 \mathbf{e}_1^H \boldsymbol{\Omega}_o + \boldsymbol{\Omega}_o \mathbf{e}_1 \mathbf{e}_1^H \Delta\boldsymbol{\Omega}_{o1}] \mathbf{a}(\theta_k)\} \\ & + 2\mathbf{a}^{(1)}(\theta_k)^H [\Delta\boldsymbol{\Omega}_{o1} \mathbf{e}_1 \mathbf{e}_1^H \boldsymbol{\Omega}_o + \boldsymbol{\Omega}_o \mathbf{e}_1 \mathbf{e}_1^H \Delta\boldsymbol{\Omega}_{o1}] \mathbf{a}^{(1)}(\theta_k)\}. \end{aligned} \tag{129}$$

Substitute $\Delta\boldsymbol{\Omega}_{o1}$ into Eq.(129) and use Eq.(8) to obtain

$$\begin{aligned} \Delta D_1 = & \frac{1}{\|\mathbf{e}_1^H \boldsymbol{\Omega}_o\|^4} \{2\Re\{\mathbf{a}^{(2)}(\theta_k)^H \boldsymbol{\Omega}_o \mathbf{e}_1 \mathbf{e}_1^H \Delta\boldsymbol{\Omega}_{o1} \mathbf{a}(\theta_k)\} \\ & + 4\Re\{\mathbf{a}^{(1)}(\theta_k)^H \Delta\boldsymbol{\Omega}_{o1} \mathbf{e}_1 \mathbf{e}_1^H \boldsymbol{\Omega}_o \mathbf{a}^{(1)}(\theta_k)\}\} \\ = & \frac{1}{\|\mathbf{e}_1^H \boldsymbol{\Omega}_o\|^4} \{2\Re\{\mathbf{a}^{(2)}(\theta_k)^H \boldsymbol{\Omega}_o \mathbf{e}_1 \mathbf{e}_1^H [\Delta\mathbf{U}_{o1} \mathbf{U}_o^H + \mathbf{U}_o \Delta\mathbf{U}_{o1}^H] \mathbf{a}(\theta_k)\} \\ & + 4\Re\{\mathbf{a}^{(1)}(\theta_k)^H [\Delta\mathbf{U}_{o1} \mathbf{U}_o^H + \mathbf{U}_o \Delta\mathbf{U}_{o1}^H] \mathbf{e}_1 \mathbf{e}_1^H \boldsymbol{\Omega}_o \mathbf{a}^{(1)}(\theta_k)\}\} \\ = & \frac{1}{\|\mathbf{e}_1^H \boldsymbol{\Omega}_o\|^4} \{2\Re\{\mathbf{a}^{(2)}(\theta_k)^H \boldsymbol{\Omega}_o \mathbf{e}_1 \mathbf{e}_1^H \mathbf{U}_o \Delta\mathbf{U}_{o1}^H \mathbf{a}(\theta_k)\} \\ & + 4\Re\{\mathbf{a}^{(1)}(\theta_k)^H \Delta\mathbf{U}_{o1} \mathbf{U}_o^H \mathbf{e}_1 \mathbf{e}_1^H \boldsymbol{\Omega}_o \mathbf{a}^{(1)}(\theta_k)\} \\ & + 4\Re\{\mathbf{a}^{(1)}(\theta_k)^H \mathbf{U}_o \Delta\mathbf{U}_{o1}^H \mathbf{e}_1 \mathbf{e}_1^H \boldsymbol{\Omega}_o \mathbf{a}^{(1)}(\theta_k)\}\}. \end{aligned} \tag{130}$$

By definition Eq.(62),

$$\begin{aligned} & B + \Delta B \\ \stackrel{\text{def}}{=} & \frac{\partial^3 P(\theta_k, \tilde{\mathbf{U}}_o)}{\partial \theta^3} \\ = & \frac{1}{\|\mathbf{e}_1^H \tilde{\boldsymbol{\Omega}}_o\|^4} \{\mathbf{a}^{(3)}(\theta_k)^H \tilde{\boldsymbol{\Omega}}_o \mathbf{e}_1 \mathbf{e}_1^H \tilde{\boldsymbol{\Omega}}_o \mathbf{a}(\theta_k) + \mathbf{a}^H(\theta_k) \tilde{\boldsymbol{\Omega}}_o \mathbf{e}_1 \mathbf{e}_1^H \tilde{\boldsymbol{\Omega}}_o \mathbf{a}^{(3)}(\theta_k) \\ & + 3\mathbf{a}^{(2)}(\theta_k)^H \tilde{\boldsymbol{\Omega}}_o \mathbf{e}_1 \mathbf{e}_1^H \tilde{\boldsymbol{\Omega}}_o \mathbf{a}^{(1)}(\theta_k) + 3\mathbf{a}^{(1)}(\theta_k)^H \tilde{\boldsymbol{\Omega}}_o \mathbf{e}_1 \mathbf{e}_1^H \tilde{\boldsymbol{\Omega}}_o \mathbf{a}^{(2)}(\theta_k)\}. \end{aligned}$$

Keep the zeroth-order term to obtain B, i.e.,

$$\begin{aligned} B = & \frac{1}{\|\mathbf{e}_1^H \boldsymbol{\Omega}_o\|^4} \{2\Re\{\mathbf{a}^{(3)}(\theta_k)^H \boldsymbol{\Omega}_o \mathbf{e}_1 \mathbf{e}_1^H \boldsymbol{\Omega}_o \mathbf{a}(\theta_k)\} \\ & + 6\Re\{\mathbf{a}^{(2)}(\theta_k)^H \boldsymbol{\Omega}_o \mathbf{e}_1 \mathbf{e}_1^H \boldsymbol{\Omega}_o \mathbf{a}^{(1)}(\theta_k)\}\}. \end{aligned} \tag{131}$$

Using Eq.(8) to obtain Eq.(75). Substitute Eq.(54) into Eq.(127) and take expectation on both sides of the equation. Using $\mathbf{a}^H(\theta_k)\mathbf{U}_o = 0$ as well as the circularity of the noise, we have

$$\begin{aligned} E(\Delta N_2) &= \frac{2}{\|\mathbf{e}_1^H \boldsymbol{\Omega}_o\|^4} \Re\{E[\mathbf{a}^{(1)}(\theta_k)^H \boldsymbol{\Omega}_o \mathbf{e}_1 \mathbf{e}_1^H \Delta\mathbf{U}_{o1} \Delta\mathbf{U}_{o1}^H \mathbf{a}(\theta_k)] \\ &\quad + E[\mathbf{a}^{(1)}(\theta_k)^H \Delta\mathbf{U}_{o1} \mathbf{U}_o^H \mathbf{e}_1 \mathbf{e}_1^H \mathbf{U}_o \Delta\mathbf{U}_{o1}^H \mathbf{a}(\theta_k)]\}. \end{aligned}$$

Substitute Eq.(42) and Eq.(53) back into the equation

$$\begin{aligned} & E(\Delta N_2) \\ = & \frac{2}{\|\mathbf{e}_1^H \boldsymbol{\Omega}_o\|^4} \Re\{E[\mathbf{a}^{(1)}(\theta_k)^H \boldsymbol{\Omega}_o \mathbf{e}_1 \mathbf{e}_1^H \mathbf{U}_s \boldsymbol{\Sigma}_s^{-1} \mathbf{V}_s^H \mathbf{N}^H \mathbf{U}_o \mathbf{U}_o^H \mathbf{N} \mathbf{V}_s \boldsymbol{\Sigma}_s^{-1} \mathbf{U}_s^H \mathbf{a}(\theta_k)] \\ & + E[\mathbf{a}^{(1)}(\theta_k)^H \mathbf{U}_s \boldsymbol{\Sigma}_s^{-1} \mathbf{V}_s^H \mathbf{N}^H \mathbf{U}_o \mathbf{U}_o^H \mathbf{e}_1 \mathbf{e}_1^H \mathbf{U}_o \mathbf{U}_o^H \mathbf{N} \mathbf{V}_s \boldsymbol{\Sigma}_s^{-1} \mathbf{U}_s^H \mathbf{a}(\theta_k)]\} \\ = & \frac{2\sigma_n^2}{\|\mathbf{e}_1^H \boldsymbol{\Omega}_o\|^4} \Re\{\mathbf{a}^{(1)}(\theta_k)^H \boldsymbol{\Omega}_o \mathbf{e}_1 \mathbf{e}_1^H \mathbf{U}_s \boldsymbol{\Sigma}_s^{-1} \mathbf{V}_s^H \mathbf{V}_s \boldsymbol{\Sigma}_s^{-1} \mathbf{U}_s^H \mathbf{a}(\theta_k) Tr(\mathbf{U}_o \mathbf{U}_o^H) \\ & + \mathbf{a}^{(1)}(\theta_k)^H \mathbf{U}_s \boldsymbol{\Sigma}_s^{-1} \mathbf{V}_s^H \mathbf{V}_s \boldsymbol{\Sigma}_s^{-1} \mathbf{U}_s^H \mathbf{a}(\theta_k) Tr(\mathbf{U}_o \mathbf{U}_o^H \mathbf{e}_1 \mathbf{e}_1^H \mathbf{U}_o \mathbf{U}_o^H)\} \\ = & \frac{2\sigma_n^2}{\|\mathbf{e}_1^H \boldsymbol{\Omega}_o\|^4} \Re\{(L-P) \mathbf{a}^{(1)}(\theta_k)^H \boldsymbol{\Omega}_o \mathbf{e}_1 \mathbf{e}_1^H \mathbf{U}_s \boldsymbol{\Sigma}_s^{-2} \mathbf{U}_s^H \mathbf{a}(\theta_k) \\ & + \|\mathbf{e}_1 \boldsymbol{\Omega}_o\|^2 \mathbf{a}^{(1)}(\theta_k)^H \mathbf{U}_s \boldsymbol{\Sigma}_s^{-2} \mathbf{U}_s^H \mathbf{a}(\theta_k)\}. \end{aligned} \tag{132}$$

Substitute Eq.(108) into Eq.(132) to obtain Eq.(76). Now let us calculate $E(\Delta N_1 \Delta D_1)$ by using Eqs.(125) and (130),

$$\begin{aligned} & E(\Delta N_1 \Delta D_1) \\ = & \frac{4}{\|\mathbf{e}_1^H \boldsymbol{\Omega}_o\|^8} \{E\{\Re\{\mathbf{a}^{(1)}(\theta_k)^H \boldsymbol{\Omega}_o \mathbf{e}_1 \mathbf{e}_1^H \mathbf{U}_o \Delta\mathbf{U}_{o1}^H \mathbf{a}(\theta_k)\} \\ & \Re\{\mathbf{a}^{(2)}(\theta_k)^H \boldsymbol{\Omega}_o \mathbf{e}_1 \mathbf{e}_1^H \mathbf{U}_o \Delta\mathbf{U}_{o1}^H \mathbf{a}(\theta_k)\}\} \\ & + 2E\{\Re\{\mathbf{a}^{(1)}(\theta_k)^H \boldsymbol{\Omega}_o \mathbf{e}_1 \mathbf{e}_1^H \mathbf{U}_o \Delta\mathbf{U}_{o1}^H \mathbf{a}(\theta_k)\} \\ & \Re\{\mathbf{a}^{(1)}(\theta_k)^H \Delta\mathbf{U}_{o1} \mathbf{U}_o^H \mathbf{e}_1 \mathbf{e}_1^H \boldsymbol{\Omega}_o \mathbf{a}^{(1)}(\theta_k)\}\} \\ & + 2E\{\Re\{\mathbf{a}^{(1)}(\theta_k)^H \boldsymbol{\Omega}_o \mathbf{e}_1 \mathbf{e}_1^H \mathbf{U}_o \Delta\mathbf{U}_{o1}^H \mathbf{a}(\theta_k)\} \\ & \Re\{\mathbf{a}^{(1)}(\theta_k)^H \mathbf{U}_o \Delta\mathbf{U}_{o1}^H \mathbf{e}_1 \mathbf{e}_1^H \boldsymbol{\Omega}_o \mathbf{a}^{(1)}(\theta_k)\}\}\}. \end{aligned}$$

Use the circular property of the noise to simplify

$$\begin{aligned} & E(\Delta N_1 \Delta D_1) \\ = & \frac{2}{\|\mathbf{e}_1^H \boldsymbol{\Omega}_o\|^8} \{\Re\{\mathbf{a}^{(1)}(\theta_k)^H \boldsymbol{\Omega}_o \mathbf{e}_1 \mathbf{e}_1^H \mathbf{U}_o \Delta\mathbf{U}_{o1}^H \mathbf{a}(\theta_k) \\ & \mathbf{a}^H(\theta_k) \Delta\mathbf{U}_{o1} \mathbf{U}_o^H \mathbf{e}_1 \mathbf{e}_1^H \boldsymbol{\Omega}_o \mathbf{a}^{(2)}(\theta_k)\} \\ & + 2\Re\{\mathbf{a}^{(1)}(\theta_k)^H \boldsymbol{\Omega}_o \mathbf{e}_1 \mathbf{e}_1^H \mathbf{U}_o \Delta\mathbf{U}_{o1}^H \mathbf{a}(\theta_k)\} \end{aligned}$$

$$\mathbf{a}^{(1)}(\theta_k)^H \Delta \mathbf{U}_{o1} \mathbf{U}_o^H \mathbf{e}_1 \mathbf{e}_1^H \boldsymbol{\Omega}_o \mathbf{a}^{(1)}(\theta_k)\}$$
$$+2\Re\{\mathbf{a}^{(1)}(\theta_k)^H \boldsymbol{\Omega}_o \mathbf{e}_1 \mathbf{e}_1^H \mathbf{U}_o \Delta \mathbf{U}_{o1}^H \mathbf{a}(\theta_k)$$
$$\mathbf{a}^{(1)}(\theta_k)^H \boldsymbol{\Omega}_o \mathbf{e}_1 \mathbf{e}_1^H \Delta \mathbf{U}_{o1} \mathbf{U}_o^H \mathbf{a}^{(1)}(\theta_k)\}\}.$$

Substitute Eqs.(42) and (53) back into the equation

$$
\begin{aligned}
& E(\Delta N_1 \Delta D_1) \\
= \quad & \frac{2}{\|\mathbf{e}_1^H \boldsymbol{\Omega}_o\|^8}\{\Re\{\mathbf{a}^{(1)}(\theta_k)^H \boldsymbol{\Omega}_o \mathbf{e}_1 \mathbf{e}_1^H \mathbf{U}_o \mathbf{U}_o^H \mathbf{N} \mathbf{V}_s \boldsymbol{\Sigma}_s^{-1} \mathbf{U}_s^H \mathbf{a}(\theta_k) \\
& \mathbf{a}^H(\theta_k) \mathbf{U}_s \boldsymbol{\Sigma}_s^{-1} \mathbf{V}_s^H \mathbf{N}^H \mathbf{U}_o \mathbf{U}_o^H \mathbf{e}_1 \mathbf{e}_1^H \boldsymbol{\Omega}_o \mathbf{a}^{(2)}(\theta_k)\} \\
& +2\Re\{\mathbf{a}^{(1)}(\theta_k)^H \boldsymbol{\Omega}_o \mathbf{e}_1 \mathbf{e}_1^H \mathbf{U}_o \mathbf{U}_o^H \mathbf{N} \mathbf{V}_s \boldsymbol{\Sigma}_s^{-1} \mathbf{U}_s^H \mathbf{a}(\theta_k) \\
& \mathbf{a}^{(1)}(\theta_k)^H \mathbf{U}_s \boldsymbol{\Sigma}_s^{-1} \mathbf{V}_s^H \mathbf{N}^H \mathbf{U}_o \mathbf{U}_o^H \mathbf{e}_1 \mathbf{e}_1^H \boldsymbol{\Omega}_o \mathbf{a}^{(1)}(\theta_k)\} \\
& +2\Re\{\mathbf{a}^{(1)}(\theta_k)^H \boldsymbol{\Omega}_o \mathbf{e}_1 \mathbf{e}_1^H \mathbf{U}_o \mathbf{U}_o^H \mathbf{N} \mathbf{V}_s \boldsymbol{\Sigma}_s^{-1} \mathbf{U}_s^H \mathbf{a}(\theta_k) \\
& \mathbf{a}^{(1)}(\theta_k)^H \boldsymbol{\Omega}_o \mathbf{e}_1 \mathbf{e}_1^H \mathbf{U}_s \boldsymbol{\Sigma}_s^{-1} \mathbf{V}_s^H \mathbf{N}^H \mathbf{U}_o \mathbf{U}_o^H \mathbf{a}^{(1)}(\theta_k)\}\} \\
= \quad & \frac{2\sigma_n^2}{\|\mathbf{e}_1^H \boldsymbol{\Omega}_o\|^8}\{\Re\{\mathbf{a}^{(1)}(\theta_k)^H \boldsymbol{\Omega}_o \mathbf{e}_1 \mathbf{e}_1^H \mathbf{U}_o \mathbf{U}_o^H \mathbf{U}_o \mathbf{U}_o^H \mathbf{e}_1 \mathbf{e}_1^H \boldsymbol{\Omega}_o \mathbf{a}^{(2)}(\theta_k) \\
& Tr[\mathbf{a}^H(\theta_k) \mathbf{U}_s \boldsymbol{\Sigma}_s^{-1} \mathbf{V}_s^H \mathbf{V}_s \boldsymbol{\Sigma}_s^{-1} \mathbf{U}_s^H \mathbf{a}(\theta_k)]\} \\
& +2\Re\{\mathbf{a}^{(1)}(\theta_k)^H \boldsymbol{\Omega}_o \mathbf{e}_1 \mathbf{e}_1^H \mathbf{U}_o \mathbf{U}_o^H \mathbf{U}_o \mathbf{U}_o^H \mathbf{e}_1 \mathbf{e}_1^H \boldsymbol{\Omega}_o \mathbf{a}^{(1)}(\theta_k) \\
& Tr[\mathbf{a}^{(1)}(\theta_k)^H \mathbf{U}_s \boldsymbol{\Sigma}_s^{-1} \mathbf{V}_s^H \mathbf{V}_s \boldsymbol{\Sigma}_s^{-1} \mathbf{U}_s^H \mathbf{a}(\theta_k)]\} \\
& +2\Re\{\mathbf{a}^{(1)}(\theta_k)^H \boldsymbol{\Omega}_o \mathbf{e}_1 \mathbf{e}_1^H \mathbf{U}_o \mathbf{U}_o^H \mathbf{U}_o \mathbf{U}_o^H \mathbf{a}^{(1)}(\theta_k) \\
& Tr[\mathbf{a}^{(1)}(\theta_k)^H \boldsymbol{\Omega}_o \mathbf{e}_1 \mathbf{e}_1^H \mathbf{U}_s \boldsymbol{\Sigma}_s^{-1} \mathbf{V}_s^H \mathbf{V}_s \boldsymbol{\Sigma}_s^{-1} \mathbf{U}_s^H \mathbf{a}(\theta_k)]\}\} \\
= \quad & \frac{2\sigma_n^2}{\|\mathbf{e}_1^H \boldsymbol{\Omega}_o\|^8}\{\|\mathbf{e}_1^H \boldsymbol{\Omega}_o\|^2 \Re\{\mathbf{a}^{(1)}(\theta_k)^H \boldsymbol{\Omega}_o \mathbf{e}_1 \mathbf{e}_1^H \boldsymbol{\Omega}_o \mathbf{a}^{(2)}(\theta_k) \\
& Tr[\mathbf{a}^H(\theta_k) \mathbf{U}_s \boldsymbol{\Sigma}_s^{-2} \mathbf{U}_s^H \mathbf{a}(\theta_k)]\} \\
& +2|\mathbf{a}^{(1)}(\theta_k)^H \boldsymbol{\Omega}_o \mathbf{e}_1\|^2 \|\mathbf{e}_1^H \boldsymbol{\Omega}_o|^2 \Re\{\mathbf{a}^{(1)}(\theta_k)^H \mathbf{U}_s \boldsymbol{\Sigma}_s^{-2} \mathbf{U}_s^H \mathbf{a}(\theta_k)\} \\
& +2|\mathbf{a}^{(1)}(\theta_k)^H \boldsymbol{\Omega}_o \mathbf{e}_1|^2 \Re\{\mathbf{a}^{(1)}(\theta_k)^H \boldsymbol{\Omega}_o \mathbf{e}_1 \mathbf{e}_1^H \mathbf{U}_s \boldsymbol{\Sigma}_s^{-2} \mathbf{U}_s^H \mathbf{a}(\theta_k)\}\}.
\end{aligned}
$$

Use Eq.(108) to obtain Eq.(77).

APPENDIX II. BIAS DERIVATION FOR POLYNOMIAL ROOTING ALGORITHMS

In this section, we derive the formulas Eqs.(91)-(95) for Root MUSIC and Eqs.(97)-(95) for Root Min-Norm. Like extrema searching algorithms, the derivation here can be applicable to other polynomial root algorithms with slight difference in perturbed weighting matrix.

A. ROOT MUSIC: For Root MUSIC, the spectral polynomial is

$$P(r_k, \tilde{\mathbf{U}}_o) = \mathbf{a}(r_k^{-1})^T \tilde{\boldsymbol{\Omega}}_o \mathbf{a}(r_k).$$

By definition Eq.(84),

$$\begin{aligned} N + \Delta N &\stackrel{\text{def}}{=} \frac{\partial P(r_k, \tilde{\mathbf{U}}_o)}{\partial z} \\ &= -r_k^{-2}\mathbf{a}^{(1)}(r_k^{-1})^T \tilde{\boldsymbol{\Omega}}_o \mathbf{a}(r_k) + \mathbf{a}(r_k^{-1})^T \tilde{\boldsymbol{\Omega}}_o \mathbf{a}^{(1)}(r_k) \\ &= -2jr_k^{-1}\Im\{r_k^{-1}\mathbf{a}^{(1)}(r_k^{-1})^T \tilde{\boldsymbol{\Omega}}_o \mathbf{a}(r_k)\}. \end{aligned} \tag{133}$$

Keep the zeroth-order term of Eq.(133) to obtain N,

$$N = -2jr_k^{-1}\Im\{r_k^{-1}\mathbf{a}^{(1)}(r_k^{-1})^T \boldsymbol{\Omega}_o \mathbf{a}(r_k)\}.$$

Use Eq.(8) to obtain Eq.(91). Clearly ΔN_1 is the first-order term of ΔN,

$$\Delta N_1 = -2jr_k^{-1}\Im\{r_k^{-1}\mathbf{a}^{(1)}(r_k^{-1})^T \Delta\boldsymbol{\Omega}_{o1} \mathbf{a}(r_k)\}. \tag{134}$$

Substitute $\Delta\boldsymbol{\Omega}_{o1}$ into Eq.(134) and use Eq.(8) to obtain

$$\begin{aligned} \Delta N_1 &= -2jr_k^{-1}\Im\{r_k^{-1}\mathbf{a}^{(1)}(r_k^{-1})^T [\Delta\mathbf{U}_{o1}\mathbf{U}_o^H + \mathbf{U}_o\Delta\mathbf{U}_{o1}^H]\mathbf{a}(r_k)\} \\ &= -2jr_k^{-1}\Im\{r_k^{-1}\mathbf{a}^{(1)}(r_k^{-1})^T \mathbf{U}_o\Delta\mathbf{U}_{o1}^H \mathbf{a}(r_k)\}. \end{aligned} \tag{135}$$

ΔN_2 is the second-order term of ΔN,

$$\Delta N_2 = -2jr_k^{-1}\Im\{r_k^{-1}\mathbf{a}^{(1)}(r_k^{-1})^T \Delta\boldsymbol{\Omega}_{o2} \mathbf{a}(r_k)\}. \tag{136}$$

Substitute $\Delta\boldsymbol{\Omega}_{o2}$ into Eq.(136) and use Eq.(8) to obtain

$$\begin{aligned} &\Delta N_2 \\ &= -2jr_k^{-1}\Im\{r_k^{-1}\mathbf{a}^{(1)}(r_k^{-1})^T [\Delta\mathbf{U}_{o2}\mathbf{U}_o^H + \mathbf{U}_o\Delta\mathbf{U}_{o2}^H + \Delta\mathbf{U}_{o1}\Delta\mathbf{U}_{o1}^H]\mathbf{a}(r_k)\} \\ &= -2jr_k^{-1}\Im\{r_k^{-1}\mathbf{a}^{(1)}(r_k^{-1})^T [\mathbf{U}_o\Delta\mathbf{U}_{o2}^H + \Delta\mathbf{U}_{o1}\Delta\mathbf{U}_{o1}^H]\mathbf{a}(r_k)\}. \end{aligned} \tag{137}$$

By definition Eq.(84),

$$D + \Delta D \stackrel{\text{def}}{=} \frac{\partial^2 P(r_k, \tilde{\mathbf{U}}_o)}{\partial z^2}$$

$$\begin{aligned} &= 2r_k^{-3}\mathbf{a}^{(1)}(r_k^{-1})^T\tilde{\boldsymbol{\Omega}}_o\mathbf{a}(r_k) + r_k^{-4}\mathbf{a}^{(2)}(r_k^{-1})^T\tilde{\boldsymbol{\Omega}}_o\mathbf{a}(r_k) \\ &\quad -2r_k^{-2}\mathbf{a}^{(1)}(r_k^{-1})^T\tilde{\boldsymbol{\Omega}}_o\mathbf{a}^{(1)}(r_k) + \mathbf{a}(r_k^{-1})^T\tilde{\boldsymbol{\Omega}}_o\mathbf{a}^{(2)}(r_k) \\ &= 2r_k^{-2}\Re\{r_k^{-2}\mathbf{a}^{(2)}(r_k^{-1})^T\tilde{\boldsymbol{\Omega}}_o\mathbf{a}(r_k)\} \\ &\quad +2r_k^{-3}\mathbf{a}^{(1)}(r_k^{-1})^T\tilde{\boldsymbol{\Omega}}_o\mathbf{a}(r_k) - 2r_k^{-2}\mathbf{a}^{(1)}(r_k^{-1})^T\tilde{\boldsymbol{\Omega}}_o\mathbf{a}^{(1)}(r_k). \end{aligned} \tag{138}$$

Hence approximate Eq.(138) to the zeroth-order term to obtain D,

$$\begin{aligned} D &= 2r_k^{-2}\Re\{r_k^{-2}\mathbf{a}^{(2)}(r_k^{-1})^T\boldsymbol{\Omega}_o\mathbf{a}(r_k)\} + 2r_k^{-3}\mathbf{a}^{(1)}(r_k^{-1})^T\boldsymbol{\Omega}_o\mathbf{a}(r_k) \\ &\quad -2r_k^{-2}\mathbf{a}^{(1)}(r_k^{-1})^T\boldsymbol{\Omega}_o\mathbf{a}^{(1)}(r_k). \end{aligned}$$

Use Eq.(8) to obtain Eq.(92). The first-order ΔD_1 is obtained by keep Eq.(138) to the first-order and neglect the zeroth-order term,

$$\begin{aligned} \Delta D_1 &= 2r_k^{-2}\Re\{r_k^{-2}\mathbf{a}^{(2)}(r_k^{-1})^T\Delta\boldsymbol{\Omega}_{o1}\mathbf{a}(r_k)\} \\ &\quad -2r_k^{-2}\mathbf{a}^{(1)}(r_k^{-1})^T\Delta\boldsymbol{\Omega}_{o1}\mathbf{a}^{(1)}(r_k) + 2r_k^{-3}\mathbf{a}^{(1)}(r_k^{-1})^T\Delta\boldsymbol{\Omega}_{o1}\mathbf{a}(r_k). \end{aligned} \tag{139}$$

Substitute $\Delta\boldsymbol{\Omega}_{o1}$ into Eq.(139) and use Eq.(8) to obtain

$$\begin{aligned} \Delta D_1 &= 2r_k^{-3}\mathbf{a}^{(1)}(r_k^{-1})^T[\Delta\mathbf{U}_{o1}\mathbf{U}_o^H + \mathbf{U}_o\Delta\mathbf{U}_{o1}^H]\mathbf{a}(r_k) \\ &\quad -2r_k^{-2}\mathbf{a}^{(1)}(r_k^{-1})^T[\Delta\mathbf{U}_{o1}\mathbf{U}_o^H + \mathbf{U}_o\Delta\mathbf{U}_{o1}^H]\mathbf{a}^{(1)}(r_k) \\ &\quad +2r_k^{-2}\Re\{r_k^{-2}\mathbf{a}^{(2)}(r_k^{-1})^T[\Delta\mathbf{U}_{o1}\mathbf{U}_o^H + \mathbf{U}_o\Delta\mathbf{U}_{o1}^H]\mathbf{a}(r_k)\} \\ &= 2r_k^{-2}\Re\{r_k^{-2}\mathbf{a}^{(2)}(r_k^{-1})^T\mathbf{U}_o\Delta\mathbf{U}_{o1}^H\mathbf{a}(r_k)\} \\ &\quad -4r_k^{-2}\Re\{\mathbf{a}^{(1)}(r_k^{-1})^T\Delta\mathbf{U}_{o1}\mathbf{U}_o^H\mathbf{a}^{(1)}(r_k)\} \\ &\quad +2r_k^{-3}\mathbf{a}^{(1)}(r_k^{-1})^T\mathbf{U}_o\Delta\mathbf{U}_{o1}^H\mathbf{a}(r_k). \end{aligned} \tag{140}$$

By definition Eq.(84),

$$\begin{aligned} B &= \frac{\partial^3 P(r_k, \mathbf{U}_o)}{\partial z^3} \\ &= -6r_k^{-4}\mathbf{a}^{(1)}(r_k^{-1})^T\boldsymbol{\Omega}_o\mathbf{a}(r_k) - 2r_k^{-5}\mathbf{a}^{(2)}(r_k^{-1})^T\boldsymbol{\Omega}_o\mathbf{a}(r_k) \\ &\quad +2r_k^{-3}\mathbf{a}^{(1)}(r_k^{-1})^T\boldsymbol{\Omega}_o\mathbf{a}^{(1)}(r_k) - 4r_k^{-5}\mathbf{a}^{(2)}(r_k^{-1})^T\boldsymbol{\Omega}_o\mathbf{a}(r_k) \\ &\quad -r_k^{-6}\mathbf{a}^{(3)}(r_k^{-1})^T\boldsymbol{\Omega}_o\mathbf{a}(r_k) + r_k^{-4}\mathbf{a}^{(2)}(r_k^{-1})^T\boldsymbol{\Omega}_o\mathbf{a}^{(1)}(r_k) \end{aligned}$$

$$+4r_k^{-3}\mathbf{a}^{(1)}(r_k^{-1})^T\boldsymbol{\Omega}_o\mathbf{a}^{(1)}(r_k)+2r_k^{-4}\mathbf{a}^{(2)}(r_k^{-1})^T\boldsymbol{\Omega}_o\mathbf{a}^{(1)}(r_k)$$
$$-2r_k^{-2}\mathbf{a}^{(1)}(r_k^{-1})^T\boldsymbol{\Omega}_o\mathbf{a}^{(2)}(r_k)-r_k^{-2}\mathbf{a}^{(2)}(r_k^{-1})^T\boldsymbol{\Omega}_o\mathbf{a}^{(1)}(r_k)$$
$$+\mathbf{a}(r_k^{-1})^T\boldsymbol{\Omega}_o\mathbf{a}^{(3)}(r_k).$$

Use Eq.(8) to obtain Eq.(93). Substitute Eq.(54) into Eq.(136) and take expectation on both sides of the equation. Notice $\mathbf{a}^H(\theta_k)\mathbf{U}_o = 0$, we have

$$E(\Delta N_2) = -2jr_k^{-1}\Im\{E\{r_k^{-1}\mathbf{a}^{(1)}(r_k^{-1})^T\Delta\mathbf{U}_{o1}\Delta\mathbf{U}_{o1}^H\mathbf{a}(r_k)\}\}.$$

Substitute Eqs.(42) and (53) back into the equation,

$$\begin{aligned}
&E(\Delta N_2)\\
=\;& -2jr_k^{-1}\Im\{E\{r_k^{-1}\mathbf{a}^{(1)}(r_k^{-1})^T\mathbf{U}_s\boldsymbol{\Sigma}_s^{-1}\mathbf{V}_s^H\mathbf{N}^H\mathbf{U}_o\mathbf{U}_o^H\mathbf{N}\mathbf{V}_s\boldsymbol{\Sigma}_s^{-1}\mathbf{U}_s^H\mathbf{a}(r_k)\}\}\\
=\;& -2\sigma_n^2 jr_k^{-1}\Im\{r_k^{-1}\mathbf{a}^{(1)}(r_k^{-1})^T\mathbf{U}_s\boldsymbol{\Sigma}_s^{-2}\mathbf{U}_s^H\mathbf{a}(r_k)Tr(\mathbf{U}_o\mathbf{U}_o^H)\}\\
=\;& -2j(L-P)\sigma_n^2 r_k^{-1}\Im\{r_k^{-1}\mathbf{a}^{(1)}(r_k^{-1})^T\mathbf{U}_s\boldsymbol{\Sigma}_s^{-2}\mathbf{U}_s^H\mathbf{a}(r_k)\}. \qquad (141)
\end{aligned}$$

Substitute Eq.(108) into Eq.(141) to obtain Eq.(94). To calculate $E(\Delta^2 N_1)$ from Eq.(135), we need to use the circularity of the noise,

$$E(\Delta^2 N_1) = -2r_k^{-2}E\{\mathbf{a}^{(1)}(r_k^{-1})^T\mathbf{U}_o\Delta\mathbf{U}_{o1}^H\mathbf{a}(r_k)\mathbf{a}^T(r_k^{-1})\Delta\mathbf{U}_o\mathbf{U}_o^H\mathbf{a}^{(1)}(r_k)\}.$$

Substitute Eqs.(42) and (53) back into the equation,

$$\begin{aligned}
&E(\Delta^2 N_1)\\
=\;& -2r_k^{-2}E\{\mathbf{a}^{(1)}(r_k^{-1})^T\mathbf{U}_o\mathbf{U}_o^H\mathbf{N}\mathbf{V}_s\boldsymbol{\Sigma}_s^{-1}\mathbf{U}_s^H\mathbf{a}(r_k)\\
&\mathbf{a}^T(r_k^{-1})\mathbf{U}_s\boldsymbol{\Sigma}_s^{-1}\mathbf{V}_s^H\mathbf{N}^H\mathbf{U}_o\mathbf{U}_o^H\mathbf{a}^{(1)}(r_k)\}\\
=\;& -2\sigma_n^2 r_k^{-2}\mathbf{a}^{(1)}(r_k^{-1})^T\boldsymbol{\Omega}_o\mathbf{a}^{(1)}(r_k)Tr[\mathbf{a}^T(r_k^{-1})\mathbf{U}_s\boldsymbol{\Sigma}_s^{-2}\mathbf{U}_s^H\mathbf{a}(r_k)]\\
=\;& -2\sigma_n^2 r_k^{-2}\|\mathbf{a}^{(1)}(r_k^{-1})^T\boldsymbol{\Omega}_o\|^2\mathbf{a}^T(r_k^{-1})\mathbf{U}_s\boldsymbol{\Sigma}_s^{-2}\mathbf{U}_s^H\mathbf{a}(r_k).
\end{aligned}$$

Use Eq.(108) to obtain Eq.(95). Now we will use circular property of the noise to derive $E(\Delta N_1\Delta D_1)$. Then substitute Eq.(42) and (53) into the equation,

$$\begin{aligned}
&E(\Delta N_1\Delta D_1)\\
=\;& 2jr_k^{-3}\Im\{E\{r_k^{-1}\mathbf{a}^T(r_k^{-1})\Delta\mathbf{U}_{o1}\mathbf{U}_o^H\mathbf{a}^{(1)}(r_k)\mathbf{a}^{(2)}(r_k^{-1})^T\mathbf{U}_o\Delta\mathbf{U}_{o1}^H\mathbf{a}(r_k)\}\}\\
&+4jr_k^{-3}\Im\{E\{r_k^{-1}\mathbf{a}^{(1)}(r_k^{-1})^T\mathbf{U}_o\Delta\mathbf{U}_{o1}^H\mathbf{a}(r_k)\mathbf{a}^{(1)}(r_k^{-1})^T\Delta\mathbf{U}_{o1}\mathbf{U}_o^H\mathbf{a}^{(1)}(r_k)\}\}
\end{aligned}$$

$$\begin{aligned}
&+2jr_k^{-3}E\{\mathbf{a}^T(r_k^{-1})\Delta\mathbf{U}_{o1}\mathbf{U}_o^H\mathbf{a}^{(1)}(r_k)\mathbf{a}^{(1)}(r_k^{-1})^T\mathbf{U}_o\Delta\mathbf{U}_{o1}^H\mathbf{a}(r_k)\}\\
=\;& 2jr_k^{-3}\Im\{E\{r_k^{-1}\mathbf{a}^T(r_k^{-1})\mathbf{U}_s\mathbf{\Sigma}_s^{-1}\mathbf{V}_s^H\mathbf{N}^H\mathbf{U}_o\mathbf{U}_o^H\mathbf{a}^{(1)}(r_k)\\
&\mathbf{a}^{(2)}(r_k^{-1})^T\mathbf{U}_o\mathbf{U}_o^H\mathbf{N}\mathbf{V}_s\mathbf{\Sigma}_s^{-1}\mathbf{U}_s^H\mathbf{a}(r_k)\}\}\\
&+4jr_k^{-3}\Im\{E\{r_k^{-1}\mathbf{a}^{(1)}(r_k^{-1})^T\mathbf{U}_o\mathbf{U}_o^H\mathbf{N}\mathbf{V}_s\mathbf{\Sigma}_s^{-1}\mathbf{U}_s^H\mathbf{a}(r_k)\\
&\mathbf{a}^{(1)}(r_k^{-1})^T\mathbf{U}_s\mathbf{\Sigma}_s^{-1}\mathbf{V}_s^H\mathbf{N}^H\mathbf{U}_o\mathbf{U}_o^H\mathbf{a}^{(1)}(r_k)\}\}\\
&+2jr_k^{-3}E\{\mathbf{a}^T(r_k^{-1})\mathbf{U}_s\mathbf{\Sigma}_s^{-1}\mathbf{V}_s^H\mathbf{N}^H\mathbf{U}_o\mathbf{U}_o^H\mathbf{a}^{(1)}(r_k)\\
&\mathbf{a}^{(1)}(r_k^{-1})^T\mathbf{U}_o\mathbf{U}_o^H\mathbf{N}\mathbf{V}_s\mathbf{\Sigma}_s^{-1}\mathbf{U}_s^H\mathbf{a}(r_k)\}\\
=\;& 2j\sigma_n^2r_k^{-3}\Im\{r_k^{-1}\mathbf{a}^T(r_k^{-1})\mathbf{U}_s\mathbf{\Sigma}_s^{-2}\mathbf{U}_s^H\mathbf{a}(r_k)Tr[\mathbf{a}^{(2)}(r_k^{-1})^T\mathbf{\Omega}_o\mathbf{a}^{(1)}(r_k)]\}\\
&+4j\sigma_n^2r_k^{-3}\Im\{r_k^{-1}\mathbf{a}^{(1)}(r_k^{-1})^T\mathbf{\Omega}_o\mathbf{a}^{(1)}(r_k)Tr[\mathbf{a}^{(1)}(r_k^{-1})^T\mathbf{U}_s\mathbf{\Sigma}_s^{-2}\mathbf{U}_s^H\mathbf{a}(r_k)]\}\\
&+2j\sigma_n^2r_k^{-3}\mathbf{a}^T(r_k^{-1})\mathbf{U}_s\mathbf{\Sigma}_s^{-2}\mathbf{U}_s^H\mathbf{a}(r_k)Tr[\mathbf{a}^{(1)}(r_k^{-1})^T\mathbf{\Omega}_o\mathbf{a}^{(1)}(r_k)]\\
=\;& 2j\sigma_n^2r_k^{-3}\Im\{r_k^{-1}\mathbf{a}^T(r_k^{-1})\mathbf{U}_s\mathbf{\Sigma}_s^{-2}\mathbf{U}_s^H\mathbf{a}(r_k)\mathbf{a}^{(2)}(r_k^{-1})^T\mathbf{\Omega}_o\mathbf{a}^{(1)}(r_k)\}\\
&+4j\sigma_n^2r_k^{-3}\|\mathbf{a}^{(1)}(r_k^{-1})^T\mathbf{\Omega}_o\|^2\Im\{r_k^{-1}\mathbf{a}^{(1)}(r_k^{-1})^T\mathbf{U}_s\mathbf{\Sigma}_s^{-2}\mathbf{U}_s^H\mathbf{a}(r_k)\}\\
&+2j\sigma_n^2r_k^{-3}\mathbf{a}^T(r_k^{-1})\mathbf{U}_s\mathbf{\Sigma}_s^{-2}\mathbf{U}_s^H\mathbf{a}(r_k)\mathbf{a}^{(1)}(r_k^{-1})^T\mathbf{\Omega}_o\mathbf{a}^{(1)}(r_k).
\end{aligned}$$

Substitute Eq.(108) into the equation to get Eq.(96).

B. Root MIN-NORM: For Root Min-Norm, the spectral polynomial is

$$P(r_k,\tilde{\mathbf{U}}_o)=\frac{1}{\|\mathbf{e}_1^H\tilde{\mathbf{\Omega}}_o\|^4}\mathbf{a}(r_k^{-1})^T\tilde{\mathbf{\Omega}}_o\mathbf{e}_1\mathbf{e}_1^H\tilde{\mathbf{\Omega}}_o\mathbf{a}(r_k).$$

By definition Eq.(84),

$$\begin{aligned}
N+\Delta N \;&\stackrel{\text{def}}{=}\; \frac{\partial P(r_k,\tilde{\mathbf{U}}_o)}{\partial z}\\
&= \frac{1}{\|\mathbf{e}_1^H\tilde{\mathbf{\Omega}}_o\|^4}\{-r_k^{-2}\mathbf{a}^{(1)}(r_k^{-1})^T\tilde{\mathbf{\Omega}}_o\mathbf{e}_1\mathbf{e}_1^H\tilde{\mathbf{\Omega}}_o\mathbf{a}(r_k)\\
&\quad+\mathbf{a}(r_k^{-1})^T\tilde{\mathbf{\Omega}}_o\mathbf{e}_1\mathbf{e}_1^H\tilde{\mathbf{\Omega}}_o\mathbf{a}^{(1)}(r_k)\}\\
&= -\frac{2jr_k^{-1}}{\|\mathbf{e}_1^H\tilde{\mathbf{\Omega}}_o\|^4}\Im\{r_k^{-1}\mathbf{a}^{(1)}(r_k^{-1})^T\tilde{\mathbf{\Omega}}_o\mathbf{e}_1\mathbf{e}_1^H\tilde{\mathbf{\Omega}}_o\mathbf{a}(r_k)\}. \qquad (142)
\end{aligned}$$

N is the zeroth-order term of Eq.(142),

$$N=-\frac{2jr_k^{-1}}{\|\mathbf{e}_1^H\mathbf{\Omega}_o\|^4}\Im\{r_k^{-1}\mathbf{a}^{(1)}(r_k^{-1})^T\mathbf{\Omega}_o\mathbf{e}_1\mathbf{e}_1^H\mathbf{\Omega}_o\mathbf{a}(r_k)\}.$$

Use Eq.(8) to obtain Eq.(97). ΔN_1 is the first-order term of Eq.(142) excluding the zeroth-order term,

$$\Delta N_1 = -\frac{2jr_k^{-1}}{\|\mathbf{e}_1^H \boldsymbol{\Omega}_o\|^4}\Im\{r_k^{-1}\mathbf{a}^{(1)}(r_k^{-1})^T[\boldsymbol{\Omega}_o \mathbf{e}_1\mathbf{e}_1^H \Delta\boldsymbol{\Omega}_{o1}+\Delta\boldsymbol{\Omega}_{o1}\mathbf{e}_1\mathbf{e}_1^H\boldsymbol{\Omega}_o]\mathbf{a}(r_k)\}. \tag{143}$$

Substitute $\Delta\boldsymbol{\Omega}_{o1}$ into Eq.(143) and use Eq.(8) to obtain

$$\begin{aligned}\Delta N_1 &= -\frac{2jr_k^{-1}}{\|\mathbf{e}_1^H \boldsymbol{\Omega}_o\|^4}\Im\{r_k^{-1}\mathbf{a}^{(1)}(r_k^{-1})^T\boldsymbol{\Omega}_o \mathbf{e}_1\mathbf{e}_1^H \Delta\boldsymbol{\Omega}_{o1}\mathbf{a}(r_k)\}\\ &= -\frac{2jr_k^{-1}}{\|\mathbf{e}_1^H \boldsymbol{\Omega}_o\|^4}\Im\{r_k^{-1}\mathbf{a}^{(1)}(r_k^{-1})^T\boldsymbol{\Omega}_o \mathbf{e}_1\mathbf{e}_1^H [\Delta\mathbf{U}_{o1}\mathbf{U}_o^H+\mathbf{U}_o\Delta\mathbf{U}_{o1}^H]\mathbf{a}(r_k)\}\\ &= -\frac{2jr_k^{-1}}{\|\mathbf{e}_1^H \boldsymbol{\Omega}_o\|^4}\Im\{r_k^{-1}\mathbf{a}^{(1)}(r_k^{-1})^T\boldsymbol{\Omega}_o \mathbf{e}_1\mathbf{e}_1^H \mathbf{U}_o\Delta\mathbf{U}_{o1}^H\mathbf{a}(r_k)\}. \end{aligned} \tag{144}$$

ΔN_2 can be obtained by keeping Eq.(142) to the second-order and throwing away the zeroth-order term,

$$\begin{aligned}\Delta N_2 &= -\frac{2jr_k^{-1}}{\|\mathbf{e}_1^H \boldsymbol{\Omega}_o\|^4}\Im\{r_k^{-1}\mathbf{a}^{(1)}(r_k^{-1})^T[\boldsymbol{\Omega}_o \mathbf{e}_1\mathbf{e}_1^H \Delta\boldsymbol{\Omega}_{o2}+\Delta\boldsymbol{\Omega}_{o2}\mathbf{e}_1\mathbf{e}_1^H\boldsymbol{\Omega}_o\\ &\quad+\Delta\boldsymbol{\Omega}_{o1}\mathbf{e}_1\mathbf{e}_1^H\Delta\boldsymbol{\Omega}_{o1}]\mathbf{a}(r_k)\}. \end{aligned} \tag{145}$$

Substitute $\Delta\boldsymbol{\Omega}_{o2}$ and $\Delta\boldsymbol{\Omega}_{o1}$ into Eqs.(145) and use (8) to obtain

$$\begin{aligned}&\Delta N_2\\ &= -\frac{2jr_k^{-1}}{\|\mathbf{e}_1^H \boldsymbol{\Omega}_o\|^4}\Im\{r_k^{-1}\mathbf{a}^{(1)}(r_k^{-1})^T[\boldsymbol{\Omega}_o \mathbf{e}_1\mathbf{e}_1^H \Delta\boldsymbol{\Omega}_{o2}+\Delta\boldsymbol{\Omega}_{o1}\mathbf{e}_1\mathbf{e}_1^H\Delta\boldsymbol{\Omega}_{o1}]\mathbf{a}(r_k)\}\\ &= -\frac{2jr_k^{-1}}{\|\mathbf{e}_1^H \boldsymbol{\Omega}_o\|^4}\Im\{r_k^{-1}\mathbf{a}^{(1)}(r_k^{-1})^T\boldsymbol{\Omega}_o \mathbf{e}_1\mathbf{e}_1^H [\Delta\mathbf{U}_{o2}\mathbf{U}_o^H+\mathbf{U}_o\Delta\mathbf{U}_{o2}^H\\ &\quad+\Delta\mathbf{U}_{o1}\Delta\mathbf{U}_{o1}^H]\mathbf{a}(r_k)+r_k^{-1}\mathbf{a}^{(1)}(r_k^{-1})^T[\Delta\mathbf{U}_{o1}\mathbf{U}_o^H+\mathbf{U}_o\Delta\mathbf{U}_{o1}^H]\\ &\quad\mathbf{e}_1\mathbf{e}_1^H[\Delta\mathbf{U}_{o1}\mathbf{U}_o^H+\mathbf{U}_o\Delta\mathbf{U}_{o1}^H]\mathbf{a}(r_k)\}\\ &= -\frac{2jr_k^{-1}}{\|\mathbf{e}_1^H \boldsymbol{\Omega}_o\|^4}\Im\{r_k^{-1}\mathbf{a}^{(1)}(r_k^{-1})^T\boldsymbol{\Omega}_o \mathbf{e}_1\mathbf{e}_1^H [\mathbf{U}_o\Delta\mathbf{U}_{o2}^H+\Delta\mathbf{U}_{o1}\Delta\mathbf{U}_{o1}^H]\mathbf{a}(r_k)\\ &\quad+r_k^{-1}\mathbf{a}^{(1)}(r_k^{-1})^T[\Delta\mathbf{U}_{o1}\mathbf{U}_o^H+\mathbf{U}_o\Delta\mathbf{U}_{o1}^H]\mathbf{e}_1\mathbf{e}_1^H\mathbf{U}_o\Delta\mathbf{U}_{o1}^H\mathbf{a}(r_k)\}. \end{aligned} \tag{146}$$

By definition Eq.(84),

$$\begin{aligned}&D+\Delta D\\ &\stackrel{\text{def}}{=} \frac{\partial^2 P(r_k,\tilde{\mathbf{U}}_o)}{\partial z^2}\end{aligned}$$

$$
\begin{aligned}
&= \frac{1}{\|\mathbf{e}_1^H\tilde{\boldsymbol{\Omega}}_o\|^4}\{2r_k^{-3}\mathbf{a}^{(1)}(r_k^{-1})^T\tilde{\boldsymbol{\Omega}}_o\mathbf{e}_1\mathbf{e}_1^H\tilde{\boldsymbol{\Omega}}_o\mathbf{a}(r_k) + r_k^{-4}\mathbf{a}^{(2)}(r_k^{-1})^T\tilde{\boldsymbol{\Omega}}_o\mathbf{e}_1\mathbf{e}_1^H\tilde{\boldsymbol{\Omega}}_o\mathbf{a}(r_k) \\
&\quad -2r_k^{-2}\mathbf{a}^{(1)}(r_k^{-1})^T\tilde{\boldsymbol{\Omega}}_o\mathbf{e}_1\mathbf{e}_1^H\tilde{\boldsymbol{\Omega}}_o\mathbf{a}^{(1)}(r_k) + \mathbf{a}(r_k^{-1})^T\tilde{\boldsymbol{\Omega}}_o\mathbf{a}^{(2)}(r_k)\} \\
&= \frac{1}{\|\mathbf{e}_1^H\tilde{\boldsymbol{\Omega}}_o\|^4}\{2r_k^{-2}\Re\{r_k^{-2}\mathbf{a}^{(2)}(r_k^{-1})^T\tilde{\boldsymbol{\Omega}}_o\mathbf{e}_1\mathbf{e}_1^H\tilde{\boldsymbol{\Omega}}_o\mathbf{a}(r_k)\} \\
&\quad +2r_k^{-3}\mathbf{a}^{(1)}(r_k^{-1})^T\tilde{\boldsymbol{\Omega}}_o\mathbf{e}_1\mathbf{e}_1^H\tilde{\boldsymbol{\Omega}}_o\mathbf{a}(r_k) - 2r_k^{-2}\mathbf{a}^{(1)}(r_k^{-1})^T\tilde{\boldsymbol{\Omega}}_o\mathbf{e}_1\mathbf{e}_1^H\tilde{\boldsymbol{\Omega}}_o\mathbf{a}^{(1)}(r_k)\}.
\end{aligned} \tag{147}
$$

D is the zeroth-order term of the Eq.(147),

$$
\begin{aligned}
&D = \\
&\frac{1}{\|\mathbf{e}_1^H\boldsymbol{\Omega}_o\|^4}\{2r_k^{-2}\Re\{r_k^{-2}\mathbf{a}^{(2)}(r_k^{-1})^T\boldsymbol{\Omega}_o\mathbf{e}_1\mathbf{e}_1^H\boldsymbol{\Omega}_o\mathbf{a}(r_k)\} \\
&+2r_k^{-3}\mathbf{a}^{(1)}(r_k^{-1})^T\boldsymbol{\Omega}_o\mathbf{e}_1\mathbf{e}_1^H\boldsymbol{\Omega}_o\mathbf{a}(r_k) - 2r_k^{-2}\mathbf{a}^{(1)}(r_k^{-1})^T\boldsymbol{\Omega}_o\mathbf{e}_1\mathbf{e}_1^H\boldsymbol{\Omega}_o\mathbf{a}^{(1)}(r_k)\}.
\end{aligned}
$$

Use Eq.(8) to obtain Eq.(98). Expanding Eq.(147) to the first-order and neglect the zeroth-order term to obtain ΔD_1,

$$
\begin{aligned}
&\Delta D_1 \\
&= \frac{1}{\|\mathbf{e}_1^H\boldsymbol{\Omega}_o\|^4}\{2r_k^{-2}\Re\{r_k^{-2}\mathbf{a}^{(2)}(r_k^{-1})^T[\Delta\boldsymbol{\Omega}_{o1}\mathbf{e}_1\mathbf{e}_1^H\boldsymbol{\Omega}_o + \boldsymbol{\Omega}_o\mathbf{e}_1\mathbf{e}_1^H\Delta\boldsymbol{\Omega}_{o1}]\mathbf{a}(r_k)\} \\
&\quad +2r_k^{-3}\mathbf{a}^{(1)}(r_k^{-1})^T[\Delta\boldsymbol{\Omega}_{o1}\mathbf{e}_1\mathbf{e}_1^H\boldsymbol{\Omega}_o + \boldsymbol{\Omega}_o\mathbf{e}_1\mathbf{e}_1^H\Delta\boldsymbol{\Omega}_{o1}]\mathbf{a}(r_k) \\
&\quad -2r_k^{-2}\mathbf{a}^{(1)}(r_k^{-1})^T[\Delta\boldsymbol{\Omega}_{o1}\mathbf{e}_1\mathbf{e}_1^H\boldsymbol{\Omega}_o + \boldsymbol{\Omega}_o\mathbf{e}_1\mathbf{e}_1^H\Delta\boldsymbol{\Omega}_{o1}]\mathbf{a}^{(1)}(r_k)\}.
\end{aligned} \tag{148}
$$

Substitute $\Delta\boldsymbol{\Omega}_{o1}$ into Eq.(148) and use Eq.(8) to obtain

$$
\begin{aligned}
&\Delta D_1 \\
&= \frac{1}{\|\mathbf{e}_1^H\boldsymbol{\Omega}_o\|^4}\{2r_k^{-2}\Re\{r_k^{-2}\mathbf{a}^{(2)}(r_k^{-1})^T\boldsymbol{\Omega}_o\mathbf{e}_1\mathbf{e}_1^H\Delta\boldsymbol{\Omega}_{o1}\mathbf{a}(r_k)\} \\
&\quad -4r_k^{-2}\Re\{\mathbf{a}^{(1)}(r_k^{-1})^T\Delta\boldsymbol{\Omega}_{o1}\mathbf{e}_1\mathbf{e}_1^H\boldsymbol{\Omega}_o\mathbf{a}^{(1)}(r_k)\} \\
&\quad +2r_k^{-3}\mathbf{a}^{(1)}(r_k^{-1})^T\boldsymbol{\Omega}_o\mathbf{e}_1\mathbf{e}_1^H\Delta\boldsymbol{\Omega}_{o1}\mathbf{a}(r_k)\} \\
&= \frac{1}{\|\mathbf{e}_1^H\boldsymbol{\Omega}_o\|^4}\{2r_k^{-2}\Re\{r_k^{-2}\mathbf{a}^{(2)}(r_k^{-1})^T\boldsymbol{\Omega}_o\mathbf{e}_1\mathbf{e}_1^H[\Delta\mathbf{U}_{o1}\mathbf{U}_o^H + \mathbf{U}_o\Delta\mathbf{U}_{o1}^H]\mathbf{a}(r_k)\} \\
&\quad -4r_k^{-2}\Re\{\mathbf{a}^{(1)}(r_k^{-1})^T[\Delta\mathbf{U}_{o1}\mathbf{U}_o^H + \mathbf{U}_o\Delta\mathbf{U}_{o1}^H]\mathbf{e}_1\mathbf{e}_1^H\boldsymbol{\Omega}_o\mathbf{a}^{(1)}(r_k)\} \\
&\quad +2r_k^{-3}\mathbf{a}^{(1)}(r_k^{-1})^T\boldsymbol{\Omega}_o\mathbf{e}_1\mathbf{e}_1^H[\Delta\mathbf{U}_{o1}\mathbf{U}_o^H + \mathbf{U}_o\Delta\mathbf{U}_{o1}^H]\mathbf{a}(r_k)\} \\
&= \frac{1}{\|\mathbf{e}_1^H\boldsymbol{\Omega}_o\|^4}\{2r_k^{-2}\Re\{r_k^{-2}\mathbf{a}^{(2)}(r_k^{-1})^T\boldsymbol{\Omega}_o\mathbf{e}_1\mathbf{e}_1^H\mathbf{U}_o\Delta\mathbf{U}_{o1}^H\mathbf{a}(r_k)\} \\
&\quad -4r_k^{-2}\Re\{\mathbf{a}^{(1)}(r_k^{-1})^T[\Delta\mathbf{U}_{o1}\mathbf{U}_o^H + \mathbf{U}_o\Delta\mathbf{U}_{o1}^H]\mathbf{e}_1\mathbf{e}_1^H\boldsymbol{\Omega}_o\mathbf{a}^{(1)}(r_k)\} \\
&\quad +2r_k^{-3}\mathbf{a}^{(1)}(r_k^{-1})^T\boldsymbol{\Omega}_o\mathbf{e}_1\mathbf{e}_1^H\mathbf{U}_o\Delta\mathbf{U}_{o1}^H\mathbf{a}(r_k)\}.
\end{aligned} \tag{149}
$$

B, by definition, is the zeroth-order term of $\frac{\partial^3 P(r_k,\tilde{\mathbf{U}}_o)}{\partial z^3}$, i.e.,

$$\begin{aligned} B &= \frac{\partial^3 P(r_k, \mathbf{U}_o)}{\partial z^3} \\ &= \frac{1}{\|\mathbf{e}_1^H \boldsymbol{\Omega}_o\|^4}\{-6r_k^{-4}\mathbf{a}^{(1)}(r_k^{-1})^T\boldsymbol{\Omega}_o\mathbf{e}_1\mathbf{e}_1^H\boldsymbol{\Omega}_o\mathbf{a}(r_k) \\ &-2r_k^{-5}\mathbf{a}^{(2)}(r_k^{-1})^T\boldsymbol{\Omega}_o\mathbf{e}_1\mathbf{e}_1^H\boldsymbol{\Omega}_o\mathbf{a}(r_k) \\ &+2r_k^{-3}\mathbf{a}^{(1)}(r_k^{-1})^T\boldsymbol{\Omega}_o\mathbf{e}_1\mathbf{e}_1^H\boldsymbol{\Omega}_o\mathbf{a}^{(1)}(r_k) - 4r_k^{-5}\mathbf{a}^{(2)}(r_k^{-1})^T\boldsymbol{\Omega}_o\mathbf{e}_1\mathbf{e}_1^H\boldsymbol{\Omega}_o\mathbf{a}(r_k) \\ &-r_k^{-6}\mathbf{a}^{(3)}(r_k^{-1})^T\boldsymbol{\Omega}_o\mathbf{e}_1\mathbf{e}_1^H\boldsymbol{\Omega}_o\mathbf{a}(r_k) + r_k^{-4}\mathbf{a}^{(2)}(r_k^{-1})^T\boldsymbol{\Omega}_o\mathbf{e}_1\mathbf{e}_1^H\boldsymbol{\Omega}_o\mathbf{a}^{(1)}(r_k) \\ &+4r_k^{-3}\mathbf{a}^{(1)}(r_k^{-1})^T\boldsymbol{\Omega}_o\mathbf{e}_1\mathbf{e}_1^H\boldsymbol{\Omega}_o\mathbf{a}^{(1)}(r_k) + 2r_k^{-4}\mathbf{a}^{(2)}(r_k^{-1})^T\boldsymbol{\Omega}_o\mathbf{e}_1\mathbf{e}_1^H\boldsymbol{\Omega}_o\mathbf{a}^{(1)}(r_k) \\ &-2r_k^{-2}\mathbf{a}^{(1)}(r_k^{-1})^T\boldsymbol{\Omega}_o\mathbf{e}_1\mathbf{e}_1^H\boldsymbol{\Omega}_o\mathbf{a}^{(2)}(r_k) - r_k^{-2}\mathbf{a}^{(2)}(r_k^{-1})^T\boldsymbol{\Omega}_o\mathbf{e}_1\mathbf{e}_1^H\boldsymbol{\Omega}_o\mathbf{a}^{(1)}(r_k) \\ &+\mathbf{a}(r_k^{-1})^T\boldsymbol{\Omega}_o\mathbf{e}_1\mathbf{e}_1^H\boldsymbol{\Omega}_o\mathbf{a}^{(3)}(r_k)\}. \end{aligned}$$

Use Eq.(8) to obtain Eq.(99). Substitute Eq.(54) into Eq.(146) and take expectation on both sides of the equation. Using $\mathbf{a}^H(\theta_k)\mathbf{U}_o = 0$ and the circular property of the noise, we have

$$\begin{aligned} E(\Delta N_2) &= -\frac{2jr_k^{-1}}{\|\mathbf{e}_1^H\boldsymbol{\Omega}_o\|^4}E\{\Im\{r_k^{-1}\mathbf{a}^{(1)}(r_k^{-1})^T\boldsymbol{\Omega}_o\mathbf{e}_1\mathbf{e}_1^H\Delta\mathbf{U}_{o1}\Delta\mathbf{U}_{o1}^H\mathbf{a}(r_k) \\ &+r_k^{-1}\mathbf{a}^{(1)}(r_k^{-1})^T\Delta\mathbf{U}_{o1}\mathbf{U}_o^H\mathbf{e}_1\mathbf{e}_1^H\mathbf{U}_o\Delta\mathbf{U}_{o1}^H\mathbf{a}(r_k)\}\}. \end{aligned}$$

Substitute Eqs.(42) and (53) back into the equation,

$$\begin{aligned} &E(\Delta N_2) \\ &= -\frac{2jr_k^{-1}}{\|\mathbf{e}_1^H\boldsymbol{\Omega}_o\|^4}\Im\{E[r_k^{-1}\mathbf{a}^{(1)}(r_k^{-1})^T\boldsymbol{\Omega}_o\mathbf{e}_1\mathbf{e}_1^H\mathbf{U}_s\boldsymbol{\Sigma}_s^{-1}\mathbf{V}_s^H \\ &\mathbf{N}^H\mathbf{U}_o\mathbf{U}_o^H\mathbf{N}\mathbf{V}_s\boldsymbol{\Sigma}_s^{-1}\mathbf{U}_s^H\mathbf{a}(r_k)]\} \\ &+E[r_k^{-1}\mathbf{a}^{(1)}(r_k^{-1})^T\mathbf{U}_s\boldsymbol{\Sigma}_s^{-1}\mathbf{V}_s^H\mathbf{N}^H\mathbf{U}_o\mathbf{U}_o^H\mathbf{e}_1\mathbf{e}_1^H\mathbf{U}_o\mathbf{U}_o^H\mathbf{N}\mathbf{V}_s\boldsymbol{\Sigma}_s^{-1}\mathbf{U}_s^H\mathbf{a}(r_k)]\} \\ &= -\frac{2j\sigma_n^2r_k^{-1}}{\|\mathbf{e}_1^H\boldsymbol{\Omega}_o\|^4}\Im\{r_k^{-1}\mathbf{a}^{(1)}(r_k^{-1})^T\boldsymbol{\Omega}_o\mathbf{e}_1\mathbf{e}_1^H\mathbf{U}_s\boldsymbol{\Sigma}_s^{-2}\mathbf{U}_s^H\mathbf{a}(r_k)Tr[\mathbf{U}_o\mathbf{U}_o^H] \\ &+r_k^{-1}\mathbf{a}^{(1)}(r_k^{-1})^T\mathbf{U}_s\boldsymbol{\Sigma}_s^{-2}\mathbf{U}_s^H\mathbf{a}(r_k)Tr[\mathbf{e}_1^H\boldsymbol{\Omega}_o\mathbf{e}_1]\} \\ &= -\frac{2j\sigma_n^2r_k^{-1}}{\|\mathbf{e}_1^H\boldsymbol{\Omega}_o\|^4}\Im\{(L-P)r_k^{-1}\mathbf{a}^{(1)}(r_k^{-1})^T\boldsymbol{\Omega}_o\mathbf{e}_1\mathbf{e}_1^H\mathbf{U}_s\boldsymbol{\Sigma}_s^{-2}\mathbf{U}_s^H\mathbf{a}(r_k) \\ &+\|\mathbf{e}_1\boldsymbol{\Omega}_o\|^2r_k^{-1}\mathbf{a}^{(1)}(r_k^{-1})^T\mathbf{U}_s\boldsymbol{\Sigma}_s^{-2}\mathbf{U}_s^H\mathbf{a}(r_k)\}. \end{aligned} \tag{150}$$

Use Eq.(108) to obtain Eq.(100). For $E(\Delta^2 N_1)$, using the circularity of the noise and substituting Eqs.(42) and (53) into the equation, we have

$$E(\Delta^2 N_1)$$

$$
\begin{aligned}
&= \frac{-2r_k^{-2}}{\|\mathbf{e}_1^H \boldsymbol{\Omega}_o\|^8} E[\mathbf{a}^{(1)}(r_k^{-1})^T \boldsymbol{\Omega}_o \mathbf{e}_1 \mathbf{e}_1^H \mathbf{U}_o \Delta \mathbf{U}_{o1}^H \mathbf{a}(r_k) \\
&\quad \mathbf{a}^T(r_k^{-1}) \Delta \mathbf{U}_{o1} \mathbf{U}_o^H \mathbf{e}_1 \mathbf{e}_1^H \boldsymbol{\Omega}_o \mathbf{a}^{(1)}(r_k)] \\
&= \frac{-2\sigma_n^2 r_k^{-2}}{\|\mathbf{e}_1^H \boldsymbol{\Omega}_o\|^8} \mathbf{a}^{(1)}(r_k^{-1})^T \boldsymbol{\Omega}_o \mathbf{e}_1 \mathbf{e}_1^H \boldsymbol{\Omega}_o \mathbf{e}_1 \mathbf{e}_1^H \boldsymbol{\Omega}_o \mathbf{a}^{(1)}(r_k) \\
&\quad Tr[\mathbf{a}^T(r_k^{-1}) \mathbf{U}_s \boldsymbol{\Sigma}_s^{-2} \mathbf{U}_s^H \mathbf{a}(r_k)] \\
&= \frac{-2\sigma_n^2 r_k^{-2}}{\|\mathbf{e}_1^H \boldsymbol{\Omega}_o\|^8} \|\mathbf{a}^{(1)}(r_k^{-1})^T \boldsymbol{\Omega}_o \mathbf{e}_1\|^2 \|\mathbf{e}_1^H \boldsymbol{\Omega}_o\|^2 \mathbf{a}^T(r_k^{-1}) \mathbf{U}_s \boldsymbol{\Sigma}_s^{-2} \mathbf{U}_s^H \mathbf{a}(r_k).
\end{aligned}
$$

Use Eq.(108) to obtain Eq.(101). To calculate $E(\Delta N_1 \Delta D_1)$, we utilize the circular property of the noise then substitute Eqs.(42) and (53),

$$
\begin{aligned}
&E(\Delta N_1 \Delta D_1) \\
&= \frac{1}{\|\mathbf{e}_1^H \boldsymbol{\Omega}_o\|^8} \{2jr_k^{-3} \Im\{E[r_k^{-1} \mathbf{a}^T(r_k^{-1}) \Delta \mathbf{U}_{o1} \mathbf{U}_o^H \mathbf{e}_1 \mathbf{e}_1^H \boldsymbol{\Omega}_o \mathbf{a}^{(1)}(r_k) \\
&\quad \mathbf{a}^{(2)}(r_k^{-1})^T \boldsymbol{\Omega}_o \mathbf{e}_1 \mathbf{e}_1^H \mathbf{U}_o \Delta \mathbf{U}_{o1}^H \mathbf{a}(r_k)]\} \\
&\quad +4jr_k^{-3} \Im\{E[r_k^{-1} \mathbf{a}^{(1)}(r_k^{-1})^T \boldsymbol{\Omega}_o \mathbf{e}_1 \mathbf{e}_1^H \mathbf{U}_o \Delta \mathbf{U}_{o1}^H \mathbf{a}(r_k) \\
&\quad \mathbf{a}^{(1)}(r_k^{-1})^T \Delta \mathbf{U}_{o1} \mathbf{U}_o^H \mathbf{e}_1 \mathbf{e}_1^H \boldsymbol{\Omega}_o \mathbf{a}^{(1)}(r_k)]\} \\
&\quad +4jr_k^{-3} \Im\{E[r_k^{-1} \mathbf{a}^{(1)}(r_k^{-1})^T \boldsymbol{\Omega}_o \mathbf{e}_1 \mathbf{e}_1^H \mathbf{U}_o \Delta \mathbf{U}_{o1}^H \mathbf{a}(r_k) \\
&\quad \mathbf{a}^{(1)}(r_k^{-1})^T \boldsymbol{\Omega}_o \mathbf{e}_1 \mathbf{e}_1^H \Delta \mathbf{U}_{o1} \mathbf{U}_o^H \mathbf{a}^{(1)}(r_k)]\} \\
&\quad +2r_k^{-3} E[\mathbf{a}^T(r_k^{-1}) \Delta \mathbf{U}_{o1} \mathbf{U}_o^H \mathbf{e}_1 \mathbf{e}_1^H \boldsymbol{\Omega}_o \mathbf{a}^{(1)}(r_k) \\
&\quad \mathbf{a}^{(1)}(r_k^{-1})^T \boldsymbol{\Omega}_o \mathbf{e}_1 \mathbf{e}_1^H \mathbf{U}_o \Delta \mathbf{U}_{o1}^H \mathbf{a}(r_k)]\} \\
&= \frac{1}{\|\mathbf{e}_1^H \boldsymbol{\Omega}_o\|^8} \{2j\sigma_n^2 r_k^{-3} \Im\{r_k^{-1} \mathbf{a}^T(r_k^{-1}) \mathbf{U}_s \boldsymbol{\Sigma}_s^{-2} \mathbf{U}_s^H \mathbf{a}(r_k) \\
&\quad Tr[\mathbf{a}^{(2)}(r_k^{-1})^T \boldsymbol{\Omega}_o \mathbf{e}_1 \mathbf{e}_1^H \boldsymbol{\Omega}_o \mathbf{e}_1 \mathbf{e}_1^H \boldsymbol{\Omega}_o \mathbf{a}^{(1)}(r_k)]\} \\
&\quad +4j\sigma_n^2 r_k^{-3} \Im\{r_k^{-1} \mathbf{a}^{(1)}(r_k^{-1})^T \boldsymbol{\Omega}_o \mathbf{e}_1 \mathbf{e}_1^H \boldsymbol{\Omega}_o \mathbf{e}_1 \mathbf{e}_1^H \boldsymbol{\Omega}_o \mathbf{a}^{(1)}(r_k) \\
&\quad Tr[\mathbf{a}^{(1)}(r_k^{-1})^T \mathbf{U}_s \boldsymbol{\Sigma}_s^{-2} \mathbf{U}_s^H \mathbf{a}(r_k)]\} \\
&\quad +4j\sigma_n^2 r_k^{-3} \Im\{r_k^{-1} \mathbf{a}^{(1)}(r_k^{-1})^T \boldsymbol{\Omega}_o \mathbf{e}_1 \mathbf{e}_1^H \boldsymbol{\Omega}_o \mathbf{a}^{(1)}(r_k) \\
&\quad Tr[\mathbf{a}^{(1)}(r_k^{-1})^T \boldsymbol{\Omega}_o \mathbf{e}_1 \mathbf{e}_1^H \mathbf{U}_s \boldsymbol{\Sigma}_s^{-2} \mathbf{U}_s^H \mathbf{a}(r_k)]\} \\
&\quad +2\sigma_n^2 r_k^{-3} \mathbf{a}^T(r_k^{-1}) \mathbf{U}_s \boldsymbol{\Sigma}_s^{-2} \mathbf{U}_s^H \mathbf{a}(r_k) \\
&\quad Tr[\mathbf{a}^{(1)}(r_k^{-1})^T \boldsymbol{\Omega}_o \mathbf{e}_1 \mathbf{e}_1^H \boldsymbol{\Omega}_o \mathbf{e}_1 \mathbf{e}_1^H \boldsymbol{\Omega}_o \mathbf{a}^{(1)}(r_k)]\}.
\end{aligned}
$$

Use Eq.(108) to obtain Eq.(102).

References

[1] M. Kaveh and A. J. Barabell, "The statistical performance of the MUSIC and the Minimum-Norm algorithms in resolving plane-waves in noise," *IEEE Transactions on Acoustics, Speech, and Signal Processing* **ASSP-34**, pp. 331–340, (1986).

[2] F. Li and R. J. Vaccaro, "Unified analysis for DOA estimation algorithms in array signal processing," *Signal Processing*, **22**, pp. 147–169, (1991).

[3] H. Wang and G. H. Wakefield, "Non-asymptotic performance analysis of eigenstructure spectral methods," *Proc. IEEE ICASSP'89*, pp. 2591–2594, Glasgow, UK (1989).

[4] X.-L. Xu and K. M. Buckley, "Bias and variance analysis of MUSIC location estimates," *Proc. 5th IEEE ASSP Workshop on Spectrum Estimation & Modeling*, pp. 332–336, Rochester, NY (1990).

[5] F. Li and Y. Lu, "Bias analysis for ESPRIT-type estimation algorithms," *IEEE Transactions on Antenna and Propagation*, **AP-42**, pp. 418–423, 1994.

[6] F. Li, *A Unified Performance Analysis of Subspace-Based DOA Estimation Algorithms.* PhD thesis, University of Rhode Island, Kingston, RI, 1990.

[7] R. O. Schmidt, "Multiple emitter location and signal parameter estimation," *Proc. RADC Spectral Estimation Workshop*, pp. 243–258, Griffiss AFB, NY (1979).

[8] R. Kumaresan and D. W. Tufts, "Estimating the angles of arrival of multiple plane waves," *IEEE Transactions on Aerospace and Electronic Systems*, **AES-19**, pp. 134–139 (1983).

[9] F. Li, R. J. Vaccaro, and D. W. Tufts, "Min-Norm Linear Prediction for arbitrary sensor array," *Proc. IEEE ICASSP'89*, pp. 2613–2616, Glasgow, UK (1989).

[10] A. J. Barabell, "Improving the resolution performance of eigenstructure-based direction-finding algorithm," *Proc. IEEE ICASSP'83*, pp. 336–339 (1983).

[11] A. Paulraj, R. Roy, and T. Kailath, "Estimation of signal parameters via rotational invariance techniques – ESPRIT," in *Proc. 19th Asilomar Conf. on Signals, Systems and Computers*, pp. 83–89, Pacific Grove, CA (1985).

[12] S. Y. Kung, K. S. Arun, and D. V. Bhaskar Rao, "State-space and singular-value decomposition-based approximation methods for the harmonic retrieval problem," *J. Opt. Soc. Am.*, **73**, pp. 1799–1811 (1983).

[13] H. Ouibrahim, D. D. Weiner, and T. K. Sarkar, "Matrix pencil approach to angle of arrival estimation," in *Proc. 20th Asilomar Conf. on Signals, Systems and Computers*, pp. 203–206, Pacific Grove, CA (1986).

[14] R. Roy, A. Paulaj, and T. Kailath, "Estimation of signal parameters via rotational invariance techniques – ESPRIT," *Proc. IEEE MILCON*, pp. 41.6.1–41.6.5 (1986).

[15] F. Li and R. J. Vaccaro, "Analysis of MUSIC and Min-Norm for arbitrary array geometry," *IEEE Transactions on Aerospace and Electronic Systems*, **AES-26**, pp. 976–985 (1990).

[16] F. Li, H. Liu, and R. J. Vaccaro, "Performance analysis for DOA estimation algorithms: unification, simplification, and Observations," *IEEE Transactions on Aerospace and Electronic Systems*, **AES-29**, pp. 1170–1184 (1993).

[17] P. Lancaster and M. Tismentsky, *The Theory of Matrices.* New York, NY: Academic Press, second ed., 1978.

Detection Algorithms for Underwater Acoustic Transients

Paul R. White

Institute of Sound and Vibration Research,
University of Southampton, Hants., U.K.

I. INTRODUCTION

The problem addressed here is that of the detection of underwater acoustic transients in a passive SONAR environment. The term 'transient' in some circles has come to describe a wide variety of acoustic events which do not necessarily conform to one's immediate image of a transient. Such misnomers are avoided, the transients to be considered are short duration events, typified by non-technical terms such as "clicks" and "bangs". The importance of such events is that they may give an initial indication of the presence of a target and possibly allow some broad form of characterisation.

The approach taken here is intended to be highly pragmatic. The resulting algorithms are implementable on real-time signal processing chips working at reasonable sampling rates. These algorithms have a strong intuitive basis and as such are often familiar, but a major aim of this discussion is to highlight the similarities between the approaches and to provide an understanding of at least where compromises have been made. A second objective is to compare the performance of these detectors using measured data, thus allowing one to gauge how well they perform in realistic scenarios.

Work on transient detection can be divided into two distinct classes. First those techniques which are based on some model of the transient signal [1,

2, 3], second those methods which require no signal model [4, 5]. One expects the former class of processors to perform better than the latter, when the transient signals comply with the assumed model, but in instances where the model is violated it is often the case that the performance of model based detectors is severely degraded.

Analysis of high Signal to Noise Ratio (SNR) recordings of underwater acoustic transient events indicates that there is no obvious model which one might use to encompass all the examples. Only a few of the transients exhibited any deterministic structure, whilst some can only be accurately modelled as highly non-stationary random processes. Modelling such a wide variety of potential events appears to be a futile objective. A detector which is applicable to a broad class of transients is an important tool. Clearly such a detector will generally perform less well than a detector optimised for a specific class of transients when transients from that class are encountered. One may consider using a set of detectors optimised for specific important classes of transients, such as narrow band FM signals, in allegiance with a general transient detector. It is only the design of these general transient detectors which is of interest here.

II. THE GENERAL PRINCIPLES

A. DETECTION THEORY

By discarding the idea of modelling the transient signals one is left with a slight dilemma. Most detection methods are based on the Generalised Likelihood Ratio (GLR) test [6, 7], which is widely optimal and can be written as :

$$\Lambda(x) = \frac{p\{ x / H_1 \}}{p\{ x / H_0 \}}$$

where $p\{ x / H_0 \}$ is the likelihood of observing the data x[1] under the null hypothesis, in this case under the assumption that there is no transient, only noise. Similarly $p\{ x / H_1 \}$ is the likelihood of observing the data x under the alternative hypothesis, here under the assumption that there is a transient present in the noise. However without some model of the transient this latter term cannot be evaluated simply.

The philosophy to be adopted is not that of looking for specific transients but merely looking for data segments which do not conform to a model of the background oceanic noise. This has the advantage of only requiring a model of the noise; such a model is also required by a GLR detector. It is implicitly assumed that the transients will be relatively rare events and so one will in fact have available relatively long time histories containing only the background noise and these can be used to accurately fit noise models. Thus in the absence of being able to evaluate the numerator in the GLR test we simply set it to a constant, say unity, and seek a detection statistic of the simpler form :

$$\Lambda(x) = \frac{1}{p\{ x / H_0 \}}$$

Clearly, as previously discussed, this method will be sub-optimal for transient signals which can be accurately modelled, but in the absence of such a model then the above represents a logical way to proceed.

---

[1] The notation for the data x is deliberately vague at this point since in the context of the present discussion it is not necessary to exactly define what form the data takes.

B. NOISE MODELS

In order to evaluate any likelihood of having observed the data one has to be able to model the noise process against which the detector is attempting to compete. In this case the major contaminating noise, at least in any well designed SONAR system, is the background oceanic noise.

The modelling of background oceanic noise has a long history [8, 9], and many authors have sought examples where the probability distribution of the noise is non-Gaussian. However these examples often occur in exceptional conditions, e.g. in the presence of close shipping [10], or only represent mild departures from Gaussianity [11]. The simplifications offered by assuming Gaussianity make it a tempting assumption to adopt. Here we are seduced by the simplifications offered by assuming Gaussianity, but acknowledge that in doing so we may have to accept a degradation in performance of the resulting detectors in certain (exceptional) circumstances. It is also noted that this degradation may be mitigated in situations where the detector acts after any beamforming has taken place. This is because the beamformer sums hydrophone measurements, the effect of which is to tend to make the beamformer output more Gaussian than the individual inputs (in the spirit of the, much abused, Central Limit Theorem [7]).

Another assumption to be made is that the background noise is not only Gaussian but also stationary over any data window considered. Clearly oceanic background noise is not stationary over arbitrary time windows. To see this, one need only realise that a major contributing factor to oceanic background noise is the surface weather [12], so variations in the background noise over periods of hours (or less) are expected. In practice the very dynamic nature of the ocean means that stationarity of the background noise statistics may only be expected on time scales of tens of seconds to minutes. It is these considerations which will in fact limit the integration times one is willing to accept in the final processors.

It is further assumed that all the processes observed are zero mean. This presents little practical difficulty since it is common practice to high-pass filter the data prior to processing to remove close to d.c. components; it is merely assumed that this pre-processing has already taken place.

Under all these assumptions one can write the probability distribution of a given data vector of L samples. The vector will be denoted $\underline{x}_n$, and is defined by

$$\underline{x}_n = [\ x(n)\ \ x(n-1)\ \ x(n-2)\ \ ....\ \ x(n-L+1)\ ]^t$$

where x(n) represents the nth sample in the input time history and ^{t} denotes simple transposition. The multivariate probability distribution for this vector is [6]

$$p\{\ \underline{x}_n\ \} = \frac{1}{(2\pi)^{L/2}\ |R|^{1/2}}\ e^{-\underline{x}_n^t\ R^{-1}\underline{x}_n/2} \qquad (1)$$

where R is the auto-correlation matrix defined by

$$R = E[\ \underline{x}_n\ \underline{x}_n^t\ ]$$

in which E[.] denotes the expectation operator.

C. SEGMENTATION

The processing strategy to be adopted can be viewed as a variation of the segmentation methods proposed by Chen [5] and others [13, 14]. The outline strategy of these techniques is discussed here. The input data stream is divided into two windows : commonly termed the test segment and the reference segment, see Figure 1. These segments are then advanced through the data, and at each new position a test statistic is calculated. One can construct various strategies for advancing the windows.

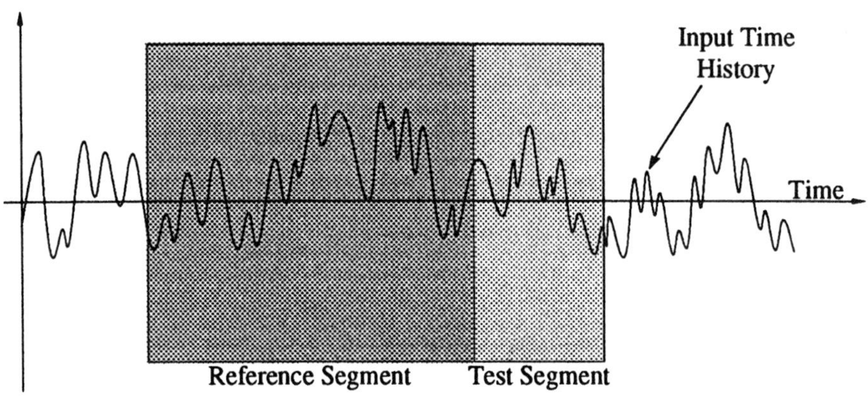

Figure 1 : The Segmentation Approach

At its most general the segmentation principle is : to compare the data in the test and reference segments to see if they conform to the same model and so to infer the absence or presence of a transient. By using test segments of variable length such an approach can be used to locate the beginning and duration of events in speech [13] so breaking up continuous speech into discrete packets prior to recognition. Whilst it would be desirable to locate the beginning and end of an underwater acoustic event, especially if the detector is to be used as a pre-cursor to an automatic classifier, it is considered that the

SNRs at which the detector is desired to work means that this goal is unrealistic. It will be sufficient merely to detect the presence of the transient, so the length of the test segment is fixed.

D. THE LIKELIHOOD TEST STATISTIC

Based on the Gaussian assumption for the noise the following test statistic is appropriate

$$\log\{ \Lambda(\underline{x}_n) \} = -\log (p\{ \underline{x}_n / H_o \}) \propto \underline{x}_n^t R^{-1} \underline{x}_n = \gamma(\underline{x}_n) \quad (2)$$

Our basic strategy is to employ a segmentation approach. The data in the reference segment is used to estimate the auto-correlation matrix R and all the data test segment is used to construct the vector $\underline{x}_n$. The principle of the approach is simple, if the test segment contains data from the same Gaussian process as the reference segment then the likelihood[2] of the vector $\underline{x}_n$ should be large, or equivalently $\gamma(\underline{x}_n)$ is small. Alternatively if the test segment is not consistent with the reference segment, for example when there is a transient contained in the test segment, then the likelihood of $\underline{x}_n$ should be small, or $\gamma(\underline{x}_n)$ should be large.

Immediately some inferences can be drawn from this discussion about the sizes of the two segments. Firstly one would like the transient size to match the length of the test segment, so that the amount of noise in the test segment is minimised and the transient makes the largest possible contribution to the test statistic. Secondly, to maximise the accuracy of the estimate of the auto-correlation matrix, the reference segment should be as large as possible. The

---

[2] The term likelihood is used (as it is in the GLR test) to denote $p\{ \underline{x}_n / H_o \}$.

limit of the size of the reference segment should either be the length of time over which one is willing to assume that the background noise is stationary or the expected interval between transients.

To accurately fix either of these lengths, so that the above conditions are met, requires an unreasonable knowledge of the environment in which the detector is to be implemented. These lengths are thus selected with the aid of considerable guess work, but as a general statement it is probably true to say that in most situations one would expect to choose a reference segment which is significantly longer than the test segment.

III. ALGORITHMS

A. THE ENERGY DETECTOR

As an illustration a detector is constructed based on the above ideas along with one further assumption, specifically it is assumed that the background noise is white. So if the noise signal is w(n) then

$$E[\, w(n)\, w(m)\,] = 0 \qquad m \neq n$$
$$= \sigma^2 \qquad m = n$$

Under this assumption the test statistic simplifies to

$$\gamma(\underline{x}_n) \;\alpha\; \underline{x}_n{}^t\, \underline{x}_n \,/\, s^2$$

where s^2 is an estimate of the energy of the signal (σ^2) generated from the reference segment. Similarly the numerator is an estimate of the signal energy generated from the test segment. This represents what may be considered as the simplest form of transient detector, simply looking for data segments which have a disproportionate amount of energy.

It is a consequence of the assumption of Gaussianity that any detector generated will only depend on the second order statistics of the data, since these statistics completely specify a Gaussian random variable. Detectors based on higher order statistics have been proposed [4].

B. INTERPRETATION OF THE LIKELIHOOD DETECTOR

Considering the definition of the likelihood test statistic (2) can shed some light on its true meaning. This is achieved by factorising the inverse of the auto-correlation matrix. Specifically it is noted that R^{-1} is positive definite so one can perform a Cholesky factorisation to yield

$$R^{-1} = C^t C$$

where C is an upper triangular matrix. Substituting this into (2) yields

$$\gamma(\underline{x}_n) = \underline{x}_n{}^t C^t C \underline{x}_n = \underline{y}_n{}^t \underline{y}_n = \sum_{j=1}^{L} y_n(j)^2$$

where $\underline{y}_n = C\underline{x}_n$ whose j^{th} component is $y_n(j)$. Further since it is easily confirmed that

$$E[\underline{y}_n \underline{y}_n{}^t] = C R C^t = I$$

then $\underline{y}_n$ is simply a pre-whitened version of the input data vector, so the $y_n(j)$ are orthonormal. The likelihood detector can be viewed as firstly whitening the data in the test segment and then applying a simple energy detector. It is of course the data in the reference segment which is used to form the basis of the whitening operation, i.e. to calculate R.

The above observations allow one to evaluate the probability distribution of the likelihood statistic itself, under the assumption that the input is truly Gaussian. Since the likelihood statistic can be written as the sum of L unit variance, independent, (orthonormal) Gaussian random variables then the likelihood test statistic must be distributed as a Chi-squared random variable with L degrees of freedom. This allows threshold levels to be calculated. By firstly specifying a required false alarm rate, i.e. probability that the detector will "detect" a transient when in fact there is only noise, and then using the Chi-squared nature of the likelihood statistic, a value (threshold) can be set such that if the test statistic exceeds that value then a detection is said to have taken place.

C. THE CHEN ALGORITHM

The standard Chen segmentation algorithm can be viewed from the stand point given above. Initially the basic algorithm is outlined here. The basis of the method is a GLR test of the generic form

$$\frac{p\{\, x / H_1 \,\}}{p\{\, x / H_0 \,\}}$$

where the hypotheses are now : -

H_0: The data in the test and reference segments are the result of Gaussian noise filtered by an Auto-Regressive (AR) filter of order k whose coefficients are the same for the test and reference segments.

H_1: The data in the test and reference segments are the result of Gaussian noise filtered by two different AR filters of order k.

The choice of an AR filter implies that one can model the data having arisen from a process of the form :

$$x(n) = \sum_{i=1}^{k} a_i\, x(n-i) + e(n)$$

where e(n) is a Gaussian white random variable. By a simple re-arrangement one immediately obtains

$$e(n) = x(n) - \sum_{i=1}^{k} a_i\, x(n-i)$$

which illustrates that under the AR assumption one can construct a Finite Impulse Response (FIR) filter of order k+1 which will exactly whiten the input sequence. The exact form of the Chen test statistic is

$$D_{ch} = (L'_{test} + L'_{ref}) \log s^2_{pool} - (L'_{ref} \log s^2_{ref} + L'_{test} \log s^2_{test})$$

where L'_{test} and L'_{ref} are the number of points in the test and reference segments, minus k, and s^2_{test}, s^2_{ref} and s^2_{pool} are the residual mean squared errors for the test, reference and 'pooled' segments. The 'pooled' segment is simply the concatenation of the test and reference segments.

For an AR model the residual mean squared error is an estimate of $E[\, e(n)^2 \,]$, i.e. an estimate of the signal's variance after it has been whitened. Conceptually one could evaluate this quantity by estimating the parameters of the optimal whitening filter, then whitening the data segment under examination and finally estimating its energy. In practice this is unnecessary since it is easier to estimate the residual mean squared errors using :

$$s^2 = \hat{r}(0) - \underline{r}^t \hat{R}^{-1} \underline{r} \qquad (3)$$

in which $\underline{r}$ represents the first column of $\hat{R}$, see Section IV (A).

There are only two significant differences between this approach and the likelihood detector. Firstly the Chen algorithm considers both the test segments as having been generated by an AR model. Whitening filters are then constructed for both segments individually and for the entire data length. The likelihood method only ever constructs one whitening filter from the reference segment. Secondly the Chen algorithm assumes an AR model of order k, which is generally significantly less than the length of the test segment, whereas the likelihood detector could be said to use a model order which is the same as the length of the test segment.

It is worth noting that the AR model may be appropriate for data segments where there is background noise only, but is inappropriate for any segment in which a transient is present. The model used is of a stationary random process driving a fixed filter; this can never model a non-stationary signal such as a transient in noise. The Chen algorithm employs an alternative hypothesis (H_1) which assumes an AR model and so is inappropriate.

The choice of a model order k in the Chen algorithm adds another variable which the user has to select. Chen in [5] suggests using a model order of 2. In our experience the use of larger model orders may increase the detection threshold, so that an individual transient is more likely to be detected. However this increase in detection threshold is generally accompanied by an increase in false alarm rates, to the detriment of the algorithm.

IV. COMPUTATIONAL ISSUES

A. CALCULATION OF THE CHEN TEST STATISTIC

Initially the computational cost associated with the Chen algorithm may seem daunting, requiring the estimation and inversion of three auto-correlation matrices. Bearing in mind that typically one may expect to have to perform $O(k^3)$ operations to invert each matrix as well as estimating the auto-correlation function, there are two ways of reducing this load.

If it is assumed that each auto-correlation matrix has a symmetric Toeplitz structure, then an estimate of R is

$$\hat{R} = \begin{bmatrix} \hat{r}(0) & \hat{r}(1) & \hat{r}(2) & \dots & \hat{r}(k-1) \\ \hat{r}(1) & \hat{r}(0) & \hat{r}(1) & \dots & \hat{r}(k-2) \\ \hat{r}(2) & \hat{r}(1) & \hat{r}(0) & \dots & \hat{r}(k-3) \\ \vdots & \vdots & \vdots & \ddots & \vdots \\ \hat{r}(k-1) & \hat{r}(k-2) & \hat{r}(k-3) & \dots & \hat{r}(0) \end{bmatrix} \quad (4)$$

where

$$\hat{r}(i) = \frac{1}{L} \sum_{j=n-L+1}^{n-i} x(j)x(j+i)$$

Using such an approximation one can apply a Levinson-Durbin recursion [16] to directly estimate the residual squared errors, using (3). This requires only $O(k^2)$ operations, and one need only store the first row of the matrix $\hat{R}$.

The second strategy for reducing the computation of a Chen test statistic is based on a series of three Adaptive Lattice filters and was proposed by Brandt [17]. This technique is based on a set of sliding window exact least squares algorithms. It requires only O(k) operations per update, but does require updating on a sample by sample basis, so one generates a new test statistic upon the arrival of each new data sample. This ensures that one never misses the point at which any transient is optimally placed within the test segment. However it does impose a heavy computational burden. Conversely the method based on the Levinson-Durbin algorithm is easily and efficiently translated in to a block format, so that the algorithm calculates the test statistic once a block of data is received. Thus the computation required per sample is reduced by the size of a block. The net effect is that neither algorithm necessarily requires less computation than the other. To accurately assess the relative computational loadings one needs detailed knowledge of the schemes to be considered.

It is worth noting that the adaptive filter approach implicitly uses a different estimate of the auto-correlation matrix, specifically

$$\hat{R} = \frac{1}{L} \sum_{j=n-L+1}^{j=n} \underline{x}_j \underline{x}_j^t \tag{5}$$

The difference between these two approximations reduces, for a stationary input, as the number of points in the averages increases.

B. CALCULATION OF THE LIKELIHOOD STATISTIC

In this section methods of calculating the likelihood test statistic are discussed. There are two basic strategies mimicking those discussed for the Chen algorithm: the first based on a simple extension of the Levinson-Durbin algorithm and the second using an exact least squares adaptive filter algorithm.

If we assume a Toeplitz structure for our estimate of the autocorrelation matrix, i.e. the estimator described by (2), then one has to calculate

$$\gamma(\underline{x}_n) = \underline{x}_n^t \hat{R}^{-1} \underline{x}_n$$

where $\underline{x}_n$ contains all the data in the test segment. Once again direct application of this formula imposes a prohibitively large computational burden, because of the requirement to invert the auto-correlation matrix estimate. This burden can be partly relieved by use of a variation on the Levinson-Durbin algorithm, specifically the use of the following recursive algorithm :

Given $\underline{r}$ *and* $\underline{x}_n$

Initialise

$$\underline{f}_0 = 1, \quad E_0 = 1/r(0), \quad \gamma_{n,1} = x(1)^2/r(0)$$

Repeat for $m = 1,2, \ldots. L-1$

$$\psi = \underline{f}_{m-1}^{\Diamond t}\, \underline{r}_{m-1}\, E_{m-1}$$

$$\underline{f}_m = \begin{bmatrix} \underline{f}_{m-1} \\ 0 \end{bmatrix} - \psi \begin{bmatrix} 0 \\ \underline{f}_{m-1}^{\Diamond} \end{bmatrix}$$

$$E_m = \frac{E_{m-1}}{1-\psi^2}$$

$$e = \underline{f}_m^{\Diamond t}\, \underline{x}_{n,m}$$

$$\gamma_{n,m+1} = \gamma_{n,m} + e^2 E_m$$

where the notation $\diamond$ is used to denote the reversing of the order of the elements in a vector, $\underline{r}_m$ denotes the first m elements in $\underline{r}$ and similarly $\underline{x}_{n,m}$ denotes the first m elements in $\underline{x}_n$. The notation $\gamma_{n,m}$ is used to denote the likelihood variable at time sample n based on the first m elements of $\underline{x}_n$. So the final value $\gamma_{n,L}$ is the same as $\gamma(\underline{x}_n)$.

This allows the computation of the likelihood statistic using a computational loading of $O(k^2)$. This again is a method suited to block processing.

The likelihood test statistic in fact plays an important role in many exact least squares adaptive filter algorithms. Sliding window forms of these algorithms can be exploited to reduce the computational load per update to O(k) operations. With the use of modern QR forms of adaptive least squares algorithms [18] the problems associated with numerical instability are also eliminated.

C. EXPONENTIAL WINDOWING

The algorithms discussed so far have used a fixed length sliding window for the reference segment, so that the estimation of the auto-correlation matrix is achieved using all the data samples in the reference segment and each is given equal weight in the final estimator. Such a method has two drawbacks. Firstly one needs to store all the samples in the reference window, and it has already been established that it would be advantageous to employ a large reference segment, which imposes a large memory requirement. Secondly the use of a sliding window means that all the data within the window have the same effect on the final estimator. For long windows this will mean that a relatively old data sample will be as influential as the most recent.

To counter these problems a sliding window approach with an exponential window is often used. The likelihood test statistic is well suited to calculation using such a window. This means that the approximation (5) is replaced by

$$\hat{R}_n = (1-\lambda) \sum_{j=1}^{n} \lambda^{n-j} \underline{x}_j \underline{x}_j^t$$

where λ (chosen to satisfy $0 < \lambda < 1$) is a user selected constant which controls the integration time of the exponential window; specifically the effective integration time is proportional to $1/(1-\lambda)$. To update the estimate of the auto-correlation matrix based on an exponential window one need only use :

$$\hat{R}_n = \lambda \hat{R}_{n-1} + (1-\lambda) \underline{x}_n \underline{x}_n^t$$

Since it is necessary to retain only the current auto-correlation matrix estimate in order to advance the solution, there is no requirement to retain a record of all the data points in the reference segment. One should be aware that when using an exponential window every data point encountered retains some effect on the current estimate, so that a particularly extraordinary series of data points may distort the estimate for a relatively long period of time.

The flexibility and simplifications offered by the use of an exponential windowing approach make it an attractive, practical, choice for processing the data in the reference segment, particularly when the reference segment is required to be long. The next section describes two approaches which are suited to estimating the likelihood test statistic with an exponentially windowed reference segment.

D. ADAPTIVE FILTERS

The concept of using an exponential window has a long history in adaptive least squares methods. Indeed these methods are ideally suited to estimating the likelihood test statistic. The vast majority of exact least squares algorithms [19] all update a quantity called the likelihood variable. It is this variable which motivated both the notation and terminology used here. The likelihood variable of adaptive filter theory is often denoted γ_n and is equivalent to

$$\gamma_n = 1 - \underline{x}_n^t \left(\sum_{j=1}^{n} \lambda^{n-j} \underline{x}_j \underline{x}_j^t \right)^{-1} \underline{x}_n \qquad (6)$$

So an approximation to the likelihood test statistic defined in (2) is

$$\gamma(\underline{x}_n) = \frac{1 - \gamma_n}{1 - \lambda}$$

Thus one need only use an exact least adaptive filter algorithm which employs the likelihood variable and monitor the behaviour of this statistic. Again one of the numerically stable forms of exact least squares methods, such as a QR adaptive filter, can be employed These algorithms actually update the square root of the likelihood variable, but this is of no consequence. The QR algorithms are numerically efficient but at present, for reasonable filter lengths and sampling rates, require too many computations for implementation on a single serial DSP processor, at the sample rates of interest.

To reduce the computational load one may seek to use simpler adaptive filter algorithms such as the gradient class of adaptive filters, of which the Least

Means Squares (LMS) algorithm is the most well known. By employing a gradient method one is implicitly making an approximation and so one expects a corresponding reduction in performance. The most suitable form of gradient algorithm for this approach is to use a Gradient Adaptive Lattice (GAL) [20]. Unfortunately GALs do not explicitly estimate the likelihood statistic, indeed they can be derived starting with an exact least squares lattice and assuming the likelihood variable is always unity. One has to take a slightly circuitous route in order to employ these algorithms. The logic behind this approach will be explained briefly.

Reconsidering the definition of the likelihood statistic (2) and factorising the auto-correlation matrix as the product of a unit lower triangular matrix L and a diagonal matrix D so that

$$R = L\,D\,L^t$$

the inverse of the auto-correlation matrix can be written

$$R^{-1} = U\,D^{-1}\,U^t$$

where U is a unit upper triangular matrix and $U = (L^{-1})^t$. Using this the expression for the likelihood statistic becomes

$$\gamma(\underline{x}_n) = \underline{x}_n{}^t\,U\,D^{-1}\,U^t\,\underline{x}_n = \underline{b}_n{}^t\,D^{-1}\,\underline{b}_n = \sum_{j=1}^{L}\frac{b_n(j)^2}{d(j)} \tag{7}$$

where $\underline{b}_n = U^t\,\underline{x}_n$ whose j^{th} element is denoted $b_n(j)$ and similarly $d(j)$ denotes the jth element on the leading diagonal of D. It is also straight forward to show that

$$E[\underline{b}_n \underline{b}_n^t] = D$$

This illustrates that the elements of the transformed data vector $\underline{b}_n$ are mutually uncorrelated. From this one can recognise that

$$d(j) = E[b_n(j)^2]$$

By considering the structure of U it is clear that $b_n(j)$ is constructed by forming a linear combination of the first j elements of $\underline{x}_n$. From this we should realise that $\underline{b}_n$ is simply the data vector $\underline{x}_n$ transformed via a Gram-Schmidt orthogonalisation process [16]. Translating this into filter terminology the $b_n(j)$'s are the backward prediction errors, and the $d(j)$'s are their mean squared values. These two quantities are readily available in most GAL algorithms and so one can construct an estimate of $\gamma(\underline{x}_n)$ based on the backward prediction errors and their means squared values, estimated via a GAL algorithm, by employing (7).

E. SPECTRAL APPROXIMATIONS

A final estimator for the likelihood statistic can be constructed by utilising the positive definite nature of the auto-correlation matrix, which allows one to construct another factorisation. In this case R is written in terms of its eigen decomposition so

$$R = Q \Lambda Q^t$$

where Q is an orthonormal matrix, whose columns are the eigenvectors of R and Λ is a diagonal matrix whose non-zero elements are the eigenvalues of R. Using this we can once again re-write (2) to give

$$\gamma(\underline{x}_n) = \underline{x}_n{}^t Q \Lambda^{-1} Q^t \underline{x}_n = \underline{\nu}_n{}^t D^{-1} \underline{\nu}_n = \sum_{j=1}^{L} \frac{\nu_n(j)^2}{\lambda(j)} \tag{8}$$

where $\underline{\nu}_n = Q^t \underline{x}_n$. The elements of the transformed data vector $\underline{\nu}_n$ are also mutually orthogonal since

$$E[\,\underline{\nu}_n \underline{\nu}_n{}^t\,] = \Lambda$$

This linear mapping is a form of Karhunen-Loéve decomposition [6]. The eigen decomposition cannot simply be performed in a recursive fashion. Thus direct application of this factored formulation does not reduce computation. We introduce a simple approximation for the asymptotic form of the eigenvectors of the auto-correlation matrix, based on work discussed in [21, 22]. If we consider an auto-correlation matrix of increasing dimension, for a stochastic process of finite energy, then the eigenvectors of the matrix tend (at least in a distributional sense) to the set of complex exponentals below

$$\underline{q}_k = \left[1 \quad e^{-2\pi ik/L} \quad e^{-4\pi ik/L} \quad \ldots\ldots \quad e^{-2(L-1)\pi ik/L}\right]^t$$

These are the set of complex exponentials which are used to compute the Discrete Fourier Transform (DFT). Inner products of this vector with the data vector yield the kth DFT coefficient of the data. Thus using (8) one can write an approximation to the likelihood test statistic as

$$\gamma(\underline{x}_n) = \sum_{k=1}^{L/2} \frac{|X_n(k)|^2}{\lambda_n(k)}$$

where $X_n(k)$ is the kth DFT coefficient of $\underline{x}_n$. The only problem is now to approximate the eigenvalues $\lambda_n(n)$. It can be shown [21] that an appropriate approximation to these eigenvalues is

$$\lambda_n(k) = E[\,|X_n(k)|^2\,]$$

This introduces a simple interpretation of this approximation to the likelihood test statistic. Specifically this estimate is evaluated by calculating the DFT for each data window, then forming a ratio of the current modulus squared DFT value with its average value calculated across the reference segment and finally summing all these ratios. This operation can be viewed as looking for individual raw spectra which do not conform to an estimate of the overall spectrum. Or alternatively, from a filter bank stand point, one might consider it as splitting the data into frequency bands and using an energy detector within each band. From any of these viewpoints one might have considered using such a method for detecting transients without recourse to the likelihood test statistic. But it is interesting to note how one can take such an intuitive detector and frame it as an approximation to a detector which has a sound theoretical basis.

There is one further interesting standpoint from which this spectral detector can be viewed, which is particularly appropriate bearing in mind the SONAR context of this discussion. Specifically the spectral detector can be realised by simple operations on a conventional sonagram (or in other terminologies a spectrogram or short time Fourier transform). The sonagram is a 2 dimensional image within which every line represents the magnitude of the Fourier transform of a window of data. Each line is constructed by advancing a time window through the incoming data stream. The result is one example of a time-frequency distribution. The test statistic can be constructed from the sonagram by taking each new point in the current Fourier transform and normalising it (dividing it by an average of the preceding values of the spectrogram in that frequency bin). The test statistic is then simply calculated by summing all the normalised sonagram values within a line. Conventionally it is prudent to normalise a sonagram prior to output to ensure that the full range of the display mechanism is used. The normalisation mechanism discussed here will tend to obscure constant tonal signals. The sonagram has historically been used to detect and classify targets using these tonal lines and consequently one would not normalise the sonagram in the proposed way if one wishes to examine tonal signals. The techniques discussed can be viewed as an alternative method for extending the usefulness of the sonagram.

The exact strategy one adopts for implementing this detector is flexible. One can use a segmentation algorithm, in which the reference segment is rectangularly windowed and used to estimate the eigenvalues (the overall spectrum), or one can adopt an exponential windowing philosophy where the eigenvalues are estimated via an exponential running average. The relative merits of the two approaches have already been discussed, and for the remainder of this work an exponential windowing approach will be employed.

V. RESULTS

A. DATA

To assess the performance of the detectors discussed so far, trials were conducted using combinations of measured data sets. The data sets considered consisted of background noise measurements and recordings of underwater acoustic events at high SNRs. The objective was to create time series with controllable SNRs, by adding the background noise and transient signals together at variable levels.

The definition of SNR for transient signals is somewhat arbitrary. This problem is exacerbated by any noise on the measured transient signals, since this tends to obscure the exact start and end of an event. To allow for easy interpretation and to avoid the ambiguity in transient SNRs, a subjective form of SNR was defined. This was calibrated by adjusting the gains for each transient signal so that in a particular background noise signal the transient could "only just be heard[3]." The level at which a transient can be just discerned is obviously dependent on a wide variety of factors, including the listener, the nature of the transient and the spectral content of the background noise. This level was used to define the 0 dB point on a scale which was dubbed the "subjective dB scale" and denoted dB_{sub}. The justification for using such a scale is purely pragmatic. At a glance one can see whether a signal is audible in background noise (positive values of dB_{sub}) or inaudible (negative values of dB_{sub}). Such a measure is only useful for data sets where the transient SNR is controllable. Care must also be taken when comparing results between different transients because of the inherently variable nature of the definition.

[3] The listener in these trials was not a trained SONAR operator.

Figure 2 (a) shows the time history of one of the high SNR transients before background noise has been added. Examination of the time history reveals little, if any, deterministic structure to the event. Figure 2 (b) shows the time history of the same transient added to background noise at $0\ dB_{sub}$. This raw time history does not betray the presence of the transient.

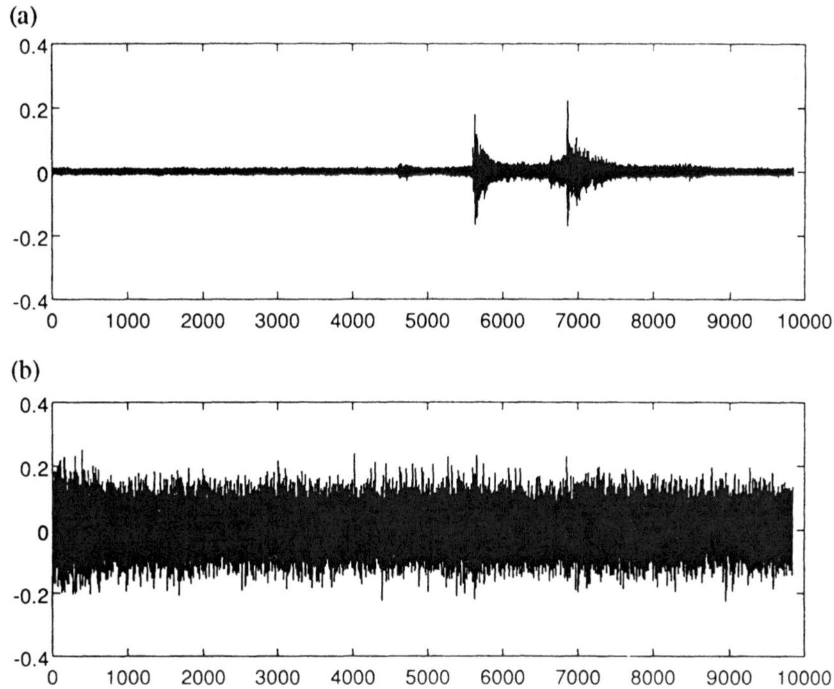

Figure 2 : (a) Transient with no noise added.
(b) Transient immersed in background noise at $0\ dB_{sub}$.

B. SIMULATION RESULTS

To examine the accuracy of the approximate methods for calculating the likelihood statistic, i.e. the spectral estimator and the GAL method, a test was conducted on using the transient depicted in Figure 2 (a) at 3 dB_{sub}. The two approximations were compared to the QR adaptive filter based method, which exactly calculates the likelihood test statistic using an exponential window. All three methods were run using the same test data and the resulting test statistics are plotted in Figure 3. An exponential window was used and care taken to ensure that all the methods used the same effective integration time, which corresponded to λ =0.9999 in (6). The test segment was 128 sample points long.

Examination of the graphs in Figure 3 reveals that all of the methods have successfully detected the transient. Further study of the plots shows a striking similarity between the results of the three algorithms. This similarity leads us to conclude that for modest test segment sizes the approximations on which the spectral and GAL methods are based are valid.

To test the relative performance of the methods a series of tests was conducted to measure the detection thresholds for the QR based algorithm and the spectral method. These tests used a set of the high SNR transients mixed with the background noise. They were conducted by firstly setting a threshold level. This was based on the Chi-squared nature of the test statistic, as discussed in Section III (B), and then employing one of the Normal approximations to the Chi-squared distribution [23]. In this case the third order approximation was used.

(a)
(b)
(c)

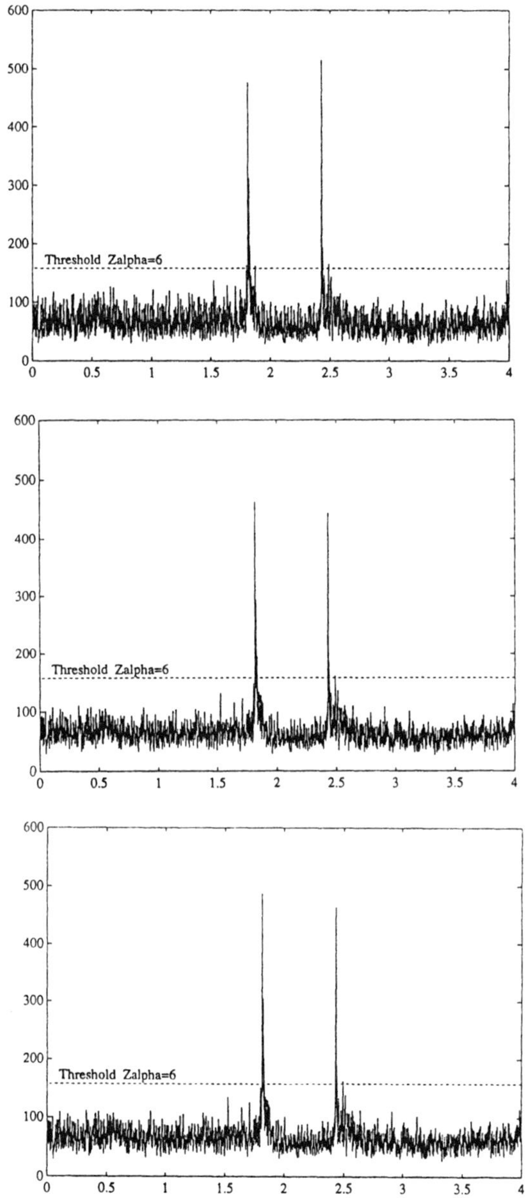

Figure 3 : Likelihood test statistic calculated using -
(a) QR Adaptive algorithm, (b) GAL Adaptive algorithm,
(c) Spectral method.

For the tests discussed in the following the threshold was selected which corresponded to a Z value in the Normal distribution of 6. Having selected a threshold, the transient signal was added to the background noise at a low SNR so that when the detector was implemented no detection occurred. Then the SNR was gradually increased until the detection algorithm produced a test statistic which exceeded the threshold level and so a detection was said to have taken place. The lowest SNR where such a detection is made is termed the detection threshold. This represents only part of a classical Receiver Operating Characteristic (ROC) curve [6]. At the time of this initial study a lack of data constrained the work so that a full ROC curve could not be performed. The above detection thresholds were calculated for all 6 transient signals, for both the QR algorithm and the spectral method, and the results are shown in Table 1. Once again a test segment of 128 points was used with $\lambda = 0.9999$.

Transient	QR	Spectral
1	-2.6	-2.6
2	-3.2	-3.7
3	-1.7	-1.8
4	0.6	-1.0
5	1.7	0.3
6	-6	-6.1

Table 1 : Detection Thresholds, in dB_{sub}, for QR and Spectral Detectors

The results in Table 1 illustrate some interesting points. One immediately sees that the majority of values in the table are negative, indicating that the detection threshold is below the hearing threshold, i.e. the detector is working at levels below which the human ear is able to detect. The two transients numbered 4 and 5 are both significantly longer in duration than the 128 point test segment used and this mismatch has caused the relatively poor performance in the results for these two examples. The results also apparently show the spectral algorithm performing better than the QR method. This is an apparent contradiction since the spectral method represents an approximation to the QR algorithm, so one would expect it to perform less well. The evaluation of the detection threshold is an incomplete performance measure since one also needs to take into account false alarm rates. The time histories available for this work were too short to construct meaningful false alarm rate measurements. We only remark that no false alarms were generated in the absence of a transient over the time history available. It is quite plausible that the spectral method achieves a reduced detection threshold at the expense of an increased false alarm rate.

VI. CONCLUSIONS

A series of transient detectors which require no signal model have been discussed within a unified framework. Several of the resulting detectors have a simple structure and are realisable within a real time system. The performance of these detectors has been examined using sets of measured data combined to create controllable, realistic, data sets. It has been shown that these methods are capable of correctly detecting signals at, and below, SNRs where the human ear can detect.

VII. ACKNOWLEDGEMENTS

The author would like to thank DRA Farnborough, for their continued financial support for this work, their technical assistance and the supply of the measured data set.

VIII. REFERENCES

1. Porat, B. and Friedlander, B. "Adaptive Detection of Transient Signals", *IEEE Trans. on ASSP*, Vol. 34, No. 6, pp 1410-1418, (1986).
2. Friedlander, B. and Porat, B. "Detection of Transient Signals by the Gabor Representation", *IEEE Trans. on ASSP*, Vol. 37, No. 2, pp 169-179, (1989).
3. Boashash, B. and O'Shea, P. "A Methodology for Detection and Classification of some Underwater Acoustic Signals using Time-Frequency Analysis Techniques", *IEEE Trans. on ASSP*, Vol. 33, No. 11, pp 1829-1841, (1988).
4. Hinich, M. "Detecting a Transient Signal by Bispectral Analysis", *IEEE Trans. on ASSP*, Vol. 38, No. 7, pp 1277-1283, (1990).
5. Chen, C. "On a Segmentation Algorithm for Seismic Signal Analysis", *Geo-exploration*, Vol. 23, pp 35-40, (1984).
6. van Trees, H. *Detection, Estimation, and Modulation Theory : Part I*, John Wiley & Sons, New York, (1969).
7. Whalen, A. *Detection of Signals in Noise*, Academic Press, New York, (1971)
8. Arase, T. and Arase, E. "Deep Sea Ambient Noise", *J. of Acoustical Society of America*, Vol. 44, pp 1679-1684, (1968).

9. Dyer, I. "Statistics of Distant Shipping Noise" *J. of Acoustical Society of America*, Vol. 53, No. 2, pp 564-570, (1973).

10. Brockett, P. *et al*, "Nonlinear and non-Gaussian Ocean Noise", *J. of Acoustical Society of America*, Vol. 82, pp 1386-1394, (1987).

11. Petit, E. *et al*, "Tests de Lois sur Bruits Preleves en Mer", *Proc. Institute of Acoustics*, Vol. 15, Part 3, pp 609-616, (1993).

12. Urick, R. *Principles of Underwater Sound*, McGraw-Hill, New York, (1983).

13. Andre-Obrecht, R. "New Statistical Approach for the Automatic Segmentation of Continuous Speech Signals", *IEEE Trans. on ASSP*, Vol. 36, No. 1, pp 29-40, (1988).

14. Basserville M. and Benvemiste, A. "Sequential Detection of Abrupt Changes in Spectral Characteristics of Digital Signals", *IEEE Trans. on Inf. Th.*, Vol. 29, No. 5, pp 709-724, (1983).

15. Srinath, M. and Rajasekaran, P. *An Introduction to Statistical Signal Processing with Applications*, John Wiley & Sons, New York, (1979).

16. Scharf, L. *Statistical Signal Processing : Detection, Estimation and Time Series Analysis*, Addison-Wesley (1990).

17. Brandt, A. "Detecting and Estimating Parameter Jumps using Ladder Algorithms and Likelihood Ratio Tests", *Proc. Int. Conf. on Acoustics Speech and Signal Processing '83*, pp 1017-1020, (1983).

18. Regalia, P. "Numerical Stability Properties of a QR-Based Fast Least Squares Algorithm", *IEEE Trans. Signal Processing*, Vol. 41, No. 6, pp 2096-2109, (1993).

19. Haykin, S. *Adaptive Filter Theory*, Prentice-Hall, New Jersey, (1986).

20. Honig, M. and Messerschmitt D. *Adaptive Filters : Structures, Algorithms and Applications*, Kulwer Academic, (1985).

21. Gray, M. "On the Asymptotic Distribution of Toeplitz Matrices", *IEEE Trans. on Inf. Th.*, Vol. 18, No. 6, pp 725- 30, (1972).

22. Grenander, U, and Szegö, G *Toeplitz Forms and Their Applications*, University of Califorina Press (1958).

23. Abramowitz, M. and Stegun, *A. Handbook of Mathematical Functions*, Dover, New York, (1972).

Constrained and Adaptive ARMA Modeling as an alternative to the DFT – with application to MRI

Jie Yang

Michael Smith

Department of Electrical and Computer Engineering,
University of Calgary, Calgary, Alberta,
Canada T2N 1N4

email: smith@enel.ucalgary.ca

1 SCOPE OF THIS ARTICLE

In many commerical and research applications, use of the discrete Fourier transform (DFT) allows the transfer of data gathered in one domain (typically spatial) into an other (frequency). This alternative representation often allows easier characterization or manipulation of the signal. For example the removal of unwanted noise components is achieved more efficiently by multiplying the frequency domain signal by the desired filter response than by a convolution operation in the original domain.

A major drawback to the DFT algorithm is that the signal resolution decreases as the data length decreases. This can be serious when there is only a finite (small) data length available, either because of experimental constraints or dynamic signal characteristics. In this article we review several methods to overcome these short comings. They involve using modeling techniques to characterize the known short data sequence. The model information is used to implicitly extrapolate the signal which permits recovery of the lost resolution.

The techniques discussed are based around constrained and adaptive variations of the auto-regressive moving average (ARMA) algorithm developed by Smith *et al.* [1, 2] as an alternative reconstruction approach for magnetic resonance imaging (MRI). This method is known as the Transient Error Reconstruction Approach (TERA). The technique is applicable to data sets that are short (truncated) in 1 or more data dimension [3].

In section 2, a background to standard DFT usage is given in the context of magnetic resonance reconstruction. In section 3, a review of the TERA

CONTROL AND DYNAMIC SYSTEMS, VOL. 77

modeling algorithm is given and its limitations examined. In section 4, some alternative MR image reconstruction methods from other authors are discussed including the Sigma and generalized series (GS) methods [4]. Then a constrained TERA (CTERA) method is presented, which attempts to combine the best features of the Sigma and TERA algorithms. In section 5, an adaptive total least squares modification schemes is introduced to overcome TERA's limitations when attempting to model the non-stationary MR data properties and for poor signal-to-noide (SNR) data sets.

In section 6, an evaluation scheme is introduced. This allows a quantitative comparision to be made of the algorithms using medical and geological MRI data sets from samples and phantoms. The areas of multi-channel analysis and neural networks are suggested in section 7 as future directions for this research. Both a qualitative and a quantitative comparison of the modeling algorithms are given in the conclusion.

2 INTRODUCTION

Magnetic resonance imaging (MRI) was first introduced in the 70's. Over the past two decades, it has become widely used for both clinical and geological imaging purposes. A major advantage of MRI is that it provides high image contrast safely [5]. Since MRI is noninvasive and uses no ionizing radiation, it does not suffer from the problems found in computer aided tomography (CAT) and positron emission tomography (PET) [6]. It also avoids difficulties from the lack of image clarity and depth of view into the body suffered by ultrasound imaging [7]. Another prominent feature of MRI is the flexibility it provides for selection of imaging planes. Image planes can be electronically rotated to any orientation without moving the object, allowing much greater flexibility than is possible with CAT.

MRI encodes spatially-resolved information about the sample as frequency and phase differences in the magnetic resonance (MR) signals which are in the low radio frequency (rf) part of the electromagnetic spectrum. Although the MR process is actually a quantum mechanical effect, it is possible to understand it using gross physical quantities in a semi-classical way. There are many excellent books and articles discussing the logistics of generating an MR image [8, 9] and of the inherent artifacts that must be overcome [10]. Despite there being variations and subtleties in the form of the MRI data, it is suitable for the purpose of this article to simplify all the approaches as producing a 2D (or 3D) complex data matrix that represents the spatial frequency (k-space) components of the MR image. The two data axes are named the frequency and phase encode directions.

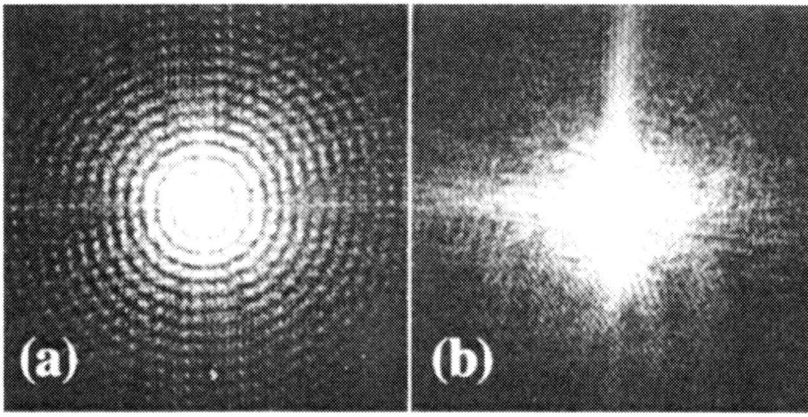

Figure 1: MR raw data "full" data set. (a) medical image phantom, (b) geological image core.

Typical data sets (medical phantom and a geological core) are shown in Figs. 1.a and 1.b respectively, where the phase encoding direction is displayed horizontally for easier conceptualization of the mathematics to be discussed later.

As is explained in the review by Liang *et al.* [4] it is neither theoretically nor experimentally desirably to gather an infinite amount of k-space MRI data. The improved resolution obtained from the longer data record would be obscured by the increased noise content in the data. However a record length of 256 points is frequently long enough that a high resolution low-noise image can be generated by directly applying a 2D inverse discrete Fourier transform (DFT) implimented using the fast Fourier transform (FFT) algorithm [11].

In other situations, time resolution, relaxation effects or other experimental considerations [8] limit one or more matrix dimension. This implied windowing leads to artifacts in any data sequence manipulated with the DFT [12]. For simplicity, we shall consider that only the horizontal (phase encoded) direction is limited in length, a common experimental situation. In this case, it is possible to reconstruct the non-truncated data direction using the DFT without introducing artifacts in that image direction. This is illustrated in Fig. 2

In the petroleum area, the images are frequently noisy because of problems associated with a fast spin-spin relaxation time of the nuclei being imaged. Improving the image SNR requires many signal averages over a

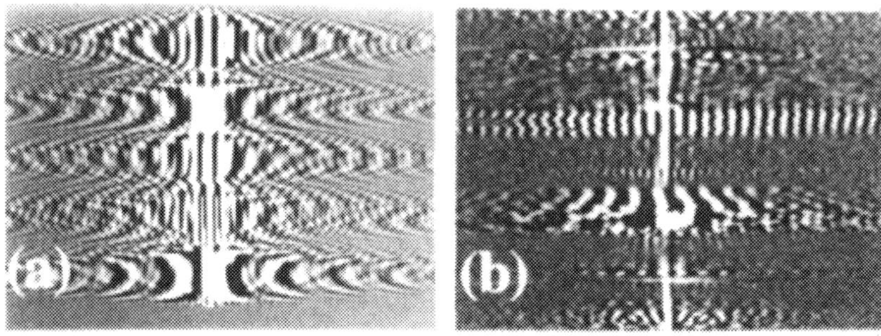

Figure 2: MR raw data set after a 1D inverse DFT in the vertical direction. (a) medical image phantom, (b) geological image core.

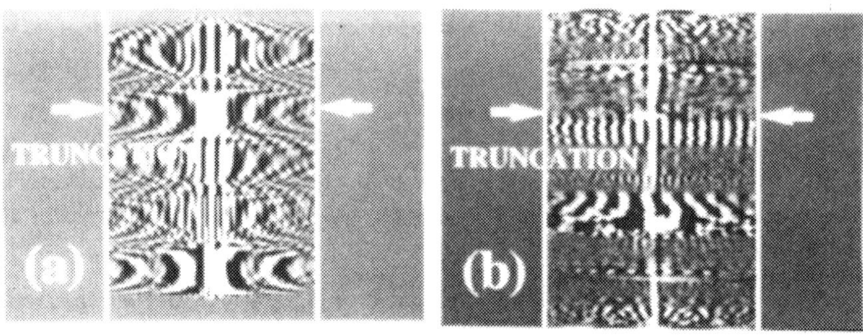

Figure 3: Truncated MR raw data, "truncated" set. (a) medical image phantom, (b) geological image core.

long period of time, sometimes days. As is shown in Fig. 2, the majority of the MR signal is in the low frequency components of the data. By not collecting the high frequency components, but performing further averaging on the lower frequency components, it is possible to improve the image SNR by using truncated data set as shown in Fig. 3.

A similar truncation effect can arise in medical imaging but for a different reason. Here the spin-lattice relaxation time of the imaged nuclei is long and effects how rapidly the data sampling in the phase encoded direction can be repeated. Requirements for fast and/or dynamic imaging again leads to the truncated data sets.

The effect of this data truncation can be seen by comparing Fig. 4 with Fig. 5. Fig. 4 shows the the 256×256 "full" or "standard" inverse DFT reconstructions from the data sets shown in Fig. 2. Fig. 5 is DFT recon-

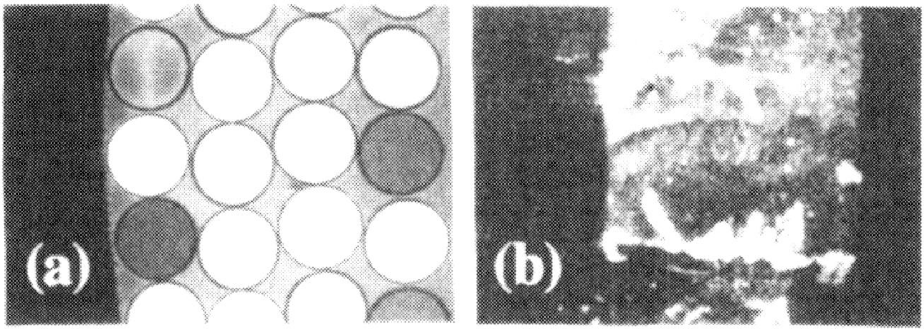

Figure 4: Image reconstructed from the "full" data sets. (a) medical image phantom, (b) geological image core.

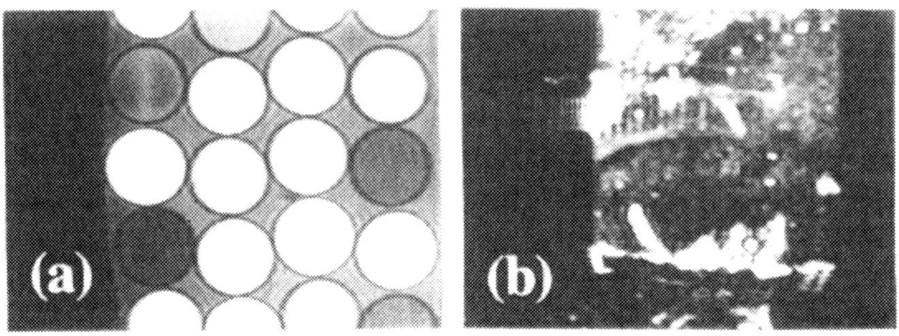

Figure 5: Image reconstructed from the "truncated" data set. (a) medical image phantom, (b) geological image core.

struction from the 256 × 128 "truncated" data sets shown in Figs. 3. The data in Fig. 3 are padded with zeroes before reconstruction to maintain image perspective, producing the ringing artifacts and resolution loss apparent in Fig. 5.

The truncation direction varies with the experimental MR technique used. Gradient echo measurement [13] can give rise to frequency encoded direction truncation. The fast echo planar imaging (EPI) [14] method can give rise to data truncation in both directions. With these approaches, and in many other areas of research, applying the standard inverse DFT method on the truncated data introduces serious artifacts and resolution loss.

2.1 Inherent Problems with the Standard DFT Method

An inverse DFT on a 2D data matrix can be divided into a series of individual 1D DFTs on first the columns and then the rows. It is therefore possible to reduce the MRI reconstruction problem to that of correctly reconstructing a series of 1D data sets.

Consider an experimentally gathered data set $s(k)$. In general, the values of $s(k)$ need to be known for all the spatial frequencies k to reconstruct exactly the image function $c(x)$ from the continuous inverse Fourier transform:

$$c(x) = \int_{-\infty}^{\infty} s(k) e^{j2\pi kx} dk \tag{1}$$

In practice the signal $s(k)$ is assumed to be uniformly sampled at $n\Delta k$, with the sampling interval Δk satisfying the Nyquist criterion to avoid aliasing. However, not all the values of $s[n\Delta k]$ are available. The approximate image function from this truncated data series is given by:

$$c_{DFT}[x] = \Delta k \sum_{n \in N_{data}} s[n\Delta k] e^{j2\pi n\Delta kx} \tag{2}$$

in which the unavailable high frequency components, *i.e.* $s[n\Delta k]$, for $n \notin N_{data}$ are treated as being zeroes.

The measured data $s[\Delta k]$ can be interpreted as part of an infinitely long set contained within a rectangular "window" $w[\Delta k]$. The inverse DFT of a windowed data set is equivalent to the inverse Fourier transform of a hypothetical infinitely long data series, convolved with the inverse Fourier transform of the window function $W[x] = FT^{-1}(w(k))$.

$$c_{DFT}[x] = \int_{-\infty}^{\infty} c[t] W[x-t] dt \tag{3}$$

or

$$c_{DFT}[x] = c[x] \otimes W[x] \tag{4}$$

where $\otimes$ signifies the convolution operation. The amplitude of the measured MR data $s[n\Delta k]$ typically decays at a rate $O(n^3)$ [4]. If the window width is large relative to the size of the data, the effect of the window can be negligible. However for truncated data, this may not be the situation.

The Fourier transform of a rectangular window is a sinc function, which is characterized by a wide centre peak and sidelobes of appreciable amplitude. The width of the centre peak determines the possible image resolution. The sidelobes, which appeared as ringing artifacts in Fig. 5, introduce uncertainty in the discrimination of anatomical detail in the MR images [10].

It is the sharp edge of the rectangular window that creates the ringing artifacts. Other windows may be used which "round off" the corners of the data set, so that the sudden truncation or discontinuity at the boundary is removed. Popular ones are the Hamming, the Blackman-Harris, and the Papoulis's optimal window [12]. All these windows reduce the discontinuity on the data boundary, so as to decrease the height of the (rippling) sidelobes of the transformed data, but (minimally) widen the central lobe. Unfortunately, these smoothing effects remove many of the high frequency components, which further reduce the resolution of the resulting image.

Another limitation of the inverse DFT method lies in its SNR inefficiency. The requirements for improved signal SNR and spatial resolution are mutually exclusive. An improvement in image resolution requires extended k-space sampling, which gathers more noise, but little additional signal, leading to an associated loss of image SNR. These limitations make the inverse DFT method less desirable for any MR applications which require high image resolution, limited data acquisition time and high SNR.

To alleviate these problems, many alternative reconstruction methods have been proposed. Without upgrading the high cost hardware system, these methods provide various means of data post-processing to achieve a better image quality. Furthermore, with the aid of these methods, the data acquisition time can be cut down without a significant image quality degradation. This decrease in data acquisition time can be used for imaging of transient effects (faster imaging) or to be traded for increased SNR by repeated sampling of a short data set.

There are a number of modeling schemes used as alternative MRI reconstruction methods. These include our transient error reconstruction approach (TERA) algorithm [1, 2, 15], the generalized series (GS) model [16] and Sigma method [17, 18]. These algorithms have been tested on clinical images, and a qualitative comparison has been given by Liang *at al.* [4].

This review article extends Liang's work to provide a quantitative comparison of the methods. An image quality comparison is especially hard for the MR image, since we lack a "standard" image. MR images are reconstructed from limited phase and frequency encoded data. The associated sampling window and instrumental errors may make it impossible to obtain a true "standard" image without error artifacts. A quantitative comparison in the frequency domain is suggested as a solution.

Modeling is only appropriate when the model matches the data. This is a major limitation with MRI data which does not present stationary signal characteristics. A number of TERA variants are presented to overcome this problem. These methods include adding pre- and post operative restraints and adaptive characteristics to the algorithm. Applications are made of recursive least square, least square lattice and total least square algorithms to determine the model coefficients. Through these modifications, we attempt to illustrate the potential of applying the TERA algorithm in any area where the standard DFT does not give a satisfactory performance.

This review article is based on an M. Sc. thesis by one of the authors (J.Y.) [19] and contains an expansion of material that has appeared in a number of publications.

3 Review of Transient Error Reconstruction Approach – TERA

Reconstruction from truncated data via the inverse DFT introduces ringing artifacts and resolution loss in the final MR image. The removal of the constraints of the measured data boundary could be achieved by modeling the data, and then using the modeling information to estimate the data beyond the given boundary. If the estimation is adequate, it can significantly reduce the ringing artifacts and possibly improve the resolution in the reconstructed image. The transient error reconstruction approach (TERA), based on an autoregressive moving average (ARMA) model, is one of these modeling algorithms invented for this purpose [1, 2, 15].

3.1 Basics of the TERA Algorithm

The MR data collected using the Fourier imaging technique [11] is usually a 2D complex matrix. After the DFT of this data matrix in the lesser truncated data direction, each row of the data in the truncated direction has the typical form shown Fig. 6. It is evident from the figure that the

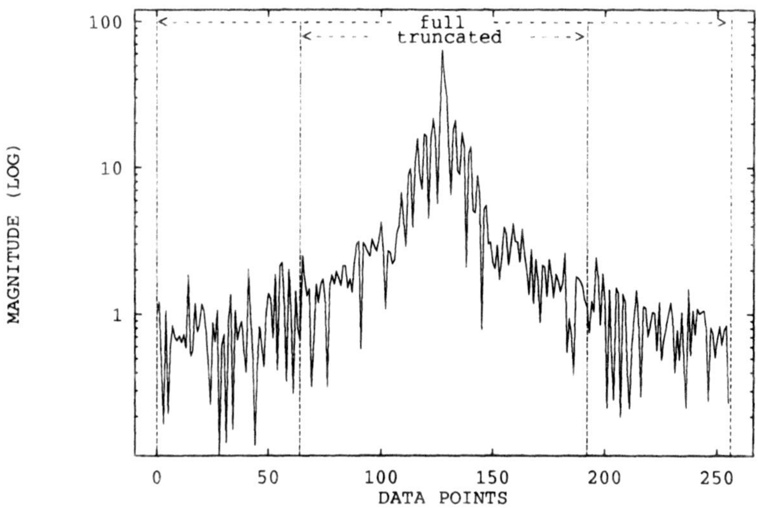

Figure 6: A typical single row from MR data array after the vertical DFT. Both the "full" and "truncated" data show a double-sided decay from the centre point and data truncation at both ends.

"truncated" data set, and to a lesser extent the "full" data set, is truncated at both ends. The TERA modeling algorithm takes a single row $mr[0], mr[1], \cdots, mr[N_{data} - 1]$ of such MR data and models it as follows.

The data $mr[n]$ decays quasi-exponentially from the central point towards both ends. Modeling such a data series is difficult as it has a non-stationary characteristic. Any modeling parameters calculated from the first (increasing) half of data will not fit for the second (decreasing) half of data. The TERA algorithm attempts to solve this problem by decomposing the data array into a Hermitian and an anti-Hermitian sub-array. For mathematical convenience, we relabel the data samples $mr[n]$ as $s[n]$ with the indexes from $-L$ to $L-1$ where $L = N_{data}/2$. The Hermitian sequence $x[n]$ and anti-Hermitian sequence $y[n]$ are defined as:

$$x[n] = (s[n] + s[-n]^*)/2 \quad (5)$$

$$y[n] = (s[n] - s[-n]^*)/2 \quad (6)$$

where $0 \leq n < L$. It is sometimes possible to correct the MR data set for experimentally introduced phase effects so that $y[n]$ is near-zero which can lead to a reduction in image reconstruction time.

The original MR data array is now reformatted into two data series, each following a single decaying train. The TERA algorithm models these two sub-array $x[n]$ and $y[n]$ separately as a subset of the infinite output of an excited infinite impulse response filter (IIR filter):

$$x[n] = -\sum_{i=1}^{p} a_i x[n-i] + \sum_{i=0}^{q} b_i e_{n-i} \tag{7}$$

$$y[n] = -\sum_{i=1}^{p'} a'_i y[n-i] + \sum_{i=0}^{q'} b'_i e'_{n-i} \tag{8}$$

where a_i, a'_i are the AR coefficients, b_i, b'_i are the MA coefficients e_i, e'_i are the excitation functions, p, p', q, q' are the orders of the AR and MA filters.

Considering the $x[n]$ sub-array, Eqn. 7 could be expressed in z-domain as:

$$X[z] = E[z]\frac{B[z]}{A[z]} \tag{9}$$

where $B[z]/A[z]$ is the transfer function of the ARMA filter, while $E[z]$ and $X[z]$ represent the excitation process and final data respectively.

This ARMA filter can be split into two cascaded filters, an AR filter and a MA filter. The MA portion is described by:

$$B[z] = \sum_{i=0}^{q} b_i z^i. \tag{10}$$

The AR portion is depicted as:

$$A[z] = 1 + \sum_{i=0}^{p} a_i z^{-i}. \tag{11}$$

The combined ARMA model has the order of (p, q). In the basic TERA algorithm, the simpliest excitation, e_n, of this ARMA filter is a Kronecker delta function, as shown in Fig. 7. In this case, Eqn. 7 could then be expressed as:

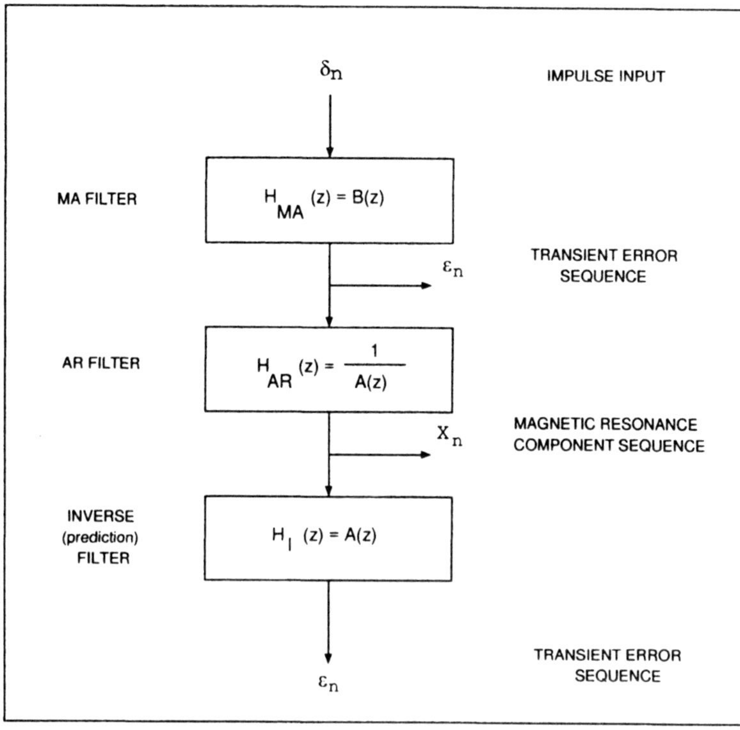

Figure 7: The three basic filter block structures of the original TERA algorithm.

$$x[n] = -\sum_{i=1}^{p} a_i x[n-i] + \sum_{i=0}^{q} b_i \delta[n-i] \tag{12}$$

The excited MA filter produces a response series $\varepsilon[n]$. The MR series, $x[n]$, is modeled as the output of a pth-order AR filter excited by this sequence. In this case, if the AR coefficients can be determined, the application of an inverse AR filter allows $\varepsilon[n]$ to be determined:

$$\varepsilon[n] = x[n] + \sum_{i=1}^{p} a_i x[n-i] \tag{13}$$

From Eqn. 12 and 13, it is clear that the MA coefficients b_i are equivalent to the terms $\varepsilon[n]$, and we end with an ARMA(p, L) model. The error stream $\varepsilon[n]$ can be divided into two sections. The first section, $n < p$, can be thought as being associated with "priming" the AR filter so that the output can follow the MR signal. The final section, $n \geq p$, is equivalent to the prediction error sequence. If the MR data is perfectly predicted by the pth order model, then the error sequence following the pth point will be zero (except the noise). Hence the name "Transient Error".

The infinite impulse response of the total ARMA filter could be calculated using the AR and MA coefficients to explicitely extrapolate the known data set. This was the approach taken in using an earlier TERA variant used to analyse multi-component exponential decays [20]. However such an approach is enherently unstable. This instability is exasperated by the low SNR of the later portions of the MRI signal.

In TERA a different, more efficient and stable approach [1] is to reconstruct the Hermitian component of the image data $c_x[x]$ directly from the transfer function (Eqn. 9) using the relation:

$$\begin{aligned} c_x[x] &= \Delta k \sum_{k=-\infty}^{+\infty} x[n\Delta k] exp(j2\pi n\Delta kx) && (14) \\ &= \Delta k X[z]|_{z=exp(-j2\pi\Delta kx)} && (15) \end{aligned}$$

where:

$$X[z] = \sum_{k=\infty}^{+\infty} x[n\Delta k] z^{-n} \qquad (16)$$

In the model we have:

$$X[z] = B[z] \;/\; A[z] \qquad (17)$$

so that the image function is given by:

$$\begin{aligned} c_x[x] &= \Delta k \frac{\sum_{i=0}^{L-1} b_i z^{-i}}{\sum_{i=0}^{p} a_i z^{-i}} |_{z=exp(-j2\pi kx)} && (18) \\ &= \Delta k \frac{\sum_{i=0}^{L-1} b_i exp(-j2\pi kx)}{\sum_{i=0}^{p} a_i exp(-j2\pi kx)} && (19) \end{aligned}$$

The final image array $c_x[x]$ can be evaluated for any value of x. However by choosing $x = m\Delta x$ and $1/(\Delta x \Delta k) = N_{image} \geq L-1$, $N_{image} = 2^k \times N_{data}$ then:

$$c_x[x] = \Delta k \frac{\sum_{i=0}^{L-1} b_i exp(j2\pi nm/N_{image})}{\sum_{i=0}^{p} a_i exp(-j2\pi nm/N_{image})} \tag{20}$$

can be calculated using the efficient FFT algorithm.

Provided the model is accurate, there are a finite number of AR coefficient and transient error (MA) terms, which can be zero padded to an appropriate length N_{image} (*i.e.* a power of two). The padding with zeroes introduces absolutely no errors since the coefficients following the highest order of (p, q) are supposed to be zero, *if the model is applicable.* The evaluation of the spectral estimate of $x[n]$ using the DFT algorithm in this fashion will therefore not re-introduce undesirable windowing effects into the final image.

The image function for the anti-Hermitian component, $c_y[x]$, can be found from a similar expression. The final MR image can be obtained by recombining the Hermitian and anti-Hermitian image data array, $c_x[x]$ and $c_y[x]$:

$$\begin{aligned} c_{MRI}[m\Delta x] &= [m\Delta x](2Re\{c_x[m\Delta x]\} - \Delta k Re\{s[0]\} \\ &\quad +j(2Im\{c_y[m\Delta x]\} - \Delta k Im\{s[0]\}) \end{aligned} \tag{21}$$

Since the TERA modeling technique successfully estimates the uncollected high frequency components (Fig. 8.a), it reduces the ringing artifacts in the image (Fig. 8.b). It is clear that there is now considerable signal beyond the truncation limit. By comparing this signal with the original datait is possible to quantitatively calculate how well the algorithm performs.

by making an . The most difficult part of the algorithm is the modeling order selection. High order modeling helps in resolution improvement, but it often introduces some undesirable side effects in the final image, such as spikes ("hot spots") (Fig. 9.a). However a low order typically means limited artifacts removal (Fig. 9.b). The data length is one of the major factors in the choosing of the model order. Normally the modeling order should not exceed one third of the given data length [1, 2, 15].

Many linear prediction algorithms can be manipulated to minimize the forward prediction errors, the backward predicting errors or both simul-

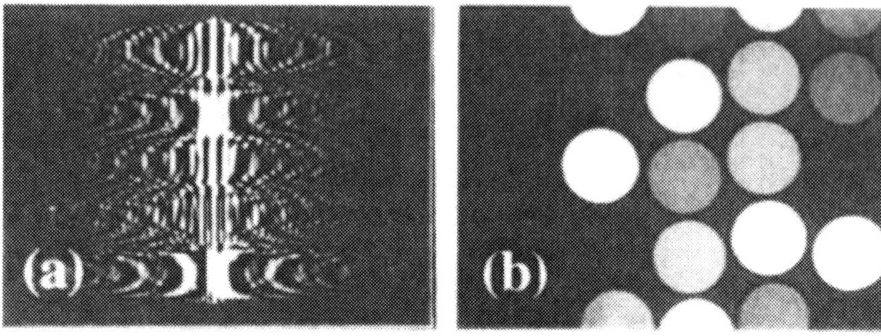

Figure 8: The TERA algorithm (ARMA (15, 64)) makes an estimation of the truncated high frequency components (a) reducing the ringing artifacts and improving the resolution (b).

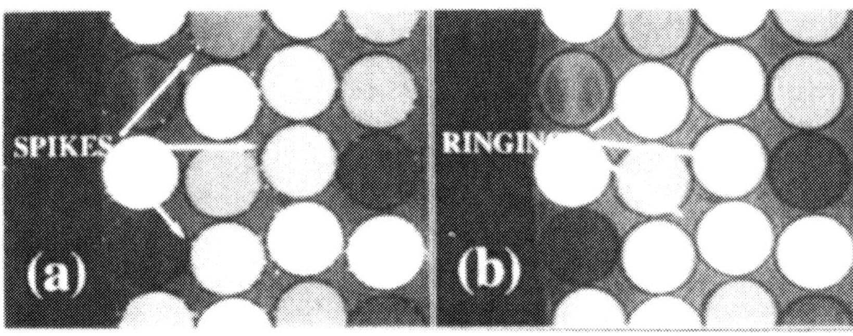

Figure 9: (a) The over-modeled ARMA(25, 64) reconstruction causes spikes in the image. (b) Under-modeling ARMA(5, 64) reintroduces the truncation artifacts.

taneously. To take into account of the decaying characteristic of the MR data, a forward prediction least square was used in the TERA. The simple Burg linear prediction algorithm, which minimizes both the forward and backward prediction errors, was found inappropriate.

3.2 Solutions to the TERA Modeling Errors

Modeling instability is one of the reasons causing the spiking error in the image. It can be decreased by adjusting the location of the poles of the model's transfer function [2]. The zeros of the $A(z)$ transfer function (or the poles of the ARMA transfer function $B(z)/A(z)$) should be moved inwards away from the unit circle, increasing the modeling stability. The adjustment is achieved by multiplying the AR coefficients with a weighting factor, α, which ensures that the poles are inside the unit circle:

$$(a_i)_{new} = (a_i)_{old}\alpha^i \tag{22}$$

where $\alpha < 1.0$. The MA coefficients are calculated from the new AR coefficients to ensure data consistency.

A proper α has to be chosen. If it is too small, all the spikes are removed, but the poles are moved so far from the unit circle that little modeling is achieved. This reintroduces the artifacts and resolution loss. Automatic adjustment of α to the data characteristics is discussed in Ref. [2].

The spikes can be further removed by adding post-operative restraints. For any given resolution improvement, it is theoretically possible to calculated the maximum difference between the modeled and truncated DFT reconstructions[15]. For the spike error, the actual difference value will be abnormally high and the pixel value can be replaced by a suitable value based on the truncated or modeled image. The spike detection threshold balances loss of resolution against spike removal. This technique was named match- it is too small, all the spikes removed, but the improved resolution by modeling is also removed. This "DFT matching" method (MTERA), combined with pole-pulling , gives good reconstruction at a reasonable computational cost. This approach is used as the standard againest which the other algorithms are compared.

3.3 An Alternative TERA Approach Involving Data Modification

The modeling difficulty, which is associated with the double-sided quasi-exponential decaying characteristics of MR raw data, is solved in the TERA algorithm by splitting the data into a Hermitian and an anti-Hermitian sub-arrays. These two sub-arrays each follows a quasi-exponential decaying train, but still have the nonstationary character. In the TERA approach, the sub-array is assumed to be represented by a discrete sum of (decaying) complex exponentials [2]:

$$x[n] = \sum_{i=1}^{p} a_i exp(-(r_i + j2\pi x_i)n\Delta k) \tag{23}$$

The validity of such a model is that it is an extension of Armstrong's method of expanding a transient response in terms of a set of orthonormal exponential function [21]. If the image is real, the model assumes that the contrast function can be represented by a train of delta functions located at x_i convolved with a magnitude squared Lorentzian function whose width is controlled by r_i. In practice, the data asymmetry makes the effective shaping function more complex.

Haacke *et al.* [22]suggested that the image can represented as a series of rectangles so that the data can be modeled as a sum of sinc function modulated complex exponentials:

$$x[n] = \sum_{i=1}^{p} a_i sinc(r_i n\Delta k) exp(-j2\pi x_i n\Delta k) \tag{24}$$

where $sinc(k) = sin(\pi k)/\pi k$. If applied to an image with a broad range of varying intensities, the images become a continuum of rectangles.

Assuming this model to be valid, then multiplying the signal $x[n]$ by $j2\pi n\Delta k$ (ramping the data) gives the modified sequence:

$$x'[n] = j2\pi n\Delta k x[n] \tag{25}$$

$$= \sum_{i=1}^{p+1} a_i exp(j2\pi n\Delta k) \tag{26}$$

which is a sum of (non-decaying) complex exponentials, which is a more stationary signal than the original. However, the original white gaussian noise, v_n, present on the data becomes modified to be proportional to nv_n. This indicates that the noise on the modified sequence is monotonically increasing and nonstationary, which contradicts the white weakly stationary noise assumption inherent in many linear prediction algorithms.

The modified stationary data makes it possible for many algorithms to be used in the AR coefficient determination in the TERA algorithm, such as the Burg forward-backward predicting least square algorithm. A variant of the TERA approach has been used to reconstruct the ramped image data, generating an image. This image is then transformed to the frequency domain, where it is inverse ramped before applying the inverse DFT to get the final image. Many problems beset this approach and are discussed in the paper [23]. The major problem is that the ramped data corresponds to consists entirely of image edge information. If the model

Figure 10: The TERA using the Burg algorithm on the original data set (a). A better extrapolation is obtained using ramped data to generate an edge image (b) which can be super-sampled and transformed back to a normal image (c)

is appropriate, "super resolution" may be achieved making it difficult to sample the edges sufficiently to avoid aliasing when transforming back to the frequency domain. Typically it is necessary to generate the edge image with 4- or 16- fold "super-sampling".

The importance of matching the modeling algorithm againest the data characteristics is clearly brought out in Fig. 10. The Burg algorithm, with its forward/back-ward predicting characteristics, is poorly matched to the original data set. Little useful extrapolation is achieved, Fig. 10.a. Ramping the data, prior to modeling, leads to considerable extrapolation of the k-space data Fig. 10.b. Super-sampling this edge image and transforming it back to k-space before "deramping" allows reconstruction of a standard image with a considerable reduction in artifacts. Comapring the k-space images in Figs. 10.a and .c shows the improved in the extrapolation achieved by the alternative data representation.

3.4 Advantages of the TERA algorithm

The conventional MR DFT reconstruction technique applied to the truncated data causes ringing artifacts and resolution loss in the final image. The TERA algorithm attempts to solve the problem by modeling the data as a subset of the infinitely long output of the ARMA filter excited by a Kronecker delta function.

The reason that TERA is very stable for a wide range of images can be explained by the manner in which the MA coefficients are calculated from the prediction errors [2]. If the model is valid, then these errors are zero.. If the model is not totally appropriate then this approach "re-introduces all data components that can't be modeled" into the MA coefficients. This guarantes that the image will never be worse than using the DFT approach. In addition, this method lessens the problems with model order determination. A moderate over-modeling (too many AR coefficients) will be counter-balanced by a change in the prediction errors, and hence an increased number of MA coefficients. These MA additional coefficients can be shown to cancel out the effect of the additioanal AR coefficients [1].

In addition, calculating the MA coefficients from the prediction errors ensures data consistancy. This permits the AR coefficients to be determined in a variety of ways but still lead to good reconstruction. In addition to the techniques discussed later in this article, it is also possible to average a number of data rows prior to modeling in order to improve the image SNR [24].

4 Iterative Sigma, Generalized Series and Constrained TERA algorithms

In this section we shall discuss some of the other modeling algorithms used to over-come the artifacts introduced by applying the DFT to short length data records. These techniques and TERA differ in their model characteristics but we were able to combine the best parts of both in a constrained TERA (CTERA) algorithm [25]. We shall show that this approach consistently outperforms the other algorithms.

4.1 Iterative Sigma Filter Method

The Sigma filter method, an edge-preserving smoothing technique, was first suggested and investigated by Constable *et. al.* [17] in the use of MR image reconstruction. Unlike the TERA algorithm, Sigma works in the image domain. The generalization of this algorithm to the more realistic complex images by Amartur and Haacke [18] is discussed here.

4.1.1 Theory of the Sigma filter

The Sigma filter is applied repeatly on the image to obtain a maximum ringing artifact reduction but retain image details. The theoretical basis of the modified 2D Sigma filter algorithm operating on the complex image can be described as follows [18].

Let $x(i,j)$ be the complex gray value of $pixel(i,j)$ in an image. The distance between $pixel(i,j)$ and its neighbour $pixel(i+k,j+l)$ within the Sigma filter $K \times L$ mask is given by:

$$d(i,j;k,l) = ||x(i+k,j+l) - x(i,j)|| \tag{27}$$

where $k = -K/2, \cdots, 0, \cdots, K/2$, $l = -L/2, \cdots, 0, \cdots, L/2$.

The Sigma filter smoothed pixel value $y(i,j)$ is given by:

$$y(i,j) = \sum_{k=-K/2}^{k=K/2} \sum_{l=-L/2}^{l=L/2} W(i,j;k,l)x(i+k,j+l) \tag{28}$$

where the weighting coefficients $W(i,j;k,l)$ of the filter are given by:

$$W(i,j;k,l) = F(i,j;k,l) / \sum_{k=-K/2}^{k=K/2} \sum_{l=-L/2}^{l=L/2} F(i,j;k,l) \tag{29}$$

and

$$F(i,j;k,l) = 1/\{1 + [d(i,j;k,l)/v]^{\alpha}\} \tag{30}$$

where v is the homogeneity threshold, a value that represents the minimum variation in the image that qualifies for heavy smoothing or averaging.

When $k, l = 0$. $F(i,j;k,l) = 1$. If the distance $d(i,j;k,l) >> v$, indicating rapid change between $pixel(i+k, j+l)$ and $pixel(i,j)$, then $F(i,j;k,l)$ is close to 0; This means that $pixel(i+k, j+l)$ will not contribute to the average value, therefore the edges are not averaged away. If $d(i,j;k,l) << v$, $F(i,j;k,l)$ approachers 1, and this pixel will contribute in the averaging. The factor α controls the amount of edge enhancement in the filtered image. If α approachs infinity, the filter is reduced to an averaging filter with the image data being evenly smoothed. If α is kept small, the edge information is enhanced with a increased magnitude, while ringing artifacts are reduced. We follow Amartur's implementation [18] and set $\alpha = 3$.

The Sigma filter is a selective averaging filter with a mask moving over the image plane. The larger jumps (edges) in the image are kept from being averaged, while the truncated ringing artifacts are supposed to be averaged. Amartur and Haacke suggested using a 2D window. However, our quantitative analysis indicated that a 1D window, customized simply for each 2D image row gave results that were closer to the "standard" image. In addition, since the algorithm can give "super-resolution" in certain image areas, we increased the sampling rate in the image domain to 4 times that suggested by Armartur to avoid aliasing problems/distortions as the Sigma algorithm switches back and forwards between the frequency and spatial domains.

4.1.2 Iterative Sigma Method Implementation

Assume we have a truncated $N \times M$ MR data set. If we reconstruct the image by padding with zeros to the size of $N' \times M$, $(N' >> N)$ and apply the inverse DFT reconstruction method, the ringing artifacts will be apparent in the image in the direction corresponding to the shorter N data length. The object of the Sigma method is to reduce these ringing artifacts, but preserve the edge details. Our 1D iterative procedures requires 7 steps.

1. Zero-pad the given N-points spatial frequency data to N' points and inverse Fourier transform to obtain a zoomed N'-point image (*i.e.* $N' = 2N$ or $4N$).
2. The initial homogeneity threshold of the sigma filter is set at 20% of the peak contrast value of the local row, corresponding to the maximum amplitude of the Gibbs ringing. The size of the mask used for the 1D Sigma filter is $2[N'/N] + 1$ to cover one period of Gibbs oscillation in the zoomed image.

3. Apply a Sigma filter to the N'-point complex image data (row by row) to smooth the ringing artifacts and noise while maintaining the edge information.

4. Fourier transform the image back to frequency domain where the data will now have the range of $-N'/2, \cdots, N'/2 - 1$. Merge the data samples in the new data set in the range of $-N/2, \cdots, N/2 - 1$ with the original data samples. The data replacement is a constraint imposed with consideration of the fact that the original data is "valid" and should not be discarded.

5. Inverse Fourier transform the new data set to obtain an N'-point image.

6. Decrease the Sigma filter homogeneity threshold to half the previous value. If the threshold value is less than $\sqrt{2}$ times the standard deviation of the image noise, terminate the process. Otherwise repeat step 3 to 6;

7. Display the final $N' \times M$ magnitude image.

Amartur and Haacke [18] set the initial threshold value proportional to the peak contrast value of the whole image. However, a large single edge in the original image produces a threshold too high for most portions of the image. We choose to set a row by row threshold according to the peak contrast value of each row.

The data merge in step 4 is imposed to preserve the original measured data. In order to reduce the possible discontinuity of the Sigma derived data and "original" data, a smoothing is necessary on the boundary [18]. A m-point cosine weighted linear smoothing is applied in the ranges $-q \leq i \leq -q + m - 1$ and $q - m \leq i \leq q - 1$, where $q = N/2$. The smoothed values $y(i)$ are given in terms of the "Sigma" data $x_{sigma}(i)$ and original data $x_{original}(i)$ as:

$$y(i) = (1 - a)x_{original}(i) + ax_{sigma}(i) \qquad ; -q \leq i \leq -q + m - 1 \quad (31)$$

and

$$y(i) = bx_{original}(i) + (1 - b)x_{sigma}(i) \qquad q - m \leq i \leq q - 1 \quad (32)$$

where $a = cos[\pi(q+i+1)/2m]$, $b = cos[\pi(q-m-j)/2m]$. The data continuity is crucial, since any discontinuity here may reintroduce the ringing to the final image. This is why the complex image method [18] works better than the magnitude one [17].

Amartur and Haacke [18] state that the iterative Sigma filter method achieves simultaneous suppression of ringing artifacts and noise. This is not accuracte as reintroducing the data (step 4) reintroduces the noise.

4.2 Generalized Series Theory Model

The generalized series (GS) model was developed by Liang and Lauterbur [16] in their attempt to establish a general mathematical framework for handling *a priori* constraints in the data-consistent way. In this model, an image function is represented as:

$$c[x] = \sum_i a_i \psi_i \left[\vec{\theta}, x\right] \tag{33}$$

where $\psi_i \left[\vec{\theta}, x\right]$ are parameterized basis functions with parameter function $\vec{\theta} = [\theta_1, \theta_2, \cdots, \theta_i]$ adaptively chosen for optimal modeling in a particular application; a_i are the series coefficients chosen to match the measured data. Of particular interest is the class of basis functions proposed by Liang [16], which are formed from the family of weighted complex sinusoids as:

$$\psi_i \left[\vec{\theta}, x\right] = \mathcal{C}\left[\vec{\theta}, x\right] e^{j2\pi i \Delta k x} \tag{34}$$

where $\mathcal{C}\left[\vec{\theta}, x\right]$ is the constraint function, which can take a variety of forms for incorporating available *a priori* information. For a stable model, it is suggested [4] that $\mathcal{C}\left[\vec{\theta}, x\right]$ be a non-negative function. Based on this set of basis functions, the GS model of the image function becomes

$$c[x] = \mathcal{C}\left[\vec{\theta}, x\right] \sum_{i \in N_{data}} a_i e^{j2\pi i \Delta k x} \tag{35}$$

The function is Eqn. 35 gives a flexible framework for incorporating different *a priori* information. It can be characterized [16] as:

1. If there is no *a priori* information available, namely, $\mathcal{C}\left[\vec{\theta}, x\right] = 1$, Eqn. 35 reduces to the Fourier representation.

2. Whenever *a priori* information is available, this GS model incorporates the information into the final reconstruction as described in Eqn. 35.

3. For the extreme case when the *a priori* function is equal to the desired image, $\mathcal{C}\left[\vec{\theta}, x\right] = c[x]$, the multiplicative Fourier series factor will be forced to unity to give an exact reconstruction.

Various available *a priori* information could be used by this GS model. In general, the basis function enables the GS model to converge faster than the Fourier series model (a_i decay faster than the original measured data) and, therefore, within a certain error bound, one can use far fewer terms to represent the spatial distributions of image information than are required by the Fourier series method, leading to a reduction of the truncation artifacts.

The GS model can use the *a priori* information of edge enhancement and artifact reduction of the Sigma filter results in the final reconstruction. More important, if the Sigma method is not appropriate or optimal, the GS model will automatically compensate by enforcing the data consistency constraints.

In the current implementation, we assume the magnitude Sigma filtered N'-points image data $x_{sig}[-N'/2], \cdots, x_{sig}[N'/2 - 1]$, as our constraint function $\mathcal{C}\left[\vec{\theta}, x\right]$. Doing a Fourier transform on both sides of Eqn. 35, we have a convolution in frequency domain given by:

$$\vec{s} = \vec{s_o} \otimes \vec{a} \tag{36}$$

where $\vec{s_o}$ is the Fourier transform of the constrained function $\mathcal{C}\left[\vec{\theta}, x\right]$, and $\vec{s}$ is the available measured data.

In contrast to the usual linear prediction formalism discussed earlier, Eqn. 36 implies an exactly matching. We must solve the set of linear equations:

$$\vec{H}\vec{a} = \vec{s} \tag{37}$$

with

$$\vec{H} = \begin{bmatrix} s_o[0] & s_o[-1] & \cdots & s_o[-N_{data}/2] \\ s_o[1] & s_o[0] & \cdots & s_0[-N_{data}/2+1] \\ \vdots & \vdots & \ddots & \vdots \\ s_o[N_{data}/2-1] & s_o[N_{data}/2-2] & \cdots & s_o[0] \end{bmatrix} \quad (38)$$

Since $\vec{H}$ is a Toelptz matrix, the linear equation can be solved efficiently.

After $\vec{a}$ is available, we can apply the DFT on the zero-padded $\vec{a}$ to the same digital resolution as $\mathcal{C}\left[\vec{\theta}, x\right]$. The image data can be then be obtained through Eqn. 35. The GS model is very flexible and effective for handling *a priori* information. The modeling data is always consistent with the original measured data.

Liang [16] states that stability is ensuredif the $\mathcal{C}\left[\vec{\theta}, x\right]$ was chosen be a non-negative function [16]. Our quantitative analysis has verified that this is true.

4.3 The Constrained TERA Model - CTERA

The Sigma method smooths the ringing artifacts while trying to preserve the edge in the MR image. This implies that in the frequency domain, it reintroduces some missing high frequency components. Fig. 11 shows the mean square error in the frequency domain between the Sigma and MTERA "modeled" data and "original" data on a row basis (for more information of the error analysis see Section 6). The Sigma and MTERA algorithm seem to work better in different parts of the data, suggesting they are modeling different data characteristics. A combination method that introduces *a priori* constraints about the edges from the Sigma model into the MTERA lagorithm might be superior to either [25]. This is the concept behind CTERA, a constrainedTERA model.

In order to combine the approaches, we start off by zero padding the truncated data set $N \times M$, reconstruct the image using standard inverse DFT to the size of $N' \times M$, $N' > N$, $(N' = 2N$ or $4N)$. The iterative Sigma filter method (1 to 3 iterations) is applied on the image. Transferring the image data back to the frequency domain, this extended data (without the merging with the original data) is used instead of the raw data as the input of the MTERA algorithm to determine the AR coefficients. The MA coefficients are determined from the original data set to ensure the data consistency constraint.

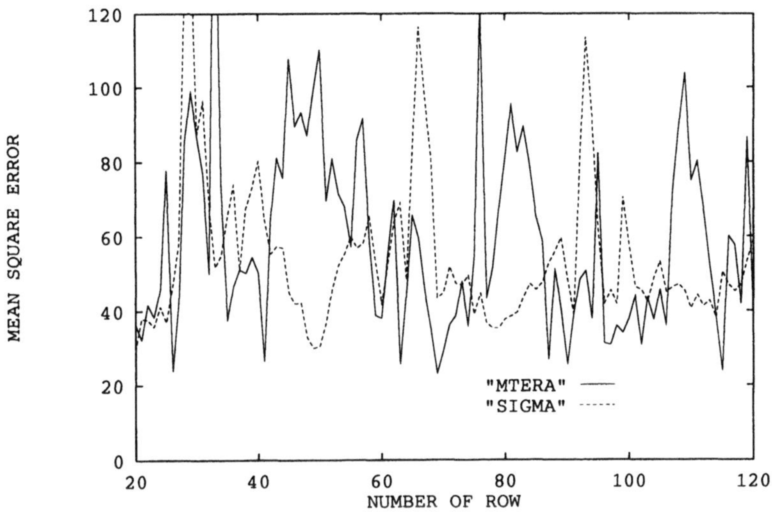

Figure 11: The mean square error between the MTERA and Sigma "modeled" data and original data in the frequency domain show these algorithms model different portions of the data.

The frequency data, transformed from the Sigma filtered image, has non-zero components beyond the original data limits. These data resulted from the edge sharpening and ringing reduction of the Sigma filter method. The increased data length allows the use of higher order modeling, which should give better representation, but with an increased computation load. In addition the AR coefficients should contain edge information and are naturally damped when determined from the long-length Sigma frequency data. This damping should ensure increased algorithm stability without heavy pole pulling.

5 An Adaptive TERA Algorithm

The TERA algorithm builds a model from the known MR data set and uses the modeling information to perform an implicit infinite extrapolation. The modeling difficulty with double-sided decaying MR signal was partially solved by independently fitting both the Hermitian and anti-Hermitian data components. By minimizing only the forward prediction errors, some attempt is give the decaying experimental data the "stationary" format required by the modeling algorithms. Further compensations are made with the MTERA variant by "pole-pulling" and "DFT-matching".

A better way to match the data characteristics is to choose an algorithm other than the forward predicting least square method, one that is suited for the non-stationary or possibly low SNR MR data. For his thesis [19] Yang developed an adaptive TERA variant to deal with non-stationary date aspects. It is implemented using both the recursive least square (RLS) and least square lattice (LSL) algorithms. In an attempt to fit high noise MR data, a total least square (TLS) implementation [26] was evaluated.

5.1 Theory behind the adaptive TERA algorithm

A forward prediction error minimizing least square algorithm [27] is employed to calculate the AR coefficients a_i in the current TERA algorithm. Given a time series $x[0], x[1], \cdots, x[n]$, $n < N_{data}$, we assume the $x[n]$ can be estimated by $\hat{x}[n]$:

$$\hat{x}[n] = -\vec{x}_p^T[n]\vec{a}^f_{p,N_{data}} \tag{39}$$

where p is the prediction order, $\vec{x_p}[n]$ is the input data vector and $\vec{a}^f_{p,N_{data}}$ is the prediction coefficient vector determined from the N_{data} points:

$$\vec{x}_p[n] = \begin{bmatrix} x[n] \\ x[n-1] \\ \vdots \\ x[n-p] \end{bmatrix}, \qquad \vec{a}^f_{p,N_{data}} = \begin{bmatrix} 1 \\ a_p[1] \\ \vdots \\ a_p[p] \end{bmatrix}. \tag{40}$$

The $\vec{a}_{p,N_{data}}$ expressed in Eqn. 39 in the linear prediction system are usually overdetermined, since in practice the number of data points N_{data} normally exceeds the model order p, *i.e.* $N_{data} >> p$. In this case, the parameters a_i do not uniquely exist. A least square method can be used to calculate a meaningful unique solution.

If we define a residual:

$$\varepsilon[n] = x[n] - \hat{x}[n], \tag{41}$$

Eqn. 39 can be rewritten as:

$$x[n] = \vec{x}_p^T[n]\vec{a}^f_{p,N_{data}} + \varepsilon[n] \tag{42}$$

This is a one step forward prediction system. The residual $\varepsilon[n]$ is called the prediction error. The prediction coefficients, $\vec{a}_{p,N_{data}}$, can be estimated by minimizing the sum of error squares $\sum_{n=p}^{N_{data}-1} |\varepsilon[n]|^2$ based on the least square criterion.

The current least square method makes an estimate of the AR coefficients based on the whole data set. It would be better to extend the least square method to adapt to the non-stationary signal. This can be achieved by assuming that the Hermitian or anti-Hermitian components are the outputs of real systems with a (slow) time varying behavior. As each new data sample is received, we update the parameter estimates to adapt to the changes in the data properties.

We implemented this apparoach to work with RLS and LSL algorithms [28] using only the forward prediction errors to better match the MRI data characteristics.

5.1.1 Recursive Least Square Method

The following analysis is based on that of Marple's [29]. For the available data series up to time t, $x[0], x[1], \cdots, x[N_t - 1]$, the pth forward prediction errors may be defined as:

$$\varepsilon^f_{p,N_t}[n] = \vec{x}_p^T[n]\vec{a}^f_{p,N_t} \tag{43}$$

where the input t vector $\vec{x}_p[n]$, forward linear prediction coefficient vector $\vec{a}^f_{p,N_t}$ has the definition:

$$\vec{x}_p[n] = \begin{bmatrix} x[n] \\ x[n-1] \\ \vdots \\ x[n-p] \end{bmatrix}, \qquad \vec{a}^f_{p,N_t} = \begin{bmatrix} 1 \\ a^f_{p,N_t}[1] \\ \vdots \\ a^f_{p,N_t}[p] \end{bmatrix}, \tag{44}$$

The sets of $\vec{a}^f_{p,N_t}$ can be determined by minimizing the forward exponentially weighted squared error, ρ^f_{p,N_t} as:

$$\rho^f_{p,N_t} = \sum_{n=0}^{N_t-1} \lambda^{N_t-n} |\varepsilon^f_{p,N_t}[n]|^2 \tag{45}$$

where the scalar λ $(0 < \lambda \leq 1)$ is called the forgetting factor.

The forgetting factor, λ, is introduced into the performance to ensure the current error $\varepsilon^f_{p,N_t}[N_t]$ has the least reduction (greatest weight), and the error in the distant past has the most reduction. Thus λ allows the prediction to better follow the statistical variation of the observable signal when the algorithm is operating in a nonstationary environment.

The forward prediction errors $\varepsilon^f_{p,N_t}[n]$ is defined over the range from $n = 0$ to $n = N_t - 1$ in Eqn. 43. N_t is the last data point introduced into the analysis. Here we have $x[n] = 0$ for $n < 0$, by definition of the Hermitian and anti-Hermitian components (Eqns. 5, 6).

The linear prediction/autoregressive coefficients that minimize ρ^f_{p,N_t} satisfy the $(p+1) \times (p+1)$ matrix normal equations

$$\vec{\Phi}_{p,N_t} \vec{a}^f_{p,N_t} = \begin{bmatrix} \rho^f_{p,N_t} \\ \vec{0}_p \end{bmatrix} \tag{46}$$

where

$$\vec{\Phi}_{p,N_t} = \sum_{n=1}^{N_t} \lambda^{N_t-n} \vec{x}^*_p[n] \vec{x}^H_p[n] \tag{47}$$

and $\vec{0}_p$ is an all-zero $p \times 1$ vector.

The vector $\vec{\Phi}_{p,N_t}$ is updated as every new sample, $x[N_t+1]$, is received. The new set of $\vec{a}^f_{p,N_t+1}$ and $\vec{a}^b_{p,N_t+1}$ are generated with the least square fit to the entire set of $N_t + 1$ points.

The traditional RLS algorithm [30] requires an exact least square recursive solution for the linear prediction coefficients as each tth sample is made available. This algorithm is very time consuming and requires a number of computations proportional to p^2 per time update. Updates from a_{p,N_t} to a_{p,N_t+1} could be avoided to explicitly solving the normal equation by a recursive least square procedure. A fast RLS algorithm used here was developed by Ljung [30] with a procedure that only requires computations of $O(p)$.

1. Time Update

The time-index update of the forward linear prediction coefficient vector for N_t to $N_t + 1$ points is given by:

$$\vec{a}^f_{p,N_t+1} = \vec{a}^f_{p,N_t} - \varepsilon^f_{p,N_t+1}[N_t + 1] \begin{bmatrix} 0 \\ \vec{c}_{p-1,N_t} \end{bmatrix} \tag{48}$$

where $\vec{c}_{p-1,N_t}$ is defined as:

$$\vec{\Phi}_{p-1,N_t}\vec{c}_{p-1,N_t} = \vec{x}_{p-1}[N_t] \tag{49}$$

The time update of the real forward linear prediction square error is expressed as:

$$\rho^f_{p,N_t+1} = \lambda\rho^f_{p,N_t} + \varepsilon^f_{p,N_t+1}[N_t + 1]\varepsilon^{f*}_{p,N_t}[N_t + 1] \tag{50}$$

The time update of the forward linear error is given by:

$$\varepsilon_{p,N_t+1}[N_t + 1] = \varepsilon_{p,N_t}[N_t + 1]/\gamma_{p-1,N_t} \tag{51}$$

where, γ_{p-1,N_t} is a real, positive scalar

$$\gamma_{p-1,N_t} = 1 + \vec{x}^T_{p-1}[N_t]\vec{c}_{p-1,N_t} \tag{52}$$

2. Order Update

The order update recursions for $\vec{c}_{p,N_t+1}$

$$\vec{c}_{p,N_t+1} = \begin{bmatrix} 0 \\ \vec{c}_{p-1,N_t} \end{bmatrix} + (\lambda\rho^f_{p,N_t})^{-1}\varepsilon^{f*}_{p,N_t}[N_t + 1]\vec{a}^f_{p,N_t} \tag{53}$$

The scalar γ_{p,N_t+1} order update is

$$\gamma_{p,N_t+1} = \gamma_{p-1,N_t} + |\varepsilon^b_{p,N_t}[N_t + 1]|^2/(\lambda\rho^f_{p,N_t}) \tag{54}$$

3. Initialization

The steady-state fast RLS algorithm for a fixed-order p filter applies only when $N_t > p$; otherwise, the least squares normal equation is underdetermined. During the interval $1 \leq N_t \leq p$, *simultaneous order and time updating* can be developed to provide a $(N_t - 1)$-order exact least squares solution to the available N_t data samples. With the arrival of time sample $x[N_t + 1]$, a switch from the initialization *order* and *time* updation procedure to the steady-state *fixed-order time-update* RLS algorithm is made in order to continue to the exact RLS solution for $N_t > p$.

Eqns. 48, 50, 51, 53 and 54 together in sequence comprise the fast RLS algorithm.

However this fast algorithm appears to have poor long term numerical stability [31]. This is due to error accumulation from finite precision computations. The instability increase with MR signals which decay in amplitude so that new points have a monotonically decreasing SNR. Several alternative solutions have been presented [32, 33] to overcome this problem. Among them the LSL algorithm is the most prominent one because its structure is successfull for conquering finite-precision errors [34].

5.1.2 Least Square Lattice Method

The history of LSL algorithms can be traced back to the pioneering work of Morf on efficient solutions for least squares predictors with a non-Toeplitz input correlation matrix [34]. The formal derivation of LSL algorithms was first presented by Morf, Vieira, and Lee [35].

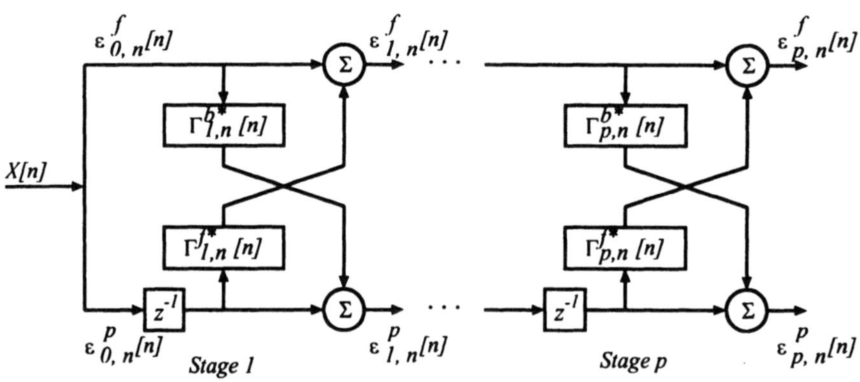

Figure 12: Multi-stage p^{th} least square lattice predictor.

The least square lattice algorithm derived by Morf [34] is composed of two main parts: order-update and time-update recursions. It has a structure shown as Fig. 12, where $x[n]$ is the input sequence, $\varepsilon^f_{p,n}[n]$ and $\varepsilon^b_{p,n}[n]$ are the forward and backward *a posteriori* prediction errors. $\Gamma^f_{p,n}[n]$ and

$\Gamma^b_{p,n}[n]$ are the forward and backward reflection coefficients, respectively.

Some Preliminaries

Before giving all the order and time update algorithms, we define both the forward and backward error predictor of order p.

The forward *a posteriori* prediction error is expressed as:

$$\varepsilon^f_{p,N_t}[n] = \vec{a}^{fH}_{p,N_t}\vec{x}_{p+1}[n] \tag{55}$$

where $0 \leq n < N_t$. $\vec{x}_{p+1}[n]$ denotes the $(p+1) \times 1$ input vector, of the filter measured at time N_t, $[x[n], x[n], \cdots, x[n-p-1]]^T$. Again for the Hermitian and anti-Hermitian array, we have $x[n] = 0$ for $n \leq 0$. the vector $\vec{a}^f_{p,N_t}$ representing the forward AR coefficients have the first element equal to unity, $1, a^f_{p,N_t}[1], \cdots, a^f_{p,N_t}[p]$. The sum of weighted forward *a posteriori* prediction-error squares equals

$$\rho^f_{p,N_t} = \sum_{n=0}^{N_t-1} \lambda^{N_t-n}|\varepsilon^f_{p,N_t}[n]|^2 \tag{56}$$

where λ is the exponential weighting factor. The pth order backward prediction *a posteriori* error is expressed as

$$\varepsilon^b_{p,N_t}[n] = \vec{a}^b_{p,N_t}[n]^H\vec{x}_{p+1}[n], \qquad 0 \leq n < N_t \tag{57}$$

where the backward prediction coefficient denoted by vector $\vec{a}^b_{p,N_t}$, and its last element equal to unity, $a^b_{p,N_t}[p], a^b_{p,N_t}[p-1], \cdots, 1$. The sum of weighted backward *a posteriori* prediction error squares equals:

$$\rho_{p,N_t}[n] = \sum_{n=0}^{N_t-1} \lambda^{N_t-n}|\varepsilon_{p,N_t}[n]|^2 \tag{58}$$

The forward prediction coefficients $\vec{a}^f_{p,N_t}$ could be determined by minimizing the sum of weighted forward *a posteriori* prediction error squares ρ^f_{p,N_t}. Let Φ_{p+1,N_t} to be $(p+1) \times (p+1)$ deterministic autocorrelation matrix of the input vector $\vec{x}_{p+1}[n]$ applied to the forward prediction error filter of order p, where $0 \leq n < N_t$. The augmented normal equation for this filter is :

$$\vec{\Phi}_{p+1,N_t}\vec{a}^f_{p,N_t} = \begin{bmatrix} \rho^f_{p,N_t} \\ \vec{0}_p \end{bmatrix} \tag{59}$$

where $\vec{0}_p$ is the $p \times 1$ null vector.

Similarly, we get the normal equation for backward prediction error filter as:

$$\vec{\Phi}_{p+1,N_t}\vec{a}^b_{p,N_t} = \begin{bmatrix} \vec{0}_p \\ \rho^b_{p,N_t} \end{bmatrix} \tag{60}$$

Order Update Recursions

By partitioning matrix Φ_{p+1,N_t} and with some simplification [28], we deduce order-update recursion equations for both forward and backward prediction error filter as:

$$\vec{a}^f_{p,N_t} = \begin{bmatrix} \vec{a}^f_{p-1,N_t} \\ 0 \end{bmatrix} - \frac{\Delta_{p-1,N_t}[N_t]}{\rho^b_{p-1,N_t-1}} \begin{bmatrix} 0 \\ \vec{a}^b_{p-1,N_t-1} \end{bmatrix} \tag{61}$$

$$\rho^f_{p,N_t} = \rho^f_{p-1,N_t} - \frac{\Delta_{p-1,N_t}[N_t]\nabla_{p-1,N_t}[N_t]}{\rho^b_{p-1,N_t-1}} \tag{62}$$

$$\vec{a}^b_{p,N_t} = \begin{bmatrix} 0 \\ \vec{a}^b_{p-1,N_t-1} \end{bmatrix} - \frac{\nabla_{p-1,N_t}[n]}{\rho^f_{p-1,N_t}} \begin{bmatrix} \vec{a}^f_{p-1,N_t} \\ 0 \end{bmatrix} \tag{63}$$

$$\rho^b_{p,N_t} = \rho_{p-1,N_t-1} - \frac{\Delta_{p-1,N_t}[N_t]\nabla_{p-1,N_t}[N_t]}{\rho^f_{p-1,N_t}} \tag{64}$$

where

$$\Delta_{p-1,N_t}[N_t] = [\sum_{n=1}^{N_t} \lambda^{N_t-n}\vec{x}_p[n]x^*[n-p]]^H \vec{a}^f_{p-1,N_t} \tag{65}$$

$$\nabla_{p-1,N_t}[N_t] = [\sum_{n=1} N_t\lambda^{N_t-n}\vec{x}_p[n-1]x^*[n]]^H \vec{a}^b_{p-1,N_t-1} \tag{66}$$

$\Delta_{p-1,N_t}[N_t]$ and $\nabla_{p-1,N_t}[N_t]$ have the following relationship:

$$\nabla_{p-1,N_t}[N_t] = \Delta^*_{p-1,N_t}[N_t] \tag{67}$$

Using Eqn. 67, we can rewrite the sum of weighted forward and backward prediction error square ρ^f_{p,N_t}, ρ^b_{p,N_t} as following

$$\rho^f_{p,N_t} = \rho^f_{p-1,N_t} - \frac{|\Delta_{p-1,N_t}[N_t]|^2}{\rho^b_{p-1,N_t-1}} \tag{68}$$

$$\rho^b_{p,N_t} = \rho^b_{p-1,N_t-1} - \frac{|\Delta_{p-1,N_t}[N_t]|^2}{\rho^f_{p-1,N_t}} \tag{69}$$

Defining the forward, backward reflection coefficients as

$$\Gamma^f_{p,N_t}[N_t] = -\frac{\Delta_{p-1,N_t}[N_t]}{\rho^b_{p-1,N_t-1}[N_t-1]} \tag{70}$$

$$\Gamma^b_{p,N_t}[N_t] = -\frac{\nabla_{p-1,N_t}[N_t]}{\rho^b_{p-1,N_t}]} \tag{71}$$

$$= -\frac{\Delta^*_{p-1,N_t}[N_t]}{\rho^f_{p-1,N_t}} \tag{72}$$

eqns. 61 and Eqn. 63 can be rewritten as

$$\vec{a}^f_{p,N_t} = \begin{bmatrix} \vec{a}^f_{p-1,N_t} \\ 0 \end{bmatrix} + \Gamma^f_{p,N_t}[N_t] \begin{bmatrix} 0 \\ \vec{a}^b_{p-1,N_t-1} \end{bmatrix} \tag{73}$$

$$\vec{a}^b_{p,N_t} = \begin{bmatrix} 0 \\ \vec{a}^b_{p-1,N_t-1} \end{bmatrix} + \Gamma^b_{p,N_t}[N] \begin{bmatrix} \vec{a}^f_{p-1,N_t} \\ 0 \end{bmatrix} \tag{74}$$

Another two order-update recursions equations are the forward and backward *a posteriori* prediction errors. They can now be expressed as following:

$$\varepsilon^f_{p,N_t}[N_t] = \varepsilon^f_{p-1,N_t}[N_t] + \Gamma^{f*}_{p,N_t}[N_t]\varepsilon^b_{p-1,N_t-1}[N_t-1] \tag{75}$$

$$\varepsilon^b_{p,N_t}[N_t] = \varepsilon^b_{p-1,N_t-1}[N_t-1] + \Gamma^{b*}_{p,N_t-1}[N_t]\varepsilon_{p-1,N_t}[N_t] \tag{76}$$

There is another parameter in this LSL algorithm required for order-update, it is called the conversion factor $\gamma_{p,N_t-1}[N_t-1]$. The order-update recursion for it is defined as:

$$\gamma_{p,N_t}[N_t] = \gamma_{p-1,N_t}[N_t] - \frac{|\varepsilon^b_{p-1,N_t}[N_t]|^2}{\rho^b_{p-1,N_t}} \tag{77}$$

The role of $\gamma_{p,N-1}[N_t - 1]$ will become apparent in the context of the least-square lattice predictor later in the time update recursion.

Time Update Recursions

To make all above order-update recursions adaptive in time, it is necessary to have a time-update recursion for the parameter $\Delta_{p-1}(N)$

$$\begin{aligned}\Delta_{p-1,N_t}[N_t] &= \lambda\Delta_{p-1,N_t-1}[N_t-1] + \\ &\quad \gamma_{p-1,N_t-1}[N_t-1]\varepsilon^b_{p-1,N_t-1}[N_t-1]\varepsilon^{f*}_{p-1,N_t}[N_t]\end{aligned} \tag{78}$$

where the conversion factor $\gamma_{p-1,N_t-1}[N_t-1]$ plays as the correction term in the time update of Eqn. 78. It enables the LSL algorithm to adapt rapidly to sudden changes in the input t [36]. After using time-update recursion for $\Delta_{p-1,N_t}(N_t)$, we can get the time-update recursion for reflection coefficients as

$$\begin{aligned}\Gamma^f_{p,N_t}[N_t] &= \Gamma^f_{p,N_t-1}[N_t-1] - \\ &\quad \gamma_{p-1,N_t-1}[N_t-1]\frac{\varepsilon^b_{p-1,N_t-1}[N_t-1]\varepsilon^{f*}_{p,N_t}[N_t]}{\rho^b_{p-1,N_t-1}}\end{aligned} \tag{79}$$

$$\begin{aligned}\Gamma^b_{p,N_t}[N_t] &= \Gamma^b_{p,N_t-1}[N_t-1] - \\ &\quad \gamma_{p-1,N_t-1}[N_t-1]\frac{\varepsilon^f_{p-1,N_t}[N_t]\varepsilon^{b*}_{p,N_t}[N_t]}{\rho^f_{p-1,N_t}}\end{aligned} \tag{80}$$

In a similar way the sum of weighted forward and backward prediction-error may be updated as follows

$$\rho^f_{p-1,N_t} = \lambda\rho^f_{p-1,N_t-1} + \gamma_{p-1,N_t-1}[N_t-1]|\varepsilon^f_{p-1,N_t}[N_t]|^2 \tag{81}$$

$$\rho^b_{p-1,N_t} = \lambda\rho^b_{p-1,N_t-1} + \gamma_{p-1,N_t}[N_t]|\varepsilon^b_{p-1,N_t}[N_t]|^2 \tag{82}$$

Equation 76, 76, 78, 79, 80, 81 and 82 together in order constitute the least-square lattice predictor algorithm.

Initialization

The steady-state LSL algorithm for a fixed-order p filter applies only when $N_t > p$. During the interval $1 \leq N_t \leq p$, *simultaneous order and time updating* can be developed to provide a $(N_t - 1)$-order exact least squares solution to the available N_t data samples. With the arrival of time sample $x[N_t+1]$, a switch from the initialization *order* and *time* updating procedure to the steady-state *fixed-order time-update* LSL algorithm is made in order to continue to the exact LSL solution for $N_t > p$.

A prominent property of lattice is that the stability checking for the algorithm is very easy. The stability criteria can be deduced as follows: According to Eqn. 62, the sum of the weighted square forward prediction-error is

$$\rho^f_{p,N_t} = \sum_{n=1}^{N_t} \lambda^{N_t-n} |\varepsilon^f_{p,N_t}[N_t]|^2 \tag{83}$$

Since $0 < \lambda \leq 1$, then ρ^f_{p,N_t} is always greater than 0. From Eqn. 68 we have

$$\rho^f_{p,N_t} = \rho^f_{p-1,N_t} - \frac{|\Delta_{p-1,N_t}[N_t]|^2}{\rho^b_{p-1,N_t-1}} \tag{84}$$

Substituting Eqn. 70 into Eqn. 84 we get

$$\rho^f_{p,N_t} = \rho^f_{p-1,N_t} + \nabla_{p-1,N_t}[N_t]\Gamma^f_{p,N_t}[N_t] \tag{85}$$

From Eqn. 72 we have:

$$\nabla_{p-1,N_t}[N_t] = -\rho^f_{p-1,N_t}\Gamma^b_{p,N_t}[N_t] \tag{86}$$

Combining Eqn. 85 and Eqn. 86, we obtain

$$\rho^f_{p,N_t} = \rho^f_{p-1,N_t} - \rho^f_{p-1,N_t}\Gamma^b_{p,N_t}[N_t]\Gamma^f_{p,N_t}[N_t] \tag{87}$$

$$= \rho^f_{p-1,N_t}[1 - \Gamma^b_{p,N_t}[N_t]\Gamma^f_{p,N_t}[N_t]] \tag{88}$$

From Eqn. 88, since both ρ^f_{p,N_t} and $\rho^f_{p-1,N_t} > 0$:

$$1 - \Gamma^b_{p,N_t}[N_t]\Gamma^f_{p,N_t}[N_t] > 0 \tag{89}$$

The above inequality always holds if the prediction is stable [28].

Determining the AR coefficients

Due to the basic structural difference between the RLS and LSL algorithms, these two algorithms present the relevant information in different ways. The RLS algorithm presents information about the input data in the form of instantaneous value of transversal filter coefficients. By contrast, the LSL algorithm provides the information in the form of a corresponding set of reflection coefficients. In the AR modeling case, we need to know the AR coefficients to identify a process. Since LSL has many nice properties, we usually choose it to calculate all the reflection coefficients, checking the stability and then converting to the needed AR coefficients using the relation found in Levinson algorithm [28]. The conversion relationship between reflection coefficient and AR coefficients is deduced as follows:

From Eqs. 81 and 82 we have:

$$\varepsilon^f_{p,N_t} = \varepsilon^f_{p-1,N_t} + \Gamma^{f*}_{p,N_t}[N_t]\varepsilon^b_{p-1,N_t-1}[N_t-1] \tag{90}$$

$$\varepsilon^b_{p,N_t} = \varepsilon^b_{p-1,N_t-1} + \Gamma^{b*}_{p,N_t}[N_t]\varepsilon^f_{p-1,N_t}[N_t] \tag{91}$$

In the z-domain, these equation can be expressed as

$$\varepsilon^f_{p,N_t}(z) = \varepsilon^f_{p-1,N_t}(z) + z^{-1}\Gamma^{f*}_{p,N_t}\varepsilon^b_{p-1,N_t}(z) \tag{92}$$

$$\varepsilon^b_{p,N_t}(z) = z^{-1}\varepsilon^b_{p-1,N_t}(z) + \Gamma^{b*}_{p,N_t}\varepsilon_{p-1,N_t}(z) \tag{93}$$

The equivalent forward and backward linear traversal filter can be expressed as

$$\varepsilon^f_{p,N_t}(z) = (1 + a^f_{1,p}z^{-1} + \cdots + a^f_{p,p}z^{-p})x(z) \tag{94}$$

$$\varepsilon^b_{p,N_t}(z) = [z^{-p}(1 + a^b_{1,p}z + \cdots + a^b_{p,p})]x(z) \tag{95}$$

where

$$a^f_{p,p} = \Gamma^{*f}_{p,N_t} \tag{96}$$

$$a^b_{p,p} = \Gamma^{*b}_{p,N_t} \tag{97}$$

By using the Levinson recursion, we determine the AR coefficients as:

$$\begin{aligned} a^f_{j,p} &= a^f_{j,p-1} + \Gamma^*_{f,p} a^b_{p-j,p-1} & (98) \\ a^b_{j,p} &= a^b_{j,p-1} + \Gamma^*_{f,p} a^f_{p-j,p-1} & (99) \end{aligned}$$

LSL has the same adaptive mechanism as the RLS algorithm and assumes that the MRI data are the output of a real-time system. As every new sample, $x[N_t + 1]$ is received, the new set of Γ^f_{p,N_t+1} and Γ^b_{p,N_t+1} are generated with a fit to the entire set of $N_t + 1$ points. The forgetting factor λ is introduced into the performance which ensures the error at present has the least reduction and error in the distant past has the most reduction. Therefore the prediction follows to the statistical variation of the observable signal. This is suited for the non-stationary MR data.

5.2 Modified TERA Suited for Low SNR

The RLS and LSL algorithms are introduced into the TERA method to make it suited for the non-stationary aspect of the MR data. In the case of low SNR MR data, we suggest use of a singular value decomposition (SVD)-based total least square (TLS) algorithm for an accurate AR parameter estimation.

For a given N_{data} points data series $x[n]$, the pth forward error predictor could be represented as:

$$\vec{\varepsilon_p} = \vec{X}\vec{a} \tag{100}$$

where

$$\vec{\varepsilon_p} = \begin{bmatrix} \varepsilon_p[0] \\ \varepsilon_p[2] \\ \vdots \\ \varepsilon_p[N-p] \end{bmatrix}, \qquad \vec{a_p} = \begin{bmatrix} a_0 \\ a_1 \\ \vdots \\ a_p \end{bmatrix} \tag{101}$$

and

$$\vec{X} = \begin{bmatrix} x[p] & x[p-1] & \cdots & x[0] \\ x[p+1] & x[p] & \cdots & x[1] \\ \vdots & \vdots & \ddots & \vdots \\ x[N-1] & x[N-2] & \cdots & x[N-p-1] \end{bmatrix} \tag{102}$$

For a p'^{th} order linear predictable sequence corrupted by noise, we choose a higher order p ($p >> p'$). Therefore the system described as Eqn 100 is overdetermined. In the noiseless case, $\vec{X}$ is of rank p' and the extra coefficients $a_{p'+1}, \cdots, a_p$ are zero. In the presence of noise, these coefficients can help to absorb noise. One way to find a meaningful solution $\vec{a}$ from the infinite numbers of solution is to minimize the energy of ε_p. The energy, I, of the $\varepsilon_P[e^{j\omega}]$ is defined as:

$$I = \frac{1}{2\pi}\int_{-\pi}^{\pi} |\varepsilon_p[e^{j\omega}]|^2 d\omega. \tag{103}$$

The above condition is equivalent to minimizing the Euclidean norm of the vector $\vec{a}$. The significance of the minimum norm lies in the fact that it minimizes the variance of prediction errors.

$$I = \vec{a}^H \vec{a} \tag{104}$$

Hence, the linear prediction problem is transformed into a norm minimization problem. The problem is redefined to minimize the norm of vector $\vec{a}$ with the following constraints:

1. The vector $\vec{a}$ lies in the null space of $\vec{X}$.
2. $a_0 = 1$.

This is solved by resorting to the singular value decomposition (SVD) of the data matrix $\vec{X}$. We proceed as following:

1. Conduct a singular value decomposition of the data matrix $\vec{X}$, and thereby find the expression of associated right singular vectors $\vec{v}_i$, $i \leq 1 \leq p$.
2. Estimated the number of M eigenvalues matrix $\vec{X}$. The number M represents the actual number of signal components buried in noise.
3. Based on the constrains (1), (2), we can deduce the expression of $\vec{a}$ as:

$$\vec{a} = \frac{\vec{e}_0 - \sum_{i=0}^{M-1}(\vec{e}_0^T \vec{v}_i)\vec{v}_i}{1 - \sum_{i=0}^{M-1} v_i^2(0)} \tag{105}$$

where $\vec{e}_0 = [1, 0, 0, 0...]^T$. The details could be found in reference [26].

This method is referred to in the scientific literature as a Total Least Square solution. The robustness of this technique lies in its ability to improve the SNR of the data by removing noise both from the data matrix $\vec{X}$ and the observation vector $\vec{a}$ by using the SVD method. The SNR enhancement property of the TLS method makes it possible to detect the signal components at high noise levels. Unlike the iterative techniques, such as RLS and LSL, TLS requires neither *a prior* knowledge of the number of signal components nor initial parameter estimates. For instance, both the RLS and LSL algorithm need to be given a forgetting factor estimation to start with, and its value directly affects the performance of these algorithms. The parameters of the TLS algorithm are estimated directly from the original data with a minimum user input.

6 Quantitative Comparison of Algorithms

New algorithms are frequently discussed in the scientific literature in isolation using the "ugly picture before - pretty picture now" criterion of success. This is not particularly useful for other researchers or clinicians deciding which method to use. The lack of a quantitative criteria also means it is not possible to determine how selective the success of the method is on different part of the image. In this section we propose some new quantitative techniques to overcome these problems.

A reconstructon algorithm can be judged based on the following criteria:

1. The quality of the reconstruction. This is normally expressed as reduced ringing artifacts and improved resolution on the reconstructed image. In the frequency domain, it is indicated by the accurate reintroduction of the high frequency components that had been truncated.
2. The performance on different types of image, including images with poor SNR.
3. The computational efficiency and procedure complexity.
4. The numerical stability of the algorithm.

One quantitative measure for MRI reconstruction is the frequency domain "DIFFENERGY" method introduced by Smith *et al.* [2]. We extend this to produce measures that indicate whether the modeling algorithms are succesful on both a global and a local basis.

6.1 Quantitative Comparison in the Frequency Domain

The MR experimental data is normally discretely sampled and has a finite length. The effective window introduces image artifacts when using the standard DFT reconstruction. In order to overcome the limitations of the DFT reconstruction method, many alternative reconstruction methods have been suggested. However, the improvements of these methods are hard to evaluate. Since the "full" data set may itself be windowed, the "standard" image reconstructed from it may have artifacts too! This makes quantitative comparison in the image domain difficult and possible inaccuracte for many image classes of interest if the modeling produces true super-resolution.

The DIFFENERGY measure in the frequency domain [2] attempts to overcome this problem. A long-length (standard) frequency domain MRI data set is truncated. This truncated data set is modeled and the success of its extension can be compared with the standard. Thus a comparison of the success of the modeling algorithm is made prior to any reconstruction distortions being introduced.

For a given original set of $N' \times M$ MR data (after 1D DFT on the lesser truncated direction) (*i.e.* Fig. 2), one row of the "standard" or "full" data may be denoted by $s_{stand}[n][m]$, $n \in N'$, $m \in M$. To provide a comparison standard, we truncate the known $N' \times M$ data set to the size of $N \times M$ to give the "truncated" data, $s_{trunc}[n][m]$, $n \in N$, $m \in M$, where $N < N'$ (*i.e.* Fig. 3). We reconstruct the image using the alternative methods from this "truncated" data, The reconstructed "modeled" image is transferred back to the frequency domain. This "modeled" data, $s_{model}[n][m], n \in N, m \in M$, is then compared with those data of the same range in the "standard" data set. We judge the quality of one method from the estimate it makes in recovering valid "truncated" high frequency data, which may not be true in the extended data set.

An important point to note is that these modeling algorithms can produce "super-resolution" images when the model is appropriate. If the method generates an image directly, it is important to super-sample this image prior to transforming back to the frequency domain for the comparison. This affords introducing aliasing artifacts that would incorrectly distort the frequency domain error measurements.

6.2 Comparison Algorithm and Procedures

The DIFFENERGY measure [2] compares, on a complex point by complex point basis, the "modeled" or "truncated" data with the "full" data. If the DIFFENERGY is significantly smaller for modeling method A then it can be considered the more successfull. Summing the errors across the whole image can be misleading sometimes. A large localised error would distort the measure even if the image is basically correctly modeled. We have modified the measure to give a normalized DIFFENERGY which can be applied locally or globally.

Let us define $DIFF_{method}[n][n]$ as the complex difference between the frequency domain data for the reconstruction technique *method* and the standard "full" data set.

$$DIFF_{method} = s_{method}[n][m] - s_{stand}[n][m] \tag{106}$$

6.2.1 GDF and $GDFn$ – Global Normalized error measures

The global normalized DIFFENERGY GDF is defined as:

$$GDF = \frac{\sum_{n\in N}\sum_{m\in M_{com}} |DIFF_{model}[n][m]|^2}{\sum_{n\in N}\sum_{m\in M_{com}} |DIFF_{trunc}[n][m]|^2} \tag{107}$$

The global normalized error GDF represents the overall ratio of the valid energy information still lost by the model algorithm compared to the energy information lost in the "truncated" data.

Attempting to compare the success of the algorithms in regions that are essentially only noise would distort the measure since there is no guarentee that the noise in the standard and modeled data sets will or will not be correlated. Therefore m is limited to M_{com} beyond which the MRI data is essentially noise.

Another global measure $GDFn$ attempts to include the effect of the noise in the GDF calculation is given by:

$$GDFn = \frac{\sum_{n\in N}\sum_{m\in M_{com}} |DIFF_{model}[n][m]|^2 - \sum_{n\in N}\sum_{m\in M_{com}} |noise|^2}{\sum_{n\in N}\sum_{m\in M_{com}} |DIFF_{trunc}[n][m]|^2 - \sum_{n\in N}\sum_{m\in M_{com}} |noise|^2} \tag{108}$$

6.2.2 LDF and $LDFn$ – Local Normalized error measures

The global measures can be compromised if only a small area is poorly fitted. To determine if this might be a problem, we introduced a localized DIFFENERGY measure, the LDF, which generates an error measure based on the average of the DIFFENERGY ratios of each row. If GDF and LDF are the same, it indicates that the modeling is appropriate to all parts of the image. If GDF is high and LDF is low, it means that only certain parts of the image are being correctly modeled.

The LDF measure is based on the average of the ratio of the differences between the standard, truncated and modeled data sets for each row.

$$LDF = \sum_{m \in M_{com}} \frac{\sum_{n \in N} |DIFF_{model}[n][m]|^2}{\sum_{n \in N} |DIFF_{trunc}[n][m]|^2} / M_{com} \tag{109}$$

Again this measure can be influenced by noise. Therefore we introduced $LDFn$ with the noise effect taken in account.

$$LDFn = \sum_{m \in M_{com}} \{ \frac{\sum_{n \in N} |DIFF_{model}[n][m]|^2 - \sum_{n \in N} |noise|^2}{\sum_{n \in N} |DIFF_{trunc}[n][m]|^2 - \sum_{n \in N_{com}} |noise|^2} \} / M_{com} \tag{110}$$

6.3 Measure Reliability Testing

The test images used were a high SNR (SNR = 37.9) phantom images (Figs. 13.a and 13.b) and a lower SNR (SNR = 17.3) Thigh image (Fig. 13.c) .

The measures GDF, $GDFn$, LDF and $LDFn$ are intended to permit a quantitative comparison of the results of different modeling approaches. They represent the normalized DIFFENERGY as a percentage between the "modeled" data and the reference "standard" data. A 0 % difference means exact modeling and total truncated error recovery. A 100 % difference means that nothing is gained from the "truncated" data DFT reconstruction. A value above 100 % may indicates a degrade possibly due to model instability (spikes). However, "noise-only" portions of the original and modeled data can produce the same effect.

We are interested in finding out the differences between the modeled $s_{model}[n][m]$ data and the standard data $s_{stand}[n][m]$, in which the signals

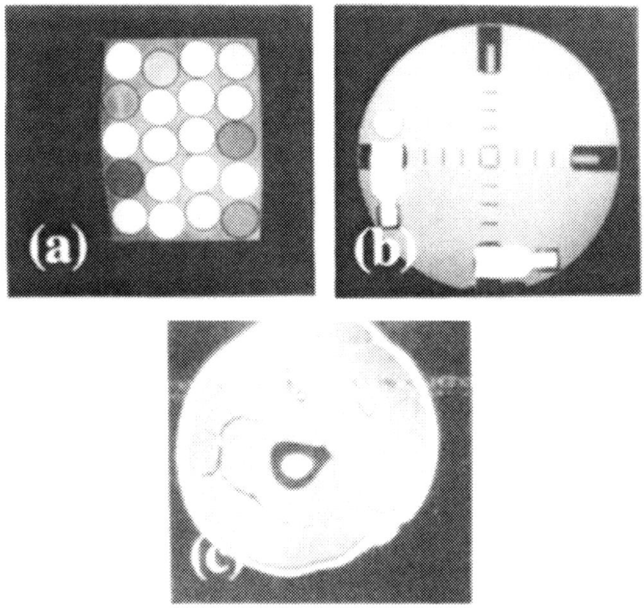

Figure 13: The test images (a) Phantom I (b) Phantom II (c) Thigh.

may contain a different level of noise. A good measure should be able to detect improvement in the image without being unduly sensitive to the presence of noise.

Fig. 14 shows the influence of noise on the global references GDF, $GDFn$ and the local references LDF and $LDFn$ for changing SNR for the human thigh image. As can be seen from Fig. 15 a similiar effect was found for both the phantom images which have very different image properties to the thigh data set. These figures indicate that the GDF and LDF measures dramatically vary with changing noise level, while $GDFn$ and $LDFn$ measures are well compensated. Experimentally we found that the $GDFn$ and $LDFn$ were essentially equivalent. This indicates the measures indicated that on average the algorithms were modeling equally across most of the images. Only a more detailed row-by-row comparision, such as that done for the CTERA method, will indicate which image details are best modeled by which algorithm.

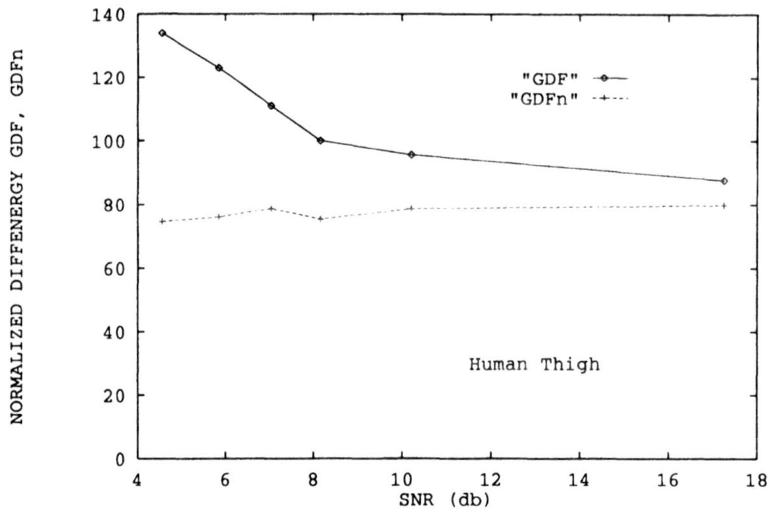

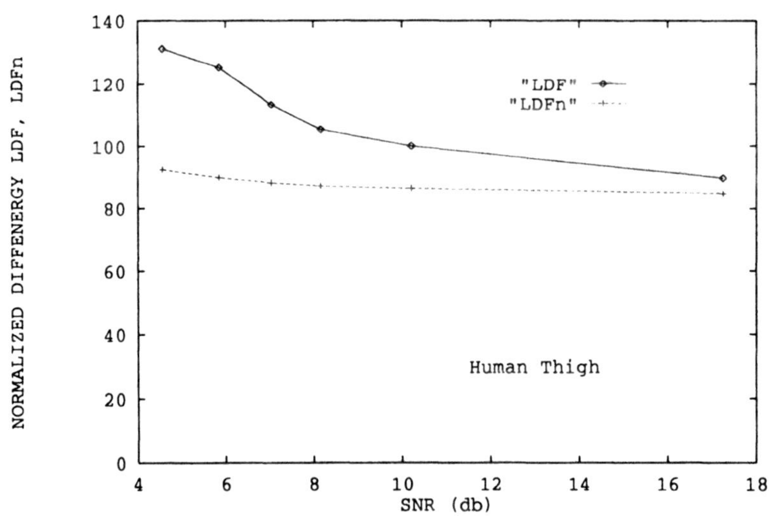

Figure 14: The influence of noise on the global references GDF, $GDFn$ and the local references LDF and $LDFn$ the changing SNR (Human thigh image).

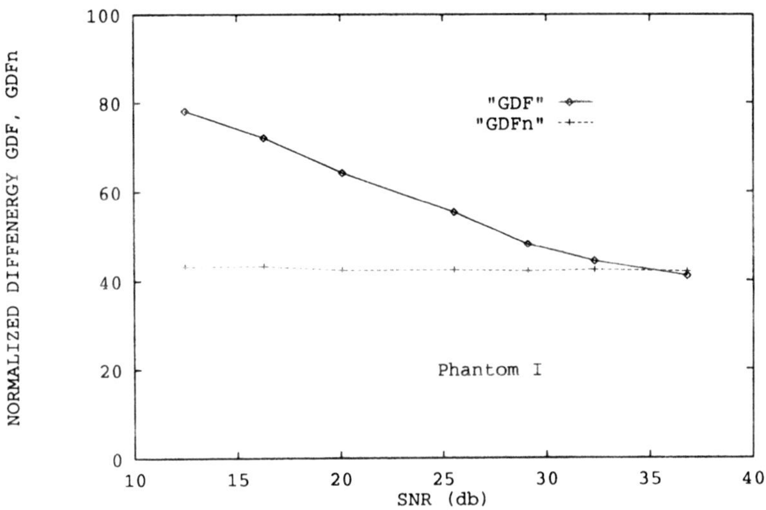

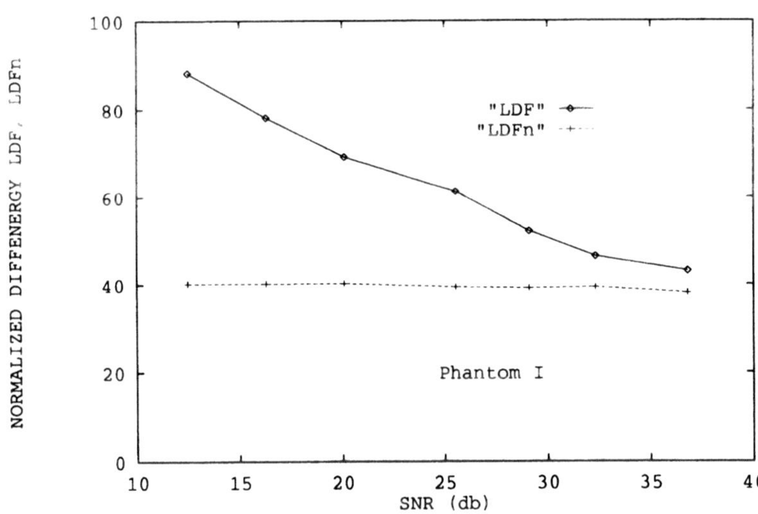

Figure 15: The influence of noise on the global references GDF, $GDFn$ and local references LDF, $LDFn$ with changing SNR (Phantom images).

Note an additional noise-related problem that must be considered. We can the evaluate the effect of noise on the modeling by truncating standard images with increasing noise levels. The noise in the central portion of the standard data may cause modeling instability which is reflected in poorer noise figures. However modeling reintroduces high frequency signal components, but no significant high frequency noise. The presence of this noise on the "standard" but not on the "modeled" image might also increase the error measure and be incorrectly interpreted as a poorer modeling fit.

6.4 A Quantitative Comparison of the Modeling Algorithms

The success of the algorithms are determined for MRI images, However, the measures should provide a valid indication of the expected relative success of the algorithms used in any situation instead of the DFT algorithm.

Phantom and human MR images have different structure and SNR; it being relatively easy to keep the phantom still while additional measurements are taken to average out the noise. We have chosen a high SNR phantom and a low SNR human thigh image for our extensive quantitative comparison, as these images are representative of the experimental situation.

All the images from the various methods are presented as a group for easy comparison at the end of the Conclusion section (Figs. 27 – 30).

6.4.1 Standard DFT

The k-space "standard" data set is used as the reference producing the image shown in Fig. 27 upper. By definition this data set has a 0 % DIFF-ENERGY measure. It is possible for the result of a modeling algorithm to exactly match this standard and generate an expolation beyond this data set. However, such a success can not be measured againest the known data set.

Note a percularity with the TERA algorithms. Because of the data consistancy, introduced on calculating the MA coefficients, only the extrapolation can be in error. The centre portion of the data set is matched exactly. In addition, since there is no need to apply a smoothing filter between the "truncated" and "extrapolated" data sets, there is no distortion introduced at the boundaries *c.f. Sigma algorithm.*

6.4.2 DFT method using zero padding

The reconstruction error using the DFT on the truncated data produces, by definition, a 100 % DIFFENERGY measure. The image is shown in the Conclusion (Fig. 27 centre).

6.4.3 TERA and MTERA methods

A forward predicting least square method [27] was used in this algorithm to calculate the set of AR coefficients. The non-stationary and noisy MR data causes the modeling to be unstable, which introduces spikes error into the final image for the standard TERA algorithm even with the stability introduced with "pole-pulling". The MTERA algorithm using "DFT matching" has significantly less spike errors.

The modeling order plays a important role in the estimation. If it is too low, the prediction is poor. If it is too high, over modeling gives a result with many spike errors. The performance of TERA and MTERA as a function of modeling order is shown at Fig. 16. MTERA gives a better fit because of the reduced spike errors resulting from the DFT-matching. The best results are found in reconstruction when the model order is about 15 to 20, corresponding to approximately 1/4 to 1/3 of the length of the Hermitian data array (64 points), which supports the suggestion made by Smith for the model order [1].

The image for the MTERA algorithm is compared to the standard and truncated images in the Conclusion (Fig. 27).

6.4.4 Burg algorithm

The computationally efficient Burg algorithm can be used on the ramped data set. The influence of modeling order is shown as Fig. 17. However, with the data transformation of ramping and de-ramping, some of the computational efficiency is lost. From the analysis, it indicates that the TERA using the Burg algorithm is better than the MTERA for the thigh image, but about the same for the phantom image. This contradicts the results in Smith and Nichols' paper [2], which indicate that the Burg algorithm was inappropriate because of the enhanced noise in the ramped data. This contradiction may be due to the fact that they did not use a normalized comparison. Alternatively, since the phantom used there differs from those used here, it may merely be further evidence that the most suitable modeling algorithm is image dependent.

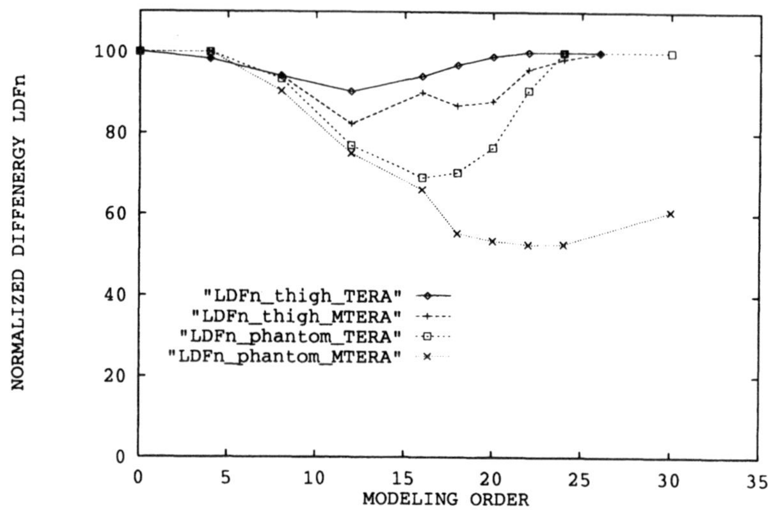

Figure 16: The modeling order influence on the TERA and MTERA algorithms. After the spikes have been reduced by MTERA, the difference between "modeled" and "original" data is decreased.

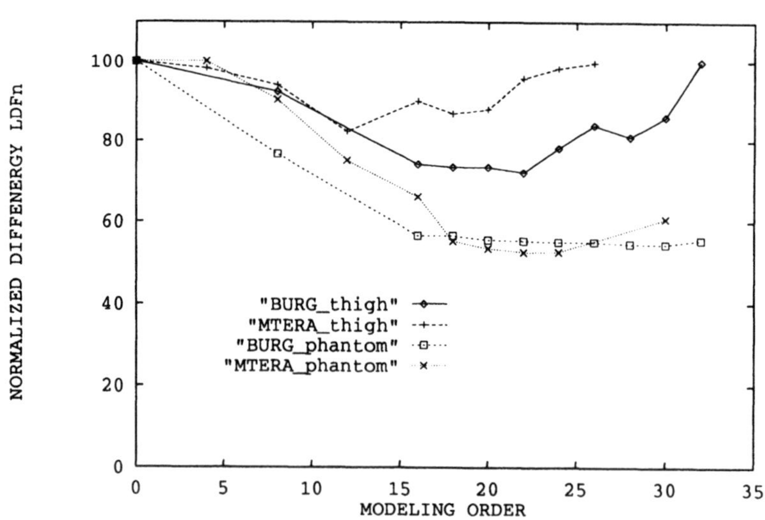

Figure 17: The modeling order influence on TERA with the Burg algorithm, compared to the MTERA results.

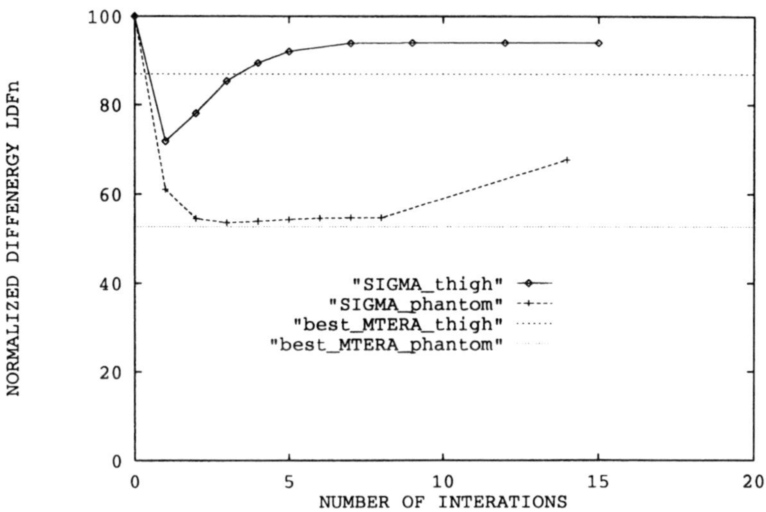

Figure 18: The results of each iteration stage of the Sigma method.

6.4.5 Iterative Sigma Filter Method

The iterative Sigma method applies a edge-enhanced average filtering on the image. It attempts to reduce the ringing artifacts while maintaining the edge information. The results of each iteration step were compared and shown in Fig. 18. Amartur and Haacke [18] reported that after 3 iterations, the fit will be stabilized. The results with our new measures show that this is true for the phantom image (mainly edges and flat surfaces). However the human thigh image gets significantly worse after the first iteration. The image for this algorithm is compared to the GS and CTERA algorithms in the Conclusion (Fig. 29).

6.4.6 GS Method

The generalized series model is designed to use available *a priori* information in the image reconstruction. Here we use the ringing reduced and edge preserved information which the magnitude Sigma images provide in GS algorithm. There are no parameters to change. The best GS fits gave a 61 % and a 77 % DIFFENERGY on the thigh and phantoms respectively which compares to the 87 % and 54 % DIFFENERGY measure with the MTERA. The image for this algorithm is compared to the Sigma and CTERA algorithms in the Conclusion (Fig. 29).

6.4.7 CTERA Method

Fig. 19 shows that the MTERA and Sigma seems to work better on different parts of the data. In the CTERA method we attempt to combine the best of both methods by using the the result of the Sigma method prior to the data merging as *a priori* information for TERA. The extended data length provided by the Sigma filter makes it possible to increase the modeling order without increasing the spike errors. The modeling order influence on the CTERA reconstruction is shown Fig. 20, which shows its high order modeling capability. The image for this algorithm is compared to the Sigma and GS algorithms in the Conclusion (Fig. 29).

6.4.8 Modified TERA Method

We modified the TERA algorithm by using different algorithms in calculating the AR coefficients, trying to account for the non-stationary characteristic and low SNR. By using the RLS, LSL and TLS algorithm instead of forward LS, we hope to obtain a reduction of the spike errors or a better estimation on a high noise image.

6.4.9 RLS method

The forward predicting fast RLS method was found to be numerically unstable on the tested MR data producing images with distortions across the full length of the image. This agrees with the comments of Cioffi, who suggested using the LSL approach[31].

6.4.10 LSL method

The LSL, which has a stable lattice structure, seems to work well in the nonstationary MR data environment. The reconstructed LSL-TERA image has far less spikes than when using LS-TERA.

The forgetting factor in LSL algorithm plays a important role in the prediction. It determines the weights of former information at the present recursion update and directly affects the accuracy of prediction. The results

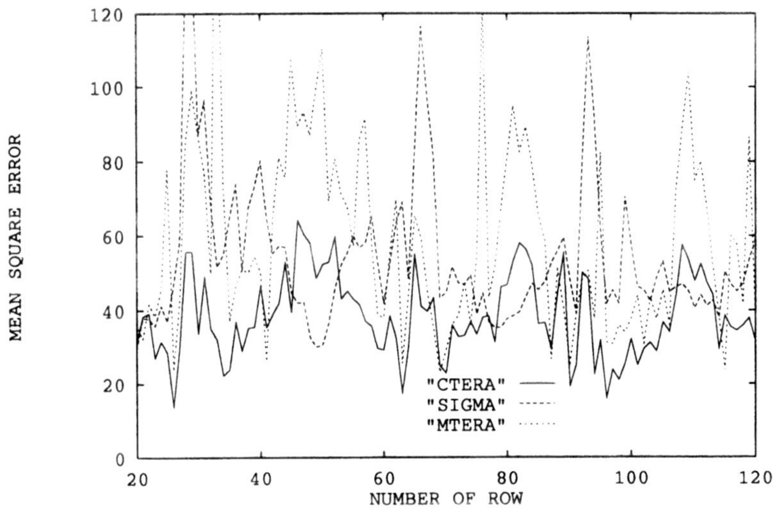

Figure 19: The mean square error between the modeling data and original data in the frequency domain. CTERA combines the best of MTERA and SIGMA.

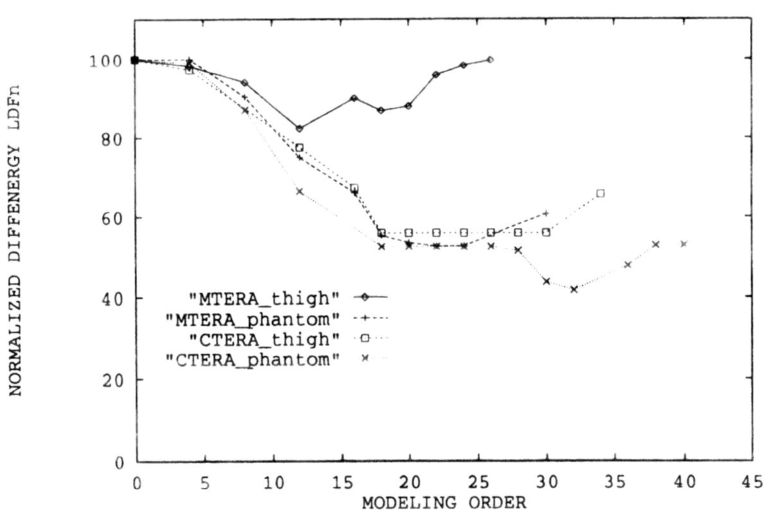

Figure 20: The modeling order influence on the CTERA algorithm, compared to the MTERA results.

of different forgetting factors along with the order are compared in Fig. 21 and 22. LSL gives the best results when the order is about the one-third of the data length.

The image for the LSL algorithm is compared TLS-TERA algorithm in the Conclusion (Fig. 28).

6.4.11 TLS method

The total least square method is not suited for the non-stationary data environment. However, with the data-ramping technique mentioned the section 3.6, it is appropriate. The influence of different model orders is shown as Fig. 23.

MTERA and TLS give equal results on the high SNR phantom image. However, the TLS method works better than TERA on the noisy thigh image, as the theory predicted [26]. The image for the TLS algorithm is compared LSL-TERA algorithm in the Conclusion (Fig. 28).

6.5 The SNR Influence on the Algorithm Performance

It appears all the algorithms work better on the phantom image than on the human image. Because of the structural complexity of the human thigh, it would require a higher model order. However choosing a high order model may not be possible, because of the limited data length. The human thigh image also has a lower SNR and the effects of this on the modeling is not clear. To investigate the effect of noise, we deliberately degraded the SNR on the phantom data. The MTERA, TERA using the TLS algorithm and CTERA were tested as a function of the SNR. The results show that when the noise level becomes high, the algorithm estimation gets poorer (Fig. 24). This could explain why our method works better on the phantom image (high SNR) than on the human thigh image (SNR). The CTERA and TERA using TLS algorithm method shows a better estimation than the MTERA with the LS algorithm. This again agrees with the theoritical prediction.

6.6 Critique of the error measures

We introduced a number of quantitative error measures in an attempt to compare various modeling algorithms on images with various SNR. The er-

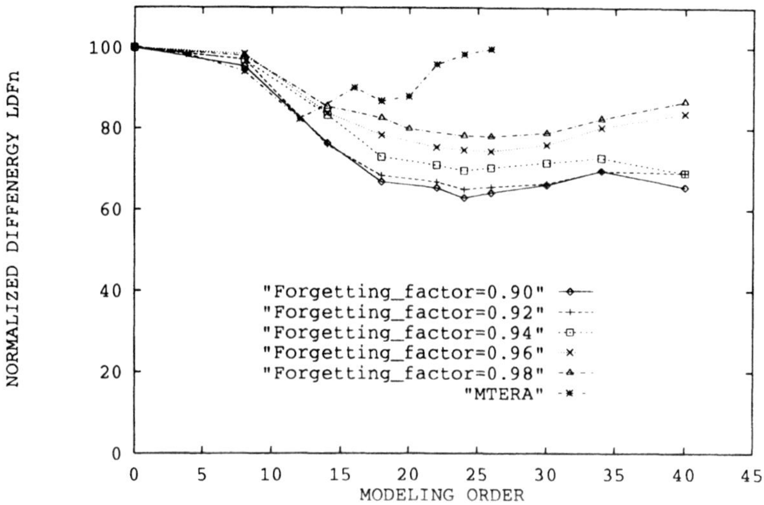

Figure 21: The modeling order and forgetting factor influences on adaptive TERA with LSL algorithm (Human thigh image), compared to the MTERA results.

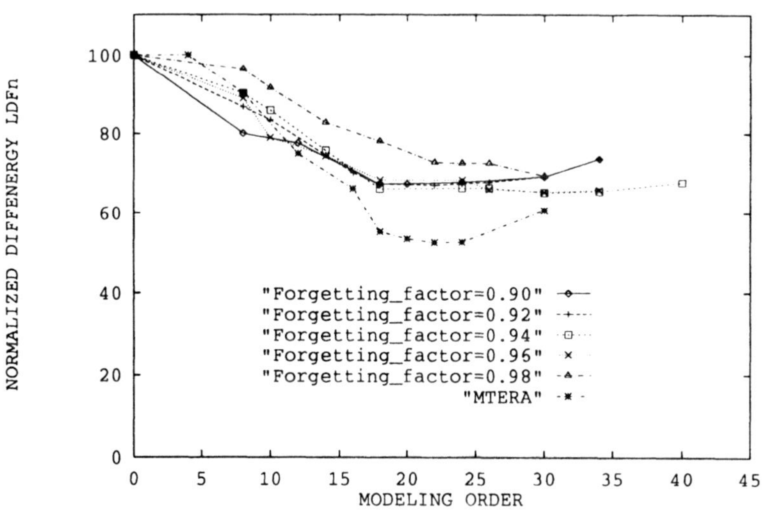

Figure 22: The modeling order and forgetting factor influences on the adaptive TERA with LSL algorithm (Phantom image), compared to the MTERA results.

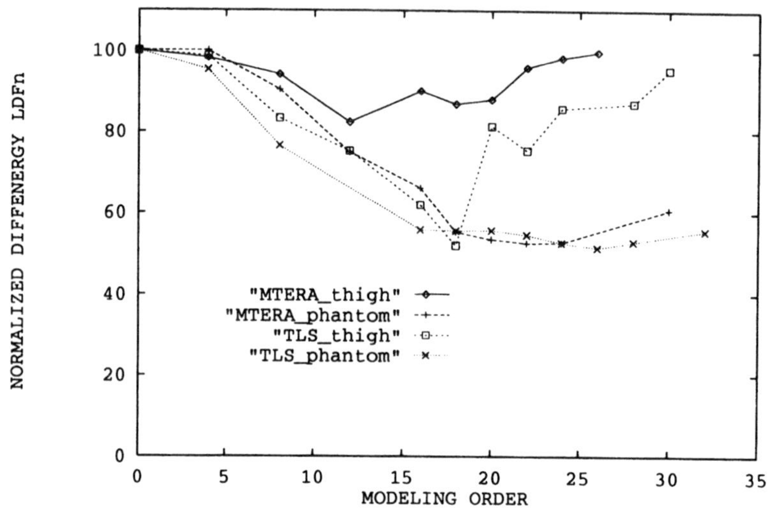

Figure 23: The modeling order influence on TERA with TLS algorithm, compared to the MTERA results.

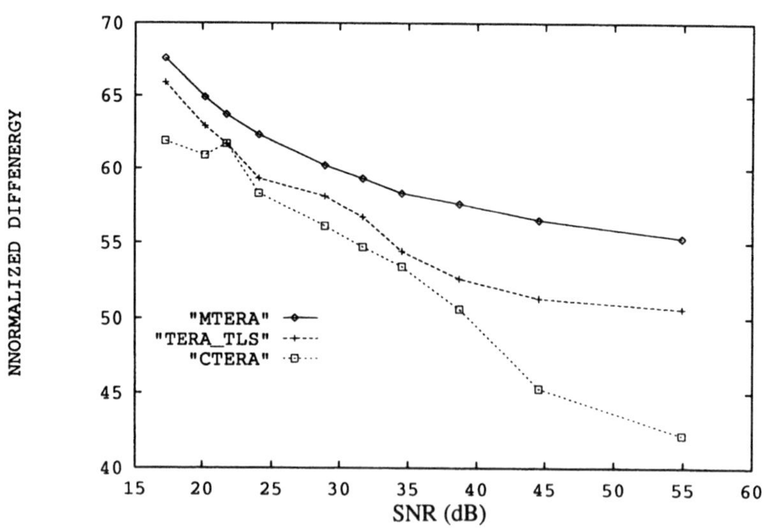

Figure 24: The SNR influence on the Phantom I reconstruction. When the SNR gets low, the performance of all the algorithms becomes poorer.

ror measures with the noise subtraction compensation gave relatively stable results without undue influences from the noise.

This quantitative comparison is however not perfect. The following areas can be justifiable criticized.

1. Some of the methods may introduce some high frequency component even beyond the cut off boundary in the "standard" data. This "super resolution" information is currently ignored, since there is no standard for comparison. This under-estimates the success of the modeling algorithm.
2. Presence of the high frequency noise in the "standard" image, but no noise in the "modeled" image can have some influence, give a larger error value than the actual difference, again underestimating the success.
3. There may be a mismatching between the phases of the high frequency component in the "standard" and "modeled" data. This may would have little effect on the reconstructed image (slight position shift) but a possible large effect on the error measures.
4. The DIFFENERGY of two modeling methods may be the same, yet the algorithms are fitting different parts of the image.
5. How does the DIFFENERGY measures relate to what the radiologist actually "sees" in the image. Is recovering 70 % of the truncated energy useful?

Nevertheless, within a certain level of error, these normalized error measures gives a consistent ranking between the methods. Large differences probably indicate a valid distinction between the approaches.

7 Future Directions

The major problem with modeling algorithms is that their success is image/feature independant. Are there methods that are succesfull with a larger numbers of image classes? Since we were approached to produce this review, we have been investigating two new areas:- multichannel analysis and neural networks. Multichannel analysis is an attempt to use the correlation between adjacent rows in an image. Neural networks are an entirely different approach to modeling the truncated data. Nets can used alone or combined with TERA.

7.1 Multichannel TERA Algorithm

Early work by Smith and Nichols attempted to use TERA on data truncated in both directions [3]. The approach was effectively repeated 2D use of the 1D TERA algorithm, with a modification to better account for the 2D data characteristics. Although "successful", it left much to be desired. A better 2D algorithm should incorporate the fact that the neighbouring lines of the MR data matrix are somewhat correlated. Two routes have been suggested to make use of this correlation. Mitchell averaged the adjacent data rows prior to the AR coefficient determination [24]. Other attempts averaged the AR and MA coefficients themselves [37, 38] to reduce the influence of noise. The major problem is that stability is gained but possible resolution improvement is lost. A more sophisticated approach is to attempt to move Scott's telecommunications work [39] on diversity to the field of MRI. This work attempts to predict a received channel on the basis of several adjacent and correlated radio channels. This is somewhat equivalent to predicting one MRI data row using information from adjacent MRI rows. However there are a number of unique problems to handle.

7.1.1 Multichannel AR Coefficients Evaluation Using MLSL

The following algorithm is from the work of Scott and Nichols [39]. The $N \times M$ MR data matrix (after 1D DFT on the lesser truncated direction) could be viewed as a collecting of N channels of data, each having M components. The neighbouring channels of data are assumed correlated. We can partition the N channels of data to N sets of L channels overlapped data set. Assume we have one of the L channels data sequences $\vec{s}[n]$:

$$\vec{s}[n] = \left[s_0[n] s_1[n] \vdots s_{L-1}[n] \right]^T \tag{111}$$

where $s_i[n]$, $0 \leq i < L$ is one channel of data. Unlike a radio channel, each MRI channel of the data has a double-sided decaying nonstationary property. As before we decompose $\vec{s}[n]$ into Hermitian $\vec{x}[n]$ and anti-Hermitian $\vec{y}[n]$ components before modeling. Each channel of data $s_i[n]$ is transformed using Eqn. 5 and 6 .

The MLSL algorithm has a similar lattice structure to that of LSL as shown in Fig. 12. The difference is that the input of MLSL is composed of a L vector input sequence, the output is a $L \times 1$ vector and the reflection coefficients of MLSL algorithm are $L \times L$ matrices. Assuming we have one of the hermitian vector arrays $\vec{x}[n]$ as the input. In a formulism similar to the single input case, the augmented normal equation for the pth multichannel forward prediction-error filter is [39]:

$$\vec{\Phi}_{p+1,N}\vec{A}^{f}_{p,N} = \left[\vec{\rho}^{f}_{p,N}\vec{0}_{Lp,L}\right]^{T} \tag{112}$$

where

$$\vec{\Phi}_{p+1,N} = \sum_{n=1}^{N} \lambda^{N-n}\vec{X}_{p+1}[n]\vec{X}^{H}_{p+1}[n] \tag{113}$$

$$\vec{X}_{p+1}[n] = \left[\vec{x}^{T}[n]\vec{x}^{T}[n-1]\cdots\vec{x}^{T}[n-p]\right]^{T} \tag{114}$$

$$\vec{A}^{f}_{p,N} = \left[\vec{I}_{L,L} - \vec{a}^{fH}_{p,N}\right]^{H} \tag{115}$$

$$\vec{\rho}^{f}_{p,N} = \sum_{n=1}^{N} \lambda|\rho^{f}_{p,N}|^{2} \tag{116}$$

$$\rho^{f}_{p,N} = \vec{A}^{fH}_{p,N}\vec{X}_{p+1}[n] \tag{117}$$

Similarly, the backward prediction-error filter has an augmented normal equation as follows:

$$\vec{\Phi}_{p+1}[n]\vec{A}^{b}_{p,N} = \begin{bmatrix} \vec{0}_{Lp,L} \\ \vec{\rho}^{b}_{p,N} \end{bmatrix} \tag{118}$$

$$\vec{A}^{b}_{p,N} = \left[-\vec{a}^{bH}_{p,N}\vec{I}_{L,L}\right]^{H} \tag{119}$$

$$\vec{\rho}^{b}_{p,N} = \sum_{n=1}^{N} \lambda^{N-n}|\rho^{b}_{p,N}|^{2} \tag{120}$$

$$\rho^{b}_{p,N} = \vec{A}^{bH}_{p,N}\vec{X}_{p+1}[n] \tag{121}$$

The recursion of MLSL algorithm consists of order-update and time-update two parts. The details of the derivation can be found in Scott's Ph.D. dissertation [39].

1. Order-Update

The order-update for the forward and backward prediction-error filters are as following:

$$\vec{A}^f_{p,N} = \begin{bmatrix} \vec{A}^f_{p-1,N} \\ \vec{0}_{L,L} \end{bmatrix} - \begin{bmatrix} \vec{0}_{L,L} \\ \vec{A}^b_{p-1,N-1} \end{bmatrix} \Delta_{p-1}[n]/\vec{\rho}^b_{p-1} \tag{122}$$

$$\vec{\rho}^f_{p,N} = \vec{\rho}^f_{p-1,N} - \nabla_{p-1}[n]\Delta_{p-1}[n-1]/\vec{\rho}^b_{p-1,N-1} \tag{123}$$

$$\vec{A}^b_{p,N} = \begin{bmatrix} \vec{0}_{L,L} \\ \vec{A}^b_{p-1,N-1} \end{bmatrix} - \begin{bmatrix} \vec{A}^f_{p-1,N} \\ \vec{0}_{L,L} \end{bmatrix} \nabla_{p-1}N/\vec{\rho}^f_{p-1,N} \tag{124}$$

$$\vec{\rho}^b_{p,N} = \vec{\rho}^b_{p-1,N} - \Delta_{p-1}[n]\nabla_{p-1}[n]/\vec{\rho}^f_{p-1,N} \tag{125}$$

where

$$\nabla_{p-1}[n] = \Delta^H_{p-1}[n] \tag{126}$$

If we define

$$\Gamma^f_p[n] = \Delta_{p-1}[n]/-\vec{\rho}^b_{p-1,N-1} \tag{127}$$

$$\Gamma^b_p[n] = \nabla_{p-1}[n]/-\vec{\rho}^f_{p-1,N} \tag{128}$$

Then we can rewrite Eqn. 122 and Eqn. 124 as:

$$\vec{A}^f_{p,N} = \begin{bmatrix} \vec{A}^f_{p-1,N} \\ \vec{0}_{L,L} \end{bmatrix} - \begin{bmatrix} \vec{0}_{L,L} \\ \vec{A}^b_{p-1,N-1} \end{bmatrix} \Gamma^f_p[n] \tag{129}$$

$$\vec{A}^b_{p,N} = -\begin{bmatrix} \vec{0}_{L,L} \\ \vec{A}^b_{p-1,N-1} \end{bmatrix} - \begin{bmatrix} \vec{A}^f_{p-1,N} \\ \vec{0}_{L,L} \end{bmatrix} \Gamma^b_p[n] \tag{130}$$

The order-update of the forward and backward *a priori* prediction errors are:

$$\vec{\varepsilon}^f_p[n] = \vec{\varepsilon}^f_{p-1}[n] + \Gamma^{fH}_p[n-1]\vec{\varepsilon}^b_{p-1}[n-1] \tag{131}$$

$$\vec{\varepsilon}^b_p[n] = \vec{\varepsilon}^b_{p-1}[n-1] + \Gamma^{bH}_p[n-1]\vec{\varepsilon}^b_{p-1}[n] \tag{132}$$

The order-update for the convention factor can be expressed as:

$$\gamma_p[n] = \gamma_{p-1}[n] - \gamma^2_{p-1}[n]\vec{\varepsilon}^{bH}_{p-1}[n]\vec{\varepsilon}^b_{p-1}[n]/\vec{\rho}^b_{p-1,N} \tag{133}$$

2. Time update

The time-update recursion for $\Delta_{p-1}[n]$ is

$$\Delta_{p-1}[n] = \lambda_{p-1}[n-1] + \gamma_{p-1}[n-1]\vec{\varepsilon}^{\,b}_{p-1}[n-1]\vec{\varepsilon}^{\,fH}_{p-1}[n] \tag{134}$$

The other time-update recursions for the reflection coefficients and the sum of weighted forward and backward prediction-error squares can be summaried as following:

$$\Gamma^f_p[n] = \Gamma^f_p[n-1] - \gamma_{p-1}[n-1]\vec{\varepsilon}^{\,b}_{p-1}[n-1]\vec{\varepsilon}^{\,fH}_p[n]/\vec{\rho}^{\,b}_{p-1,N-1} \tag{135}$$

$$\Gamma^b_p[n] = \Gamma^b_p[n-1] - \gamma_{p-1}[n-1]\vec{\varepsilon}^{\,f}_{p-1}[n-1]\vec{\varepsilon}^{\,bH}_{p-1}[n]/\vec{\rho}^{\,f}_{p-1} \tag{136}$$

$$\vec{\rho}^{\,f}_{p-1,N} = \lambda\vec{\rho}^{\,f}_{p-1,N-1} + \gamma_{p-1}[n-1]\vec{\varepsilon}^{\,f}_{p-1}[n]\vec{\varepsilon}^{\,fH}_{p-1}[n] \tag{137}$$

$$\vec{\rho}^{\,b}_{p-1,N} = \lambda\vec{\rho}^{\,b}_{p-1,N-1} + \gamma_{p-1}[n]\vec{\varepsilon}^{\,b}_{p-1}[n]\vec{\varepsilon}^{\,bH}_{p-1}[n] \tag{138}$$

Eqns. 131, 132, 134, 135, 136, 137 and 138 together, in order, constitute the MLSL algorithm.

Once the multi-channel reflection coefficient matrices have been obtained, a conversion relation similar to the single channel derivation can be used in calculating the multi-channel AR coefficients:

$$\vec{a}^{\,f}_{p,p} = \Gamma^{fH}_p[N] \tag{139}$$

$$\vec{a}^{\,b}_{p,p} = \Gamma^{bH}_p[N] \tag{140}$$

By using the Levinson recursion, we can derive the AR coefficients for the reflection coefficients:

$$\vec{a}^{\,f}_{j,p} = \vec{a}^{\,f}_{j,p-1} + \Gamma^{fH}_p[N]\vec{a}^{\,b}_{p-j,p-1} \tag{141}$$

$$\vec{a}^{\,b}_{j,p} = \vec{a}^{\,b}_{j,p-1} + \Gamma^{bH}_p[n]\vec{a}^{\,f}_{p-j,p-1} \tag{142}$$

7.1.2 Multichannel Image Reconstruction

With the AR coefficient matrices obtained, we apply the inverse AR filter to determine the transient error terms.

$$\vec{\varepsilon}_{p,N}[n] = \vec{X}[n] + \sum_{i=1}^{p} \vec{A}_{p,N}\vec{X}[n-i] \tag{143}$$

with

$$\vec{X}[n] = -\sum_{i=1}^{p} \vec{A}_{p,N}\vec{X}[n-i] + \sum_{i=1}^{q} \vec{B}^{b}_{p,N}\vec{\delta}_{n-i} \tag{144}$$

With the assumption that the filter excitations are delta functions, the MA coefficients for this algorithm are equivalent to the transient error terms.

The image data could be obtained through direct Fourier transform of these combined ARMA coefficients from the Hermitian and anti-Hermitian sub-vector array:

$$\vec{C}[p\Delta x] = (2Re(\vec{C}_x[p\Delta x]) - \Delta k Re(\vec{s}[0])) + j(2Im(\vec{C}_y[p\Delta x]) - \Delta k Im(\vec{s}[0])) \tag{145}$$

where

$$\vec{C}_x[x] = \vec{C}_y[x] = \Delta k \frac{\sum_{n=0}^{p-1} \vec{B} exp(j2\pi nm/N)}{\sum_{n=0}^{N_{data}-1} \vec{A} exp(j2\pi nm/N)} \tag{146}$$

where $x = m\Delta x$ and $1/(\Delta x \Delta k) = N_{data} \geq p-1$.

For the truncated $N \times M$ MR data set, we zero-pad $L-1$ rows before the first and after the last row and form sets of the N of L multi-channel data. The L channel input data results the L channel image data. Only the $(L/2)^{th}$ channel image data is kept at one time. Repeat this calculation N times to form the whole image plane.

Scott [39] has been very successful in implimenting this algorithm in the area of telecommunications. However our extension of the algorithm into a form suitable for MRI multi-channel fitting has been less successful. We believe that it is a problem of implimentation rather than concept leaving the algorithm a good starting point for future research.

7.2 Neural Networks

Since the time we agreed to write this article, use of truncated data sets in MR has changed. Now the focus is on dynamic imaging. The technique is to first gather a "full" data set of the "static" object. This is followed by a series of "truncated" data set taken with a small time resolution in order to follow some changing characteristic, such as contrast agent uptake.

The CTERA algorithm is appropriate for this dynamic imaging with the AR coefficients being determined (trained) from the "full" static image data set. The MA coefficients are determined from the prediction errors calculated from the truncated dynamic data sets.

An alternative approach may be to use neural networks (NN) in the training. Neural networks are relatively new information processing systems whose architectures are modeled according to the human nervous system. They have some unique capabilities which are not found in the conventional information processing techniques, such as generalization ability and parallel processing ability. Excellent reviews and tutorial on NN can be found in [40, 41, 42].

The generalization ability of NN refers to the networks producing reasonable outputs for inputs not encountered during training [42]. This makes it possible for NN to solve some complex problems that are difficult for conventional signal processing techniques. Yan and Mao [43] originally suggested that a neural network be trained on the truncated data set. The nets weights are then fixed and the network used to extrapolate the data set, which is then reconstructed using the standard DFT. Yan and Mao made use of a real-valued net where the complex MR values are independently modeled as real and imaginary components. We have recently shown that they made some errors in the analysis of the algorithm's performance. In fact it is far more efficient and stable in the presence of noise than they suggested if the network is trained on the unaliased frequency data [44].

Recently complex-valued neural networks have been discussed in the literature [45, 46]. We have had even greater success with these as there is considerable correlation between the MR real and imaginary components . In papers recently submitted we have shown successfully data extrapolation using a complex network with fewer hidden layer neurons and less training than with the real-valued nets [47]. The topology of our complex-valued network is similar to the standard real-valued backpropagation (BP) network. However, the inputs, outputs, weights, bias and activation functions are all complex-valued, and the BP algorithm itself was derived in the complex domain.

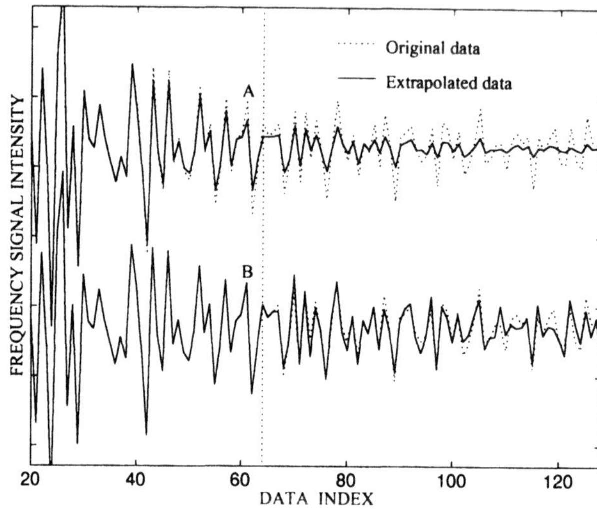

Figure 25: Comparison of the extrapolated frequency data (solid lines) and original data (dotted lines). (A) shows the real components of the extrapolated data using a real-valued network and (B) shows the result from our complex-valued network. Vertical dotted line indicates the truncation point.

Figure 25 shows one side of the real components of the extrapolated frequency data using real and complex-valued networks on a 1D simulation data set. Similiar succes was found with results with real-world MR data. The real-valued network has one hidden layer with 200 neurons as suggested in [43] and the complex one has a smaller hidden layer of 40 neurons. Training is stopped at 500 epoches in both cases. These results demonstrate the superior performance of the complex-valued NN based method over the real-valued NN based method, especially on the medical MR data. The use of fewer hidden layer neurons means less complexity and training, an important consideration as we are applying the networks to a 2D image.

Dynamic imaging has shown to be another application in which the generalization ability of NN can be fully explored [50]. In our NN based resolution enhancement method, the complex-valued network is trained on the full reference data set and adaptively restore the high frequency components based on the available low frequency components of each individual dynamic data set. This method can improve the time resolution four to eight fold without losing much spatial and contrast resolutions compared with the conventional Fourier imaging technique. Compared with existing k-space data substitution technique [48, 49] under the same time resolution, our approach can significantly enhance the spatial and contrast resolutions by reducing the edge burring, the severe intensity distortion and other artifacts (figure 26).

There are two major problems with this application of the neural networks. First one is the computation time for training the network. However, due to its parallel processing nature, the NN is ideally suitable for implementation using VLSI techniques and will become excellent in on-line processing with the advances of VLSI.

The second is the problem of stably extrapolating the MR data far enough to account for the possible "super-resolution" in the final image. We have recently attempted to combine neural networks with TERA since avoiding an explicit extrapolation was one of the main advantages of TERA. With the NN-TERA approach, the AR coefficients and prediction errors are calculated as discussed earlier. The first p terms of the prediction errors correspond to the "transient" that initializes the ARMA filter to follow the MR signal. The other terms should be close to zero if the model is totally appropriate, something that frequently is not true. The neural network is used to "extrapolate" these prediction errors to produce a new set of MA coefficients. Since much of the modeling is done via the AR coefficients, we expect that the neural network will train much faster on the MA coefficients than on the original data. The AR and new MA coefficeints can then be used in the standard TERA algorithm to avoid the explicit extrapolation. Preliminary results of this approach are to be presented [51].

The work on neural networks as an alternative to the DFT is being done as an M.Sc. disertation by Yan Hui [52].

8 Summary and Conclusions

The TERA algorithm was developed to avoid the artifacts introduced into MR images reconstructed using the standard inverse DFT method applied to truncated data sets. A number of improvements were made in response to experimental problems with the MR data characteristics. The standard TERA algorithm independently models the Hermitian and anti-Hermitian data components, using "pole-pulling" to ensure stability. Matched-TERA (MTERA) uses the known truncated DFT image as a restraint to remove any remaining instability spikes in the modeled images. This allows an increased model order (higher resolution) with out compromising stability.

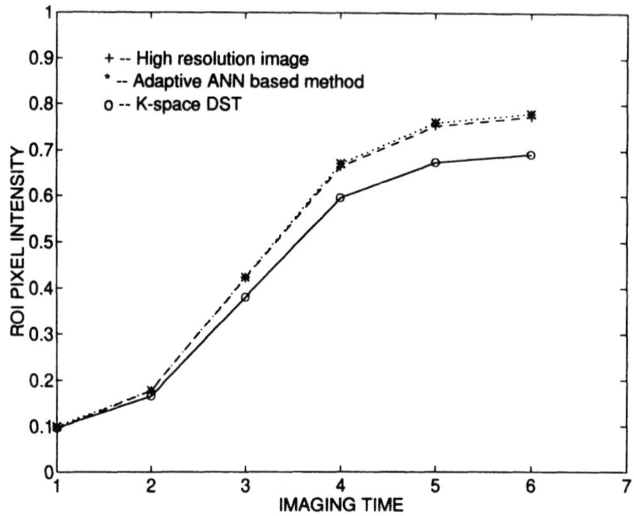

(a) 32 phase encodings

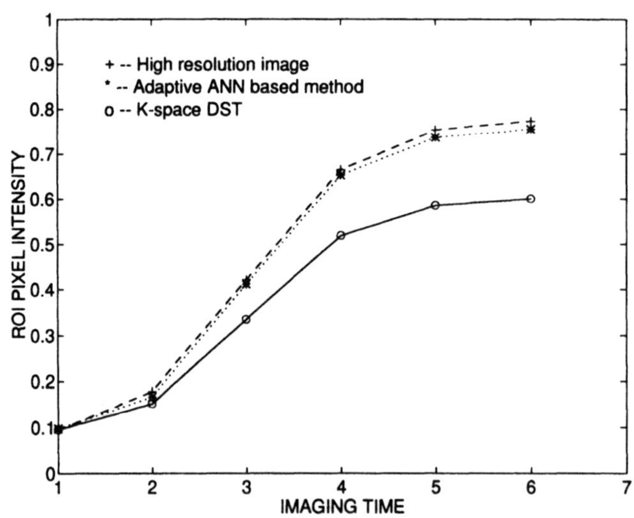

(b) 24 phase encodings

Figure 26: Average pixel intensity changes in the region of interest (ROI) versus imaging time with different methods and phase encoding numbers. The high resolution images used 256 encodings.

A number of algorithms were discussed for the first time in this article. Constrained TERA (CTERA) allowed the modeling coefficients to be determined from *a priori* information such as a data set with enhanced edges. This algorithm combined the best from the SIGMA and GS (generalized series) modeling techniques of other authors with the TERA algorithm. The fast recursive least square (RLS-TERA) method fails in the modeling because of numerical stability problems. The adaptive least square lattice (LSL-TERA) was more stable.

In terms of problems with low SNR of MR data, we suggested use of the total least square (TLS-TERA) approach. The TLS, applied to the ramped MR data, uses a SVD method allows a more accurate AR coefficient determination in the presence of noise but with a heavy calculation burden.

We used these algorithms with a number of images and evaluated their success in the recovering the energy truncated from the data using a local and global "DIFFENERGY" measure. The images included two medical phantoms (flat surfaces, high SNR), two medical images (thigh and knee) and a petroleum core [53]. The degree of truncation resulted in energy losses in the frequency domain data between 0.8 % and 4.4 %. These values may seem low, but it must be remembered that the loss is in the high-frequency detail of the image.

The best results are presented in (Table 1). The CTERA and TLS methods consistently recovered more of the truncated signal than the other modeling algorithms discussed. A qualitative comparison of the methods can be made by the reader using the images presented at the end of this section.

From the table it can be seen that the energy recovery is 60 % for the phantoms, 40 - 50 % for the thigh and petroleum core but only 7% for the thigh image. The constrained (CTERA) and the total least square variant (TLS-TERA) consistently had a better energy recovery than the other techniques.

Although the modeling algorithms discussed in this article make a substantial improvement to image quality a number of very troubling problems remain. The first is the computational complexity of the algorithms. Are they fast enough to become clinically useful? The second problem may be the more important. **Is the 60 % of the missing energy recovered by the CTERA algorithm more or less important for diagnosis using the modeled image that the 25 % energy recovery from the generalized series approach?** The answer would be obvious if the energy recovery had been close to 90 % but not with the existing imperfect algorithms. We are currently trying to set up a series of computer assisted

	PHANTOM I		PHANTOM II		THIGH		KNEE		CORE	
MISSING ENERGY	1. 1%		0.4%		4.3%		0.8%		4.4%	
SNR	37.9		29.5		17.3		29.6		15.4	
METHOD	**LDFn**	**GDFn**	**LDFn**	**GDFn**	**LDFn**	**GDFn**	**LDFn**	**GDFn**	**LDFn**	**GDFn**
CTERA	41.1●	42.3●	43.3●	42.7●	55.9	54.7	94.3	92.3	60.1●	63.1●
TLS	51.5	52.3	44.3	45.3	52.0●	53.2●	93.2●	91.7●	68.4	70.3
MTERA	52.7	51.2	47.9	48.7	86.8	95.4	97.3	97.4	70.0	65.7
SIGMA	53.6	52.5	-	-	71.9	72.1	-	-	-	-
GS	74.7	72.1	-	-	60.7	58.2	-	-	-	-
LSL	65.2	68.1	55.2	57.6	64.2	65.4	98.3	96.6	66.3	65.4
BURG	54.5	55.0	-	-	74.4	74.3	-	-	-	-
TERA	71.3	72.5	56.8	58.7	92.4	93.1	101.3	102.9	77.8	78.9

Table 1: The best results from the various methods on the value of diffenergy. The best error correction for each image is marked • .

evaluation techniques using standard images to allow researchers to evaluate their methods using a more "visual" approach [54, 55]. It is intented to mimic the response that the radiologist would have trying to identify lesions in the modeled image. It would also provide standards similar to what are used to compare micro-processors, a concept useful for researchers, manufacturers and clinicians.

Acknowledgements

The authors would like to thank the University of Calgary, the Natural Sciences and Engineering Research Council of Canada (NSERC) for financial support in the form of operating funds. We appreciate the comments from Drs. A. P. Crawley, D. Axelson and S. T. Nichols. We appreciate Y. Hui allowing us to use preliminary results for his MRI neural network algorithms.

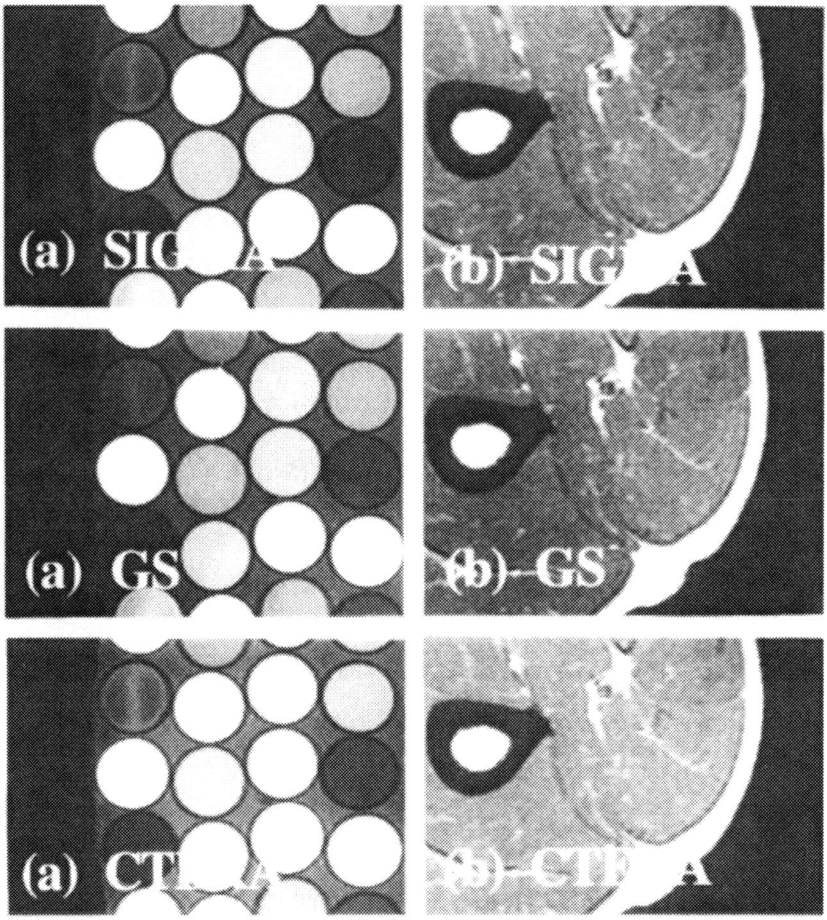

Figure 27: Phantom I (a) and thigh (b) images reconstructed using the standard DFT from the "full" (upper) and "truncated" (centre) data are compared to the MTERA reconstruction on the "truncated" data (lower).

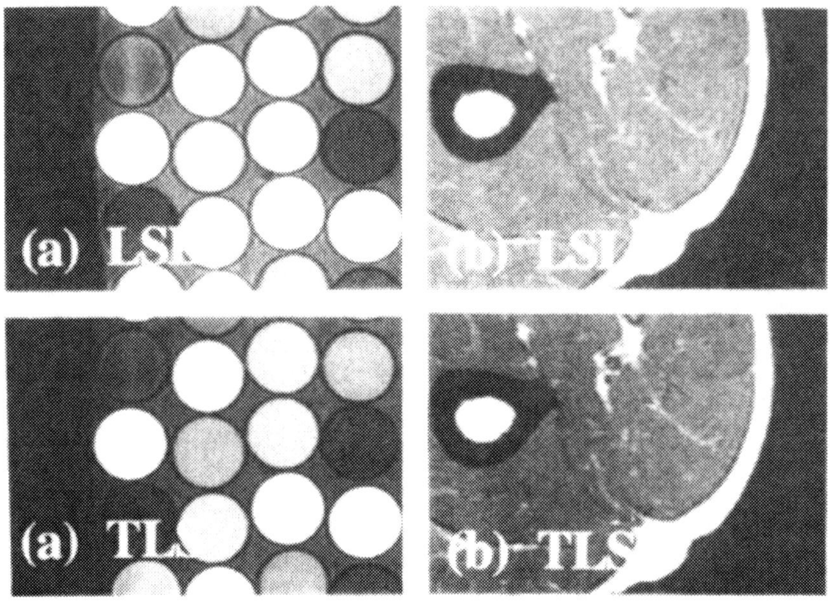

Figure 28: Phantom I (a) and thigh (b) images reconstructed from the "truncated" are compared for the adaptive-TERA algorithm using LSL (upper) and the TERA algorithm using TLS.

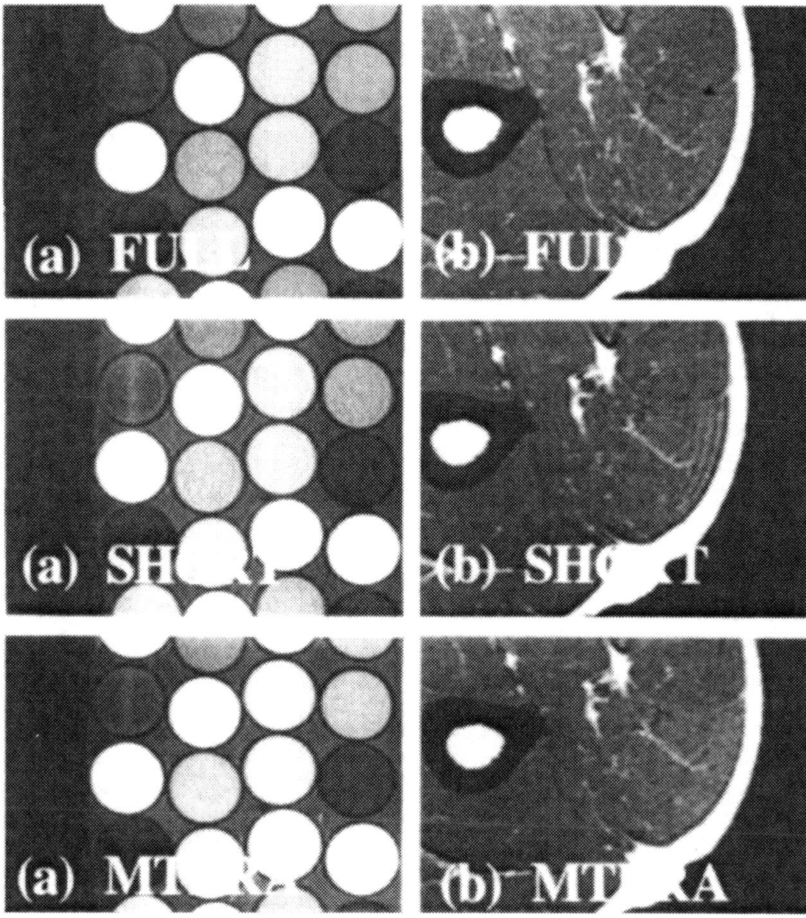

Figure 29: The "truncated" image processed by the iterative Sigma filter method (upper), the GS algorithm (centre) and constrained CTERA (lower) algorithms.

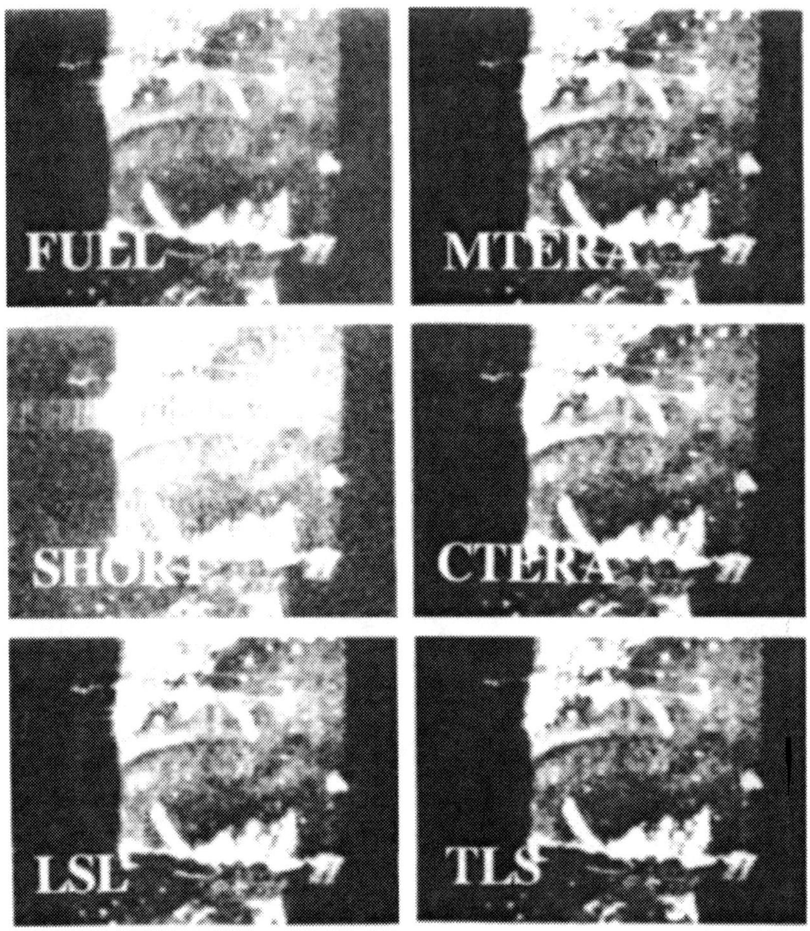

Figure 30: Comparison geology images for the DFT and TERA algorithms.

References

[1] M. R. Smith, S. T. Nichols, R. M. Henkelman and M. L. Wood, "Application of Autoregressive Moving Average Parametric Modeling in Magnetic Resonance Image Reconstruction", *IEEE Trans. Med. Imaging* M1-5:132–139, 1986.

[2] M. R. Smith, S. T. Nichols, R. T. Constable and R. M. Henkelman, "A Quantitaive Comparison of the TERA Modeling and DFT Magnetic Resonance Image Reconstrucion Techniques", *Magn. Reson. Med.* 19:1–19, 1991.

[3] M. R. Smith and S. T. Nichols, "A Two-Dimensional Modeling Reconstruction Technique for Magnetic Resonance Data", *J. Magn. Reson.* 85:573–580, 1989.

[4] Z. P. Liang, F. E. Boada, R. T. Constable, E. M. Haacke, P. C. Lauterbur and M. R. Smith, "Constrained Reconstruction Methods in MR Imaging", *Rev. Magn. Reson. Med.* 4:67–185, 1992.

[5] S. J. Gibbs and A. E. James Jr., "Bioeffects", *Magnetic Resonance Imaging*, C. L. Partain *et. al.* eds., W. B. Saunders Company, Philadelphia, 1988.

[6] M. Osbakken, J. Griffith and P. Taczanowsky, " A gross Morphologic, Histologic, Hematologic, and Blood Chemistry Study of Adults and Neonatal Mice Cronically Exposed to High Magnetic Fields", *Magn. Reson. Med.* 3:502–517, 1986.

[7] T. S. Curry, J. F. Dowdey and R. C. Murray, *Christensen's Introduction to the Physics of Diagnostic Radiology*, Lea & Febiger, Philadelphia, 1984.

[8] B. R. Friedman, J. P. Jones, G. Chaves-Munoz, A. P. Salmon and C. R. R. Merritt, *Principles of MRI*, McGraw-Hill Information Services Company, 1988.

[9] A. P. Crawley and M. R. Smith, "Magnetic Resonance Imaging", *Indian Med. Life Sci. J.*, special issue on medical imaging, vol. 2, in press 1995.

[10] R. R. Henkelman and M. J. Bronskill, "Artifacts in Magnetic Resonance Imaging", *Rev. Magn. Reson. Med.* 2:1–126, 1987.

[11] A. Kumar, D. Welti, and R. R. Ernst. "NMR Fourier Zeugmatography", *J. Magn. Reson.* 18:69–83, 1975.

[12] F. J. Harris, "On the Use of Windows for Harmonic Analysis with the Discrete Fourier Transform", *Proc. IEEE.* 66:51–84, 1978.

[13] A. Hasse, J. Frahm, D. Matthhaei, W. Haenicke and K. D. Merboldt, "FLASH imaging: rapid NMR imaging using low flip angle pulses", *J. Magn. Reson.* 67:258–266, 1986.

[14] P. Mansfield and I. L. Pykett, "Biological and medical imaging by NMR", *J. Magn. Reson.* 29:355-373, 1978.

[15] M. R. Smith and S. T. Nichols, "DFT matching - A Method of Introducing Constraints into Alternative MRI Reconstrucion Algorithm", *Proceedings of 10th SMRM Conference*, California, 2:749, 1991.

[16] Z. P. Liang and P. C. Lauterbur, "A Generalized Series Approach to MR Spectroscopic Imaging", *IEEE Trans. Med. Imaging* MI-10:132–137, 1991.

[17] R. T. Constable and R. M. Henkelman, "Data extrapolation for truncation artifact removal", *Magn. Reson. Med.* 17:108–118, 1991.

[18] S. Amartur and E. M. Haacke, "Modified Iterative Model Based on Data Extrapolation Methods to Reduce Gibbs Ringing", *J. Mag. Res. Imag.* 1:307–317, 1991.

[19] J. Yang, *Alternative MRI Reconstruction Techniques Using Modeling*, M. Sc. dissertation, Electrical and Computer Engineering, University of Calgary, Canada, 1993.

[20] M. R. Smith and S. T. Nichols, "Improved Resolution in the Analysis of Multicomponent Exponential Signals", *Nucl. Instrum. Methods.* 205:479-483, 1983.

[21] H. L. Armstrong. "On Finding an Orthonormal Basis for Representing Transients", *IRE Trans, Circ. Theory* CT-4:286, 1957.

[22] E. M. Haacke, Z. P. Liang and S. Lzen, "Superresolution Reconstruction through Object Modeling and Parameter Estimation", *IEEE Trans. Acoust., Speech, Signal Processing* 37:592–595, 1989.

[23] M. R. Smith and S. T. Nichols, "A Comparison of Models Used as Alternative Magnetic Resonance Image Reconstruction Method", *Mag. Res. Imag.* 8:173–183, 1990.

[24] D. K. Mitchell, "Image Enhancement in Magnetic Resonance Imaging", M.Sc. dissertation, Department of Electrical and Computer Engineering, University of Calgary, June, 1987.

[25] M. R. Smith, "Constrained ARMA Method for Magnetic Resonance Imaging", *IEEE S.S.A.P. Conference 92*, Victoria, Canada, Oct. 1992.

[26] J. Alam, "Multi-exponential Signal Analysis Using The Total Least Square Method", M.Sc. dissertation, Department of Electrical and Computer Engineering, University of Calgary, June, 1990.

[27] I. Barrowdale and R. E. Erickson, "Algorithm for Least Square Prediction and Maximum Entropy Spectral Analysis-Part 1: Theory," *Geophys*, 45:420–432, Mar. 1980.

[28] S. Haykin, *Adaptive Filter Theory*, Prentice-Hall, Englewood Cliffs, New Jersey.

[29] S. L. Marple, Jr, *Digital Spectral Analusis with Applications*, Prentice-Hall Inc, Englewood Cliffs, NJ, 1987.

[30] L. Ljung and T. Soderstrom, *Theory and Practice of Recursive Indentification*, The MIT Press, Cambridge, Mass., 1983.

[31] J. M. Cioffi and T. Kailath, "Fast, Recursive-Least-Square, Transversal Filters for Adaptive Filtering", *IEEE Trans. Acoust., Speech, Signal Processing* ASSP-32:304–337, 1984.

[32] D. W. Lin, "On Digital Implementation of the Fast Kalman Algorithm", *IEEE Trans. Acoust., Speech, Signal Processing* ASSP-32:998–1006, 1984.

[33] P. Fabre and C. Gueguen, "Improvement of the Fast Recursive Least-Square Algorithm via Normalization: A Comparative Study", *IEEE Trans. Acoust., Speech, Signal Processing* ASSP-34:296-308, 1986.

[34] M. Morf, "Fast Algorithm for Multivariable Systems", Ph.D. dissertation, Stanford University, Stanford, California, U.S.A., 1974.

[35] M. Morf, A. Vieira and D. T. Lee, "Ladder Forms for Identification and Speech Processign", *Proc. IEEE Conf. Decison and Control*, New Orleans, 1074–1078, 1977.

[36] M. Morf and D. T. Lee, "Recursive Least Squares Ladder Forms for Fast Parameter Tracking", *Conf. Decision and Control*, San Diego, Calif., pp. 1362–1367, 1978.

[37] W. M. Saar, "High Quality Magnetic Resonance Imaging Using Reduced Data Sets and Arma Modeling", M.Sc. dissertation, Department of Electrical and Computer Engineering, University of Calgary, 1988.

[38] A. R. Penner, M. R. Smith and S. T. Nichols, "Noise reduction in magnetic resonance images using DFT and TERA modeling reconstruction", Proc. 11th Ann. Int. Conf. IEEE Eng. Med. Biol. Soc., 642–643, 1989.

[39] K. E. Scott, "Diversity with Multichannel Equalization", Ph.D. dissertation, Department of Electrical and Computer Engineering, University of Calgary, Calgary, 1991.

[40] R. P. Lippmann, "An Introduction to Computing with Neural Nets," *IEEE Signal Processing Mag.* 4–22 1987.

[41] D. R. Hush and B. G. Horne, "Progress in Supervised Neural Networks: What's New Since Lippmann?," IEEE Signal Processing Mag. 8–39, 1993.

[42] S. Haykin, "Neural Networks: a Comprehensive Foundation," IEEE Press, 1994.

[43] H. Yan and J. Mao, "Data Truncation Artifact Reduction in MR Imaging Using a Multilayer Neural Network," *IEEE Trans. Med. Imaging* MI-12:73–77, 1993.

[44] Y. Hui and M. R. Smith, "Comment on 'Data Truncation Artifact Reduction in MR imaging Using a Multilayer Neural Network", submitted to *IEEE Trans. Med. Imaging*, Aug., 1994.

[45] N. Benvenuto and F. Piazza, "On the Complex Backpropagation Algorithm," *IEEE Trans. Signal Processing* SP-40:967–969, 1992.

[46] H. Leung and S. Haykin, "The Complex Backpropagation Algorithm," *IEEE Trans. Signal Processing*, SP-39:2101–2104, 1991.

[47] Y. Hui and M. R. Smith, "MRI Reconstruction From Truncated Data using a Complex Domain Backpropagation Neural Network," *Proc. of IEEE Pacific Rim Conf. on Communications, Computers, Visualization and Signal Processing.*, Victoria, May, 1995 – accepted.

[48] J. J. van Vaals, M. E. Brummer, W. T. Dixon, H. H. Tuithof, H. Engels, R. C. Nelson, B. M. Gerety, J. L. Chezmar, J. A. den Boer, "Keyhole Method for Accelerating Imaging of Contrast Agent Uptake," J. Magn. Reson. Imaging 3:671–675, 1993.

[49] R. A. Jones, O. Haraldseth, T. B. Müller, P. A. Rink, A. N. Øksendal, "K-space Substitution: A Novel Dynamic Imaging Technique," Magn. Reson Med. 29:830–834, 1993.

[50] Y. Hui and M. R. Smith, "An Adaptive Resolution Enhancement Method for Dynamic MRI Using Complex Domain Backpropagation Algorithm," submitted to *1995 I. Conf. Image Proc.*, October 1995, submitted February 1995.

[51] Y. Hui and M. R. Smith, "Comparing a Neural Network Based Image Enhancement Technique with Keyhole Magnetic Resonance Imaging," *1995 Int. Soc.Mag. Reson. Conf*, Sydney, Australia, July 1995, submitted April 1995.

[52] Y. Hui, "Complex Domain Backpropagation Neural Network and Its Applications in MRI Reconstruction and Dynamic Imaging," *M. Sc. thesis*, Deptartment of Electrical and Computer Engineering, Univ. of Calgary, Canada – completion date June 1995.

[53] M. R. Smith, D. Krawciw and D. Axelson, "Alternative MRI Reconstruction Methods on Geological Data", *MRI in Applied Science.*, Duke University, USA, Oct. 1992.

[54] J. Zeng, "Evaluation procedures for MRI post-processing algorithms", M. Sc. dissertation, Department of Electrical and Computer Engineering, University of Calgary, Canada – completion date May, 1995.

[55] T. Mathews, Jnr., "MRI reconstruction using Wavelets", M. Eng. dissertation, Department of Electrical and Compuer Engineering, University of Calgary, Canada – completion date September, 1995.

INTEGRATION OF NEURAL CLASSIFIERS FOR PASSIVE SONAR SIGNALS

Joydeep Ghosh and Kagan Tumer
Dept. of Elect. and Comp. Engineering,
University of Texas,
Austin, TX 78712-1084

Steven Beck and Larry Deuser
Tracor Applied Sciences
6500 Tracor Lane,
Austin, TX 78725.

Abstract

The identification and classification of underwater acoustic signals is an extremely difficult problem because of low SNRs and a high degree of variability in the signals emanated from the same type of sound source. Since different classification techniques have different inductive biases, a single method cannot give the best results for all signal types. Rather, more accurate and robust classification can obtained by combining the outputs (evidences) of multiple classifiers based on neural network and/or statistical pattern recognition techniques. In this paper, five approaches are compared for integrating the decisions made by networks using sigmoidal activation functions exhibiting global responses with those made by localized basis function networks. These methods are compared using realistic oceanic data. The first method uses an entropy-based weighting of individual classifier outputs. The second is based on combination of confidence factors in a manner similar to that used in MYCIN. The other three methods use simpler techniques of majority voting, averaging, and density estimation with little extra computational overhead. The results indicate that evidence integration provides significant gains when networks are trained on qualitatively different feature sets. Integration techniques also provide a basis for detecting outliers and "false alarms."

1 Introduction

Artificial neural network (ANN) approaches to problems in the field of pattern recognition and signal processing have led to the development of various "neural" classifiers using feed-forward networks [Lip89]. These include the Multi-Layer Perceptron (MLP) as well as kernel-based classifiers such as those employing Radial or Elliptical Basis Functions (EBFs) [BL88, MD89]. These networks can serve as non-parametric, adaptive classifiers that learn through examples [Lip89], without requiring a good *apriori* mathematical model for the underlying signal characteristics.

If the (noisy) input signals representing different classes have substantial overlap in their density distributions, then the statistically optimal expected correct classification achieved by Bayes decision theory can be significantly less than 100% [DH73]. For real-life, difficult signals such as the underwater acoustic signals considered in this paper, the apriori class distributions are not known, and thus Bayesian *a posteriori* probabilities are not immediate. However, recent results show that that multilayer feedforward networks trained by minimizing either the expected mean square error (MSE) or cross-entropy, and by using a 1 of M teaching function, yield network outputs that estimate posterior class probabilities [Gis90, RL91, SCSP91]. These estimations have been observed to be very good for low-dimensional input patterns, at least in regions where there are sufficient training patterns [RL91, SCSP91]. Detailed experiments for high-dimensional patterns have not been reported yet, but they should also give good estimates of aposteriori class probabilities if the cost function is reduced to a sufficiently low value during training, and if the training data reflect true likelihood distributions.

The idea of integrating knowledge from disparate sources has been explored in several fields. Particularly notable are the Dempster-Shaefer theory of evidence and evidence combination in machine learning literature [GLF81], and multisensor fusion in pattern recognition. In the context of feedforward networks, interpretation of network outputs as Bayesian probabilities provides a sound basis for combining the results from multiple classifiers to yield more accurate classification. This idea was first explored by us in [GBC92] wherein we examined several techniques for combining such "evidences", and compared their efficacies for classifying underwater acoustic signals. The results clearly showed that using multiple classifiers provides better and more robust classification decisions. Multiple classifier integration has also been suggested for automated cytology screening [LHDN91], using an approach based on the theory of binary decision trees [Shl90]. The concept of stacked generalization, an inductive approach to combining generalizers, has been recently introduced by Wolpert [Wol92]. Another aspect of getting more robust classification results when using a gradient-descent based network such as the MLP is explored in [KL91]. Their work uses a bank of MLPs that are identical except that they use different initial values of weights, for 3-D object classification and pose estimation. Even the direct use of a second MLP network to combine the outputs of diverse classifiers has shown statistically significant improvement in prediction of protein secondary structure [ZMW92].

In this paper, we focus again on passive sonar signals. This problem domain is of special interest because of the high dimensionality of the input patterns, and the non-trivial decision boundaries demanded. Also, they allow the study of various rejection strategies for countering outliers and false alarms. In the following section, we outline the ANN classifiers used for the experiments. Section 3 describes the database of underwater biologic signals used, and related issues of signal preprocessing and feature-extraction. Experimental results obtained by individual classifiers on this data set are presented in Section 4. The following sections introduce and evaluate various techniques to combine the evidences obtained from distinct classifiers for more accurate and robust results. The work presented in this paper is part of a larger project on the design of a detection and classification system that uses a hybrid of ANN and statistical pattern recognition techniques tailored to recognizing short-duration oceanic signals [Gho91, GBC92, GDB92, BG92].

2 Overview of ANN Classifiers Used

Our experiences, corroborated by those of several other researchers (see [NL91] for example), show that classification error rates are similar across different ANN classifiers when they are powerful enough to form minimum error decision regions, if they are properly tuned, and when sufficient training data is available. Practical characteristics such as training time, classification time and memory requirements, however, can differ by orders of magnitude [GH89]. Also, the classifiers differ in their robustness against noise, effects of small training sets, and in their ability to handle high-dimensional inputs [BG92]. A good review of probabilistic, hyperplane, kernel and exemplar-based classifiers that discusses the relative merit of various schemes within each category, is available in [Lip89, NL91]. Comparisons between these classifiers and conventional techniques such as decision trees, K nearest neighbor, Gaussian mixtures, and CART can be found in [NL91, WK91]. For this study, we chose the MLP and the EBF network, since these classifiers are comparatively insensitive to noise and to irrelevant inputs [BG92].

2.1 MLP.

The standard fully-connected MLP network that adapts weights using "backpropagation" of error is perhaps the most commonly used static ANN classifier [Lip89]. The first network used is an MLP which is trained using the back-propagation rule with momentum. A decaying learning rate is used. Though we have also experimented with various pruning techniques such as "Optimal Brain Damage" [CDS90], the results were not statistically different from a fully connected MLP of appropriate size, and hence are not reported here.

2.2 Localized Basis Function Networks.

The second network used for this study belongs to a class of single hidden-layer feedforward networks in which localized basis functions are used as the activation functions for the hidden layer units. The output of the ith output node, $f_i(\mathbf{x})$ when input vector $\mathbf{x}$ is presented, is given by :

$$f_i(\mathbf{x}) = \sum_j w_{ij} R_j(\mathbf{x}), \tag{1}$$

where $R_j(x)$ is a suitable local basis function that defines the output of the jth hidden node, $\mathbf{x}_j$ is the location of the jth centroid, where each centroid is a kernel/hidden node, and w_{ij} is the weight connecting the jth kernel/hidden node to the ith output node.

A well known example is the Radial Basis Function (RBF) network wherein a radially symmetric function is chosen for $R(x)$, i.e., $R_j(\mathbf{x}) = R(||\mathbf{x} - \mathbf{x_j}||)$ [CGDB91, MD89]. If Gaussian functions are chosen as the basis functions, we have

$$R_j(\mathbf{x}) = e^{-\frac{1}{2}\frac{||\mathbf{x}-\mathbf{x_j}||^2}{\sigma^2}}$$

where σ determines the width of the receptive field. The j hidden node has a maximum output of 1 when input $\mathbf{x} = \mathbf{x_j}$. Also, contours of constant output value are hyperspheres centered at $\mathbf{x_j}$, as is expected from a radially symmetric function. Since RBFs are equally responsive in all input dimensions, they are not so practical and efficient for high dimensional inputs. This is because a much larger number of kernel nodes are required to cover higher dimensional spaces with adequate resolution [HK91]. Moreover, they are more sensitive to noise and to irrelevant inputs [BG92].

The abovementioned drawbacks can be countered by using

Elliptical Basis Function (EBF) networks that employ Gaussian basis functions but with different variance, σ, in different dimensions. Thus, for an EBF network, Eq. 1 is used with

$$R_j(\mathbf{x}_p) = e^{-\frac{1}{2}\sum_k \frac{(x_{pk}-x_{jk})^2}{\sigma_{jk}^2}} . \tag{2}$$

The EBF network is a type of Gaussian Potential Function Network [LK88] which involves segmentation of the input domain into several potential fields in form of Gaussians. The Gaussian potential functions of this scheme need not be radially symmetric functions. Instead, the sizes of these potential fields are determined by a correlation matrix. For the EBF, the off-diagonal elements of this matrix are zero, thus reducing the number of free variables used.

The network parameters of EBF are computed by gradient descent. Consider a quadratic error function, $E = \sum_p E_p$ where $E_p = \frac{1}{2}\sum_i (t_i^p - f_i(\mathbf{x_p}))^2$. Here t_i^p is the target function for input $\mathbf{x_p}$ and f_i is as defined in Eq. 1. The mean square error is the expected value of E_p over all patterns. The update rules for various network parameters are obtained using gradient descent on E_p:

$$\Delta w_{ij} = \eta_1 (t_i^p - f_i(\mathbf{x_p}))R_j(\mathbf{x_p}) \tag{3}$$

$$\Delta x_{jk} = \eta_2 R_j(\mathbf{x_p})\frac{(x_{pk}-x_{jk})}{\sigma_{jk}^2}(\sum_i (t_i^p - f_i(\mathbf{x_p}))w_{ij}) \tag{4}$$

$$\Delta \sigma_{jk} = \eta_3 R_j(\mathbf{x_p})\frac{(x_{pk}-x_{jk})^2}{\sigma_{jk}^3}(\sum_i (t_i^p - f_i(\mathbf{x_p}))w_{ij}). \tag{5}$$

Classification experiments show that these Elliptical Basis Function Networks require considerably shorter training times compared to RBF networks and typically require a fewer number of kernel nodes. However, training time per epoch is increased since centroids and center widths are also adapted using the generalized delta rule, which is a slower procedure.

How should the widths, $\sigma_{jk}s$ be initialized? For RBF, the initial positions of the centroids are typically obtained by k-means clustering, and the width σ_j for the jth centroid is of the same order as the distance between this unit and the nearest centroid, $\mathbf{x_{j^*}}$. This suggests the initialization

$$\sigma_{jk}(init) = \sigma \times \|\mathbf{x_j} - \mathbf{x_{j^*}}\|, \quad \forall j, k.$$

However, since the spread of data is in general different in different dimensions, an initialization given by:

$$\sigma_{jk}(init) = n^{1/2}\sigma \times \|\mathbf{x_{jk}} - \mathbf{x_{j^*k}}\|, \quad \forall j, k.$$

seems more appropriate, where $\sigma = O(1)$ determines selectivity, and $n^{1/2}$ is a normalization term for n-dimensional inputs so that the average variance is σ_j^2, as before. The latter initialization strategy is used in this paper.

3 Data Description and Representation

The data used for the experiments were based on two sets of feature vectors, both of which were extracted from short–duration passive sonar signals due to four types of naturally occurring oceanic sources. The first feature set, FS1, consisted of 25–dimensional feature vectors and the second set, FS2, consisted of 24–dimensional feature vectors. The signal source type and number of training and test samples are given in Tables 1 and 2.[1]

Table 1: Description of Data for Feature Set 1 (FS1).

Class	Description	Training	Testing
1	Porpoise Sound	116	284
2	Ice	116	175
3	Whale Sound 1	116	129
4	Whale Sound 2	148	235
Total		496	823

Table 2: Description of Data for Feature Set 2 (FS2).

Class	Description	Training	Testing
1	Porpoise Sound	142	284
2	Ice	175	175
3	Whale Sound 1	129	129
4	Whale Sound 2	118	235
Total		564	823

Signal preprocessing as well as the extraction of good feature vectors is crucial to the performance of the classifiers[BDSW91]. For

[1] Note that the test sets are extracted from the same raw data.

FS1, the 25-dimensional feature vectors extracted from the raw signals consist of :

- 16 coefficients of Gabor wavelets—a multiscale representation that does not assume signal stationarity[CGE89],
- 1 value denoting signal duration, and
- 8 other temporal descriptors and spectral measurements.

For FS2, the 24–dimensional feature vectors extracted from the raw signals consist of:

- 10 reflection coefficients corresponding to the maximum broad-band energy segment, using a short term correlation estimate in the decision time–window,
- 10 reflection coefficients, obtained using a long time window spanning the entire time-window,
- 3 temporal descriptors derived from different noise thresholds, and
- 1 value denoting signal duration.

The preprocessing techniques and choice of the feature vectors reflect the following two goals:

- removal or normalization of all effects that vary, but not as a result of the events of interest, to the greatest possible extent; and

- the presence of each type of signal of interest should result in a measurable difference in the observable features.

4 Performance of Individual Classifiers

For each of the two feature sets, two different classifiers were used. These were

- C1 = MLP with a single hidden layer of 50 nodes.
- C2 = EBF with 60 centroids.

Several different sizes for each classifier were tested before settling on the appropriate ones. Each of the networks were "fully trained," i.e. were trained until the classification rate on the test set reached its maximum. The training samples are indicated in the first column of Tables 1 and 2. The networks were then tested using the samples from the second column in each of those Tables. The classification results and the confidence intervals (CI) for each individual classifiers on both feature sets (averaged over 10 runs, with different random initial weights) are presented in Table 3. These results point to three quick conclusions:

- FS1 represents the actual data better than FS2.
- C1 outperforms C2.
- The FS2-C2 combination has difficulties capturing the essential properties of the data.

Table 3: Results for Individual Classifiers.

Classifier/ Feature set	% Correctly Classified	Standart Deviation	95% Confidence Interval
C1/FS1	92.66	.63	92.21–93.11
C1/FS2	88.60	.73	88.08–89.12
C2/FS1	91.30	.64	90.84–91.76
C2/FS2	82.02	2.22	80.43–83.61

Table 4: Confusion Matrix for C1 with FS1.

	decision			
	1	2	3	4
Class				
1	268	5	6	5
2	2	157	3	13
3	2	4	117	6
4	5	0	6	224

While the overall performance is comparable among the two classifiers for FS1, there is a significant difference for FS2. This can be explained by the fact that FS2 may have less local information, and therefore not lend itself to a classifier that takes advantage of the local aspects of the data. This type of difficulty may manifest itself during the placement of the centroids, and may prevent the centroids from representing the input space with sufficient precision.

Table 5: Confusion Matrix for C1 with FS2.

	decision			
	1	2	3	4
Class				
1	254	18	12	0
2	7	158	5	5
3	3	8	115	3
4	8	12	7	208

Table 6: Confusion Matrix for C2 with FS1.

	decision			
	1	2	3	4
Class				
1	274	4	3	3
2	2	150	5	18
3	2	5	115	7
4	6	4	11	214

Table 7: Confusion Matrix for C2 with FS2.

	decision			
	1	2	3	4
Class				
1	216	37	17	14
2	5	123	11	36
3	2	7	111	9
4	4	13	6	212

The classification results on the test set are displayed in Tables 4–7 using confusion matrices.[2] For each technique, the correct class is given by the horizontal row label and the misclassification (or confused) class assigned is given by the vertical column label. The numerical entries give the number of vectors in each class that is classified as belonging to each decision region. Elements on the diagonal represent correct classifications, whereas elements off the diagonal represent classification errors. For example, in Table 4 there were 13 vectors belonging to class 2 that were erroneously classified as belonging to class 4. Such a display format exposes which signals cause the greatest problems for a given technique. Table 7, for example points to the potential source of the problem for using C2 with FS2: There is a disproportionate amount of patterns from class 1 that are classified as belonging to class 2, and a similar number of patterns

[2]The results in Table 3 are based on averages, whereas the confusion matrices in Tables 4–7 are based on sample runs. Therefore there may be slight differences in classification percentages.

from class 2 that are classified as belonging to class 4. That information can be used in directing new training strategies, by either focusing on those centroids or classes that cause the most problems.

5 Evidence Integration and Decision Making

It has been recently shown that training multilayer feedforward networks by minimizing the expected mean square error (MSE) at the outputs and using a 0/1 teaching function yields network outputs that approximate posterior class probabilities [Gis90, RL91, SCSP91]. In particular, the MSE is equivalent to

$$MSE = K_1 + \sum_c \int_x D_c(x)(P(c/x) - f_c(x))^2 dx \tag{6}$$

where K_1 and $D_c(x)$ depend on the class distributions only, $f_c(x)$ is the output of the node representing class c given an input x, $P(c/x)$ denotes the posterior probability, and the summation is over all classes. Thus, minimizing the (expected) MSE corresponds to a weighted least squares fit of the network outputs to the posterior probabilities.

This result gives a sound mathematical basis for interpreting the network outputs as probabilities, and for using an integrator to combine the outputs from multiple classifiers to yield a more accurate classification. For very low values of MSE, $f_c(x)$ approximates

$P(c/x)$ according to Eq. 6. Let $f_{c,i}(x)$ be the output of the node that denotes membership in class c in the ith neural classifier. We expect that, for all i, x [Mak91]:

$$\sum_c f_{c,i} \simeq 1. \tag{7}$$

Similarly, if the posteriori estimate is very good, one would expect for all c, i:

$$\frac{1}{N}\sum_{j=1}^{N} f_{c,i}(\mathbf{x}_j) \simeq P(c), \tag{8}$$

where j indexes the N training data samples, and P(c) is obtained by counting the frequency of class c in the training data. Indeed, both Eq. 7 and Eq. 8 are observed to be quite accurate for signals that are classified with high confidence by all the three classifiers.

Note that $f_{c,i}(\mathbf{x}_j)$ can be interpreted as the confidence of membership for a specific input sample, $\mathbf{x}_j$. The *class* confidences for each class can be derived for different classifiers by the following rule [Shl90]:

$$g_{c,i} = \frac{n_{c/c,i} + 1}{(\sum_k n_{k/c,i}) + 2}, \tag{9}$$

where $n_{k/c,i}$ is the number of input samples in the training set that belong to class c but are labeled as class k by the ith classifier. By adding 1 and 2 in the numerator and denominator respectively, we ensure that $0 < g_{c,i} < 1$, and is particularly helpful if the training set is small.

By combining the concept of class confidence with the interpretation of the outputs as *aposteriori* class probabilities, an overall classification confidence, $y_{c,i}(x)$ is obtained for each sample and for each classifier:

$$y_{c,i}(x) = g_{c,i} \times f_{c,i}(x) \tag{10}$$

Our goal is now to combine the $y_{c,i}(x)$s from individual classifiers to obtain an overall confidence across all classifiers. Two methods for such evidence combination are proposed:

5.1 Entropy-Based Integrator (ENT).

In this method, a weighted average of the outputs of n different classifiers is first performed, with a larger entropy resulting in a smaller weight. The integrator then selects the class corresponding to the maximum value, providing this value is above a threshold. Otherwise, the input is considered as a false alarm, since there is no strong evidence that it belongs to any of the n classes.

First, for every classifier, the entropy is calculated using normalized outputs $y'_{c,i} = \frac{y_{c,i}}{\sum_c y_{c,i}}$. The weight given to each classifier (normalized) output differs from sample to sample according to the (approximated) entropy at the output of that classifier, as follows:

$$H(c) = \frac{1}{n} \sum_{i=1}^{n} \frac{y_{c,i}}{-\sum_c y'_{c,i}\ ln\ y'_{c,i}};$$

$$\text{assigned class label} = \text{c} : max\ H(c). \tag{11}$$

In this way, the outputs of a classifier with several similar values get a lower weighting as opposed to classifiers which strongly hypothesize a particular class membership.

5.2 Heuristic Combination of Confidence Factors (CF).

This approach is inspired by techniques for parallel combination of rules in expert systems. Certainty factors were introduced in the MYCIN expert system for reasoning in expert systems under uncertainty, and reflect the confidence in a given rule [SB75]. The original method of rule combination in MYCIN was later expressed in a more probabilistic framework by Heckerman [Hec86], and serves as the basis for the method proposed below:

First, the outputs, which are in the range [0,1], are mapped into confidence factors (CFs) in the range [-1,1] using a log transformation. Then, a Mycin-type rule is used to combine the CFs for each class. The advantage of this combination rule is that it makes the result invariant to the sequence in which the different network outputs are combined.

The individual CFs are first obtained using:

$$CF_{c,i} = \log_n((n - 1/n)y_{c,i} + 1/n). \tag{12}$$

For each class c, all positive CFs and all negative CFs are combined separately. The resultant positive and negative CFs are combined in the final step, to obtain a combined confidence CF_c, for each class c. The classification decision is:

$$\text{assigned class label} = \text{c} : max\ CF_c.$$

The equations used for combining the CFs are similar to those used in the original MYCIN [SB75], namely:

$$\begin{aligned} CF(a,b) &= 1-(1-a)(1-b) \text{ if } a,b>0; \\ &= -CF(-a,-b) \quad \text{if} \quad a,b<0; \\ &= a+b, \qquad\qquad \text{otherwise.} \end{aligned} \tag{13}$$

Again, the values of $max\ CF_c$ were much lower for the deviant signals, yielding another metric for detecting false alarms and unknown signals. Indeed, by varying the threshold for the minimum acceptable value for $max\ H(c)$ or $max\ CF_c$, one can obtain a range of classification accuracy versus false alarm rates, and be able to choose a suitable trade-off point.

The other three techniques are straightforward:

5.3 Geometric Mean (GM).

This approach is based on the assumption that the network outputs represent aposteriori probabilities, and that the outputs of the different classifiers are independent. The geometric mean of each classifier is used to determine the class of the vectors. Therefore, the maximum of the geometric means:

$$\overline{y_c} = \sqrt[n]{\prod_i y_{c,i}}$$

is chosen as the correct class, where n is the number of classifiers.

5.4 Averaging (AVE).

This approach simply averages the individual classifier outputs. The maximum of the averaged values:

$$\overline{y_c} = \frac{1}{n} \sum_i y_{c,i}$$

is chosen as the correct class, where n is the number of classifiers.

5.5 Majority Vote (MV).

The last approach is the majority vote, where the class which yields the maximum value in at least two of the three classifiers is chosen. If all three classifiers indicate different classes, then the one with overall maximum output value is used to indicate the correct class.

6 Integration Results

6.1 Combining Fully Trained Classifiers.

First the different classifiers were combined using data from a single feature set, then two feature sets were combined with data from a single classifier and finally both feature sets were combined with both classifiers. All classifier outputs were obtained after the classifiers were fully trained. The combination strategies for partially trained classifiers is addressed in the next subsection.

The combination of two classifiers on a single feature set provides only minimal improvements. Table 8 shows the results of com-

Table 8: Combining classifiers with a single feature set.

	C1	C2	AVE	CF	ENT	GM	MV
FS1	92.66	91.30	93.03	93.03	92.98	92.88	93.35
FS2	88.60	82.02	89.14	88.70	88.97	88.77	88.60

bining C1 and C2 for a single feature set. The results were combined using the arithmetic mean (AVE), confidence factors (CF), entropy (ENT), the geometric mean (GM) and finally a majority vote (MV). [3] The different combining techniques provided almost identical results for FS1, and only slightly varying results for FS2. The averaging method provided a 5% reduction in the error rate for both feature sets, when compared to the results of using C1 alone. In both cases however the improvements were marginal at best, and at the limit of statistical significance (The values were either within, or barely outside the 95% confidence interval).

Table 9: Combining features sets with a single classifier.

	FS1	FS2	AVE	CF	ENT	GM	MV
C1	92.66	88.60	95.24	94.78	95.07	94.63	93.46
C2	91.30	82.02	93.34	93.05	92.59	92.86	91.76

Table 9 shows the results for combining FS1 and FS2 for each classifier. The improvements are much more pronounced when two feature sets are combined, and clearly statistically significant. For C1

[3] For the case with only two combiners, the majority vote is reduced to picking the combiner which provides the highest value.

there was a 35% reduction in the error rate for the AVE combiner, and for C2 there was a more modest, but still highly significant reduction of 23%.

Finally we combined all four combinations of feature sets and classifiers. The results are presented in Table 10. The error rates were reduced by 39% for the average combiner. Table 11 presents the improvements in the error rates in more detail. The improvements are measured against the best results obtained by a single classifier on a single feature set. For example in the first row of Table 11, improvements are measured against 92.66% correct classification, the result obtained by C1 on FS1.

Table 10: Combining both features sets and classifiers.

AVE	CF	ENT	GM	MV
95.53	94.92	95.12	94.56	94.48

Table 11: Reductions in Error Rates

Combinations	AVE	CF	ENT	GM	MV
C1 and C2 (FS1)	5.0	5.0	4.4	3.0	9.4
C1 and C2 (FS2)	4.7	0.9	3.3	1.5	0.0
FS1 and FS2 (C1)	35.2	28.9	32.5	26.8	10.6
FS1 and FS2 (C2)	23.5	20.1	14.8	17.9	5.3
C1, C2, FS1 and FS2	39.1	30.8	33.5	25.9	24.8

Tables 12–14 present confusion matrices for the AVE combiner. An important observation here is that the errors are not con-

Table 12: Confusion Matrix for C1-C2 combination for FS1 using AVE.

	decision			
	1	2	3	4
Class				
1	277	1	3	3
2	2	164	3	6
3	0	5	124	0
4	3	2	3	227

Table 13: Confusion Matrix for FS1-FS2 combination for C1 using AVE.

	decision			
	1	2	3	4
Class				
1	272	4	4	4
2	2	156	4	13
3	0	5	119	5
4	6	3	7	219

Table 14: Confusion Matrix for FS1-FS2 and C1-C2 combination using AVE.

	decision			
	1	2	3	4
Class				
1	275	2	3	4
2	1	165	3	6
3	0	3	125	1
4	2	3	3	227

centrated in any particular region for the cases where more than one feature set is used. Using more than one feature set improves the performance by eliminating systematic errors.

6.2 Combining Partially Trained Classifiers.

The results using fully trained networks confirm the notion that using two or more qualitatively different feature extractors helps to obtain significant improvements on the quality of the solutions. For the application presented in this article, combining two classifiers on a single feature set provided improvements too modest to warrant the increased computation. This situation raises an interesting point, namely the computational overhead of training more than one classifier and then combining the results. This motivates another set of experiments with classifiers trained for only a fifth of the time they were trained previously.

Table 15: Combining partially trained classifiers with a single feature set.

	C1	C2	AVE	CF	ENT	GM	MV
FS1	90.88	85.54	91.01	91.13	91.01	91.86	90.89
FS2	85.66	73.50	87.24	87.00	86.39	86.69	87.12

Table 16: Combining features sets with a single partially trained classifier.

	FS1	FS2	AVE	CF	ENT	GM	MV
C1	90.88	85.66	94.65	93.20	94.29	93.92	92.47
C2	85.54	73.50	85.66	85.05	87.97	87.24	83.96

Tables 15–17 show the results on combining such partially trained classifiers. The time required for combining results is negligible compared to the training time of each network. Therefore, the results in Tables 15 and 16 were obtained in only 40 % of the time required for obtaining results in Table 3. We conclude from these results that for a global classifier (C1), and in the presence of more than one feature set, the combination of partially trained networks yields results that are superior to a fully trained network tuned to one fea-

Table 17: Combining both features sets and partially trained classifiers.

AVE	CF	ENT	GM	MV
93.80	93.56	93.68	93.44	92.71

ture set. The same however does not hold true for a local classifier, where a partially trained network is not sufficiently "tuned" to the feature vectors to provide data for combinations. Consequently the combination results fall short of the classification rate attained with a single fully trained network. If only a single feature set is available, the accuracy of the combination of partially trained networks does not reach the accuracy of the best fully trained classifier.

The most important result is conveyed by Table 17. Here the total time for training and combining is about the same as the average time for "fully" training a single classifier on a single data set, but the results are significantly superior, i.e. are well outside the confidence intervals provided in Table 3.

6.3 Classification Accuracy vs. False Alarms.

For the results reported in sections 6.1 and 6.2, no thresholding was done, so that a classification decision had to be made for every test vector. Since low output values represent low confidences, this information can be used to reject some of the test samples. A simple technique is to check whether the strongest combined confidence for a given sample is greater than a threshold θ. For example, for the AVE technique, we reject a sample if $max(\overline{y_c}) \leq \theta$, while for ENT, a sample is rejected if $max(\overline{H(c)}) \leq \theta$, where $\overline{H(c)}$ is the normalized value of $H(c)$. This type of thresholding can also be performed on the outputs of single classifiers, by simply comparing the largest output to the threshold, and only assigning a class if the output value exceeds this threshold. Otherwise, we say the the pattern has "missed detection."

Table 18: False Alarms: Missed Detection Percentages for AVE combiner.

θ	C1/ FS1	C1/ FS2	C2/ FS1	C2/ FS2	C1/ FS1–FS2	C1–C2/ FS1	C1–C2/ FS1–FS2
0.0	0	0	0	0	0	0	0
0.2	.85	4.98	.49	13.60	.49	.24	0
0.4	1.58	6.68	.97	23.21	1.72	1.81	4.74
0.6	3.28	8.38	3.28	34.51	13.72	7.29	18.11

Table 19: Misclassification Among Detected Vectors for AVE combiner.

θ	C1/ FS1	C1/ FS2	C2/ FS1	C2/ FS2	C1/ FS1–FS2	C1–C2/ FS1	C1–C2/ FS1–FS2
0.0	7.34	11.40	8.70	17.98	4.76	6.97	4.47
0.2	6.74	8.37	8.48	16.32	4.62	6.97	4.47
0.4	6.28	8.11	8.38	14.34	3.74	6.43	2.56
0.6	5.82	8.12	7.71	11.64	1.60	5.13	.69

Tables 18–19 show the percentage of samples rejected and the classification accuracy among the accepted samples for different values of θ, for the AVE combiner. Tables 20–21 show similar results for the ENT combiner. In these tables the second through fifth columns represent the results for individual classifiers on single data sets. For

Table 20: False Alarms: Missed Detection Percentages for ENT combiner.

θ	C1/ FS1	C1/ FS2	C2/ FS1	C2/ FS2	C1/ FS1–FS2	C1–C2/ FS1	C1–C2/ FS1–FS2
0.0	0	0	0	0	0	0	0
0.2	.85	4.98	.49	13.60	1.46	.24	1.74
0.4	1.58	6.68	.97	23.21	2.67	2.19	3.28
0.6	3.28	8.38	3.28	34.51	4.13	5.13	5.47

Table 21: Misclassification Among Detected Vectors for ENT combiner.

θ	C1/ FS1	C1/ FS2	C2/ FS1	C2/ FS2	C1/ FS1–FS2	C1–C2/ FS1	C1–C2/ FS1–FS2
0.0	7.34	11.40	8.70	17.98	4.93	7.02	4.88
0.2	6.74	8.37	8.48	16.32	4.43	6.91	4.67
0.4	6.28	8.11	8.38	14.34	3.77	6.42	3.41
0.6	5.82	8.12	7.71	11.64	3.23	5.16	2.88

example, for $\theta = .4$, C1 rejects 1.58% of the samples from FS1, and has an accuracy of 93.72% among detected samples, as opposed to 92.66% for all samples. Similarly combining results from FS1 and FS2 obtained using C1, with a threshold of $\theta = .4$, rejects 1.72% of the samples and increases the classification accuracy of accepted samples to 96.26% (instead of 95.24% for all the samples).

Studying these tables provides an interesting conclusion: One can tune the classifiers to reach a pre–set classification rate, as long as one is allowed to reject certain samples. Therefore, increased classification accuracy can be achieved at the expense of increased rejection rates.

6.4 Limits on Improvements due to Combination.

An issue that has been so far neglected is the potential limits on the expected improvements. While combining results from different classifiers and/or feature sets, we are combining evidence that stems from a single raw data set. Therefore, certain errors may be too deeply rooted in the data to be eliminated with these methods. To analyze this further, consider the case where a pair of classifiers are used, and thus the classification errors can divided into three categories:[4]

- Only one classifier fails to produces the correct class.
- Both classifiers fail to produce the correct class, but select different (incorrect) classes.
- Both classifiers produce the same incorrect class.

The first type of error is the most common, and can often be corrected by the combiner. The second type of error is more problematic, yet correctable. As long as the output value of the

[4]The method also applies for combining more than two sources, but the specific cases have to be modified.

correct class is sufficiently high in both classifiers, a correct decision can be made. It is however, unrealistic to expect the correct class to be have the second highest output value in every case. Therefore, this error count gives a realistic bound on the expected number of errors after combination, i.e. the number of errors that can not be corrected without special conditions. The third type of error is not rectifiable by a combination method, since all the evidence points to a particular erroneous class. Therefore, the number of errors in the third category provides an absolute lower bound on the number of errors after combination—or an upper bound on the correct classification rate.

Table 22: Combination Results and Upper Bounds.

	AVE	CF	ENT	GM	MV	Realistic Bound	Absolute Bound
C1	95.14	94.78	95.07	94.63	93.46	97.21	98.66
C2	93.34	93.05	92.59	92.86	91.76	96.96	98.78
FS1	93.03	93.03	92.98	92.88	93.35	95.02	95.14
FS2	89.14	88.70	88.97	88.77	88.60	92.71	94.78

Table 22 shows the result obtained by the different combiners, and compares them to the mentioned upper bounds [5]. The difference between the AVE combiner and the realistic bound varies between 2–3.6%, even though the actual values are significantly different. Therefore an inspection of errors of different type can provide

[5]Here the first column gives the classifier used if two feature sets were combined, and the feature set used if two classifiers were combined.

a guide to the expected combiner error rates. Furthermore, the number of errors of type 2 can be used in the selection of the feature sets, since significantly different feature sets are expected reduce this type of errors.

7 Concluding Remarks

The results presented in this paper highlight the advantage of using evidence combination techniques for improved classification in a situation when the capabilities of a single classifier are fundamentally limited. They also yield a simple metric for expressing the strength of confidence in a given classification decision. Decisions with low confidence measures indicate either improper feature vector selection, inadequate representation by the networks, or that the corresponding signals are outliers or false alarms.

The performance is relatively similar for all five combination techniques. Simple averaging and using the entropy generally provide slightly better results for the data used in this study. Interestingly, the arithmetic mean provides better results than the geometric mean, even though the geometric mean is qualitatively closer to combining conditional probabilities. One explanation for this discrepancy is the dominating effect of low values. For example, if one network has a very low (say 0.01) value for the correct output, then even if the second network has a high (say 0.8) output, the geometrically combined score is less than 0.09. Thus a class which had a value of .1 in both networks will be erroneously chosen.

The results obtained in this study support the theory that more than one feature set is needed to provide significant gains in the classification accuracy. Using different classifiers on a single feature set did not provide improvements that were statistically significant, and hence the added computational expense was unjustified. The results obtained from partially trained classifiers enforced this view. In the presence of at least two feature sets, partial training and subsequent combination does lead to results superior to that of any single classifier. However, in the absence of different feature sets, the results of combining two partially trained classifiers fall short of that of the best fully trained single classifier.

Finally, the percentage of correct classifiers can be significantly boosted, if a modest percentage of samples are rejected. With the proposed type of thresholding, and in the presence of at least two feature sets, the classification error percentage can be reduced by as much as 30%, by only rejecting 2–3% of the samples. Under similar rejection circumstances, but in the absence of two feature sets, only a 5–10% reduction in the error rate is achieved.

Acknowledgements: This research was supported in part by ONR contracts N00014-89-C-0298 and N00014-92-C-0232. We thank Jim Whiteley, and Russell Still of Tracor Applied Sciences for their contributions at various stages of this project.

References

[BDSW91] S. Beck, L. Deuser, R. Still, and J. Whiteley. A hybrid neural network classifier of short duration acoustic signals. In *Proc. IJCNN*, pages I:119–124, July 1991.

[BG92] S. Beck and J. Ghosh. Noise sensitivity of static neural classifiers. In *SPIE Conf. on Applications of Artificial Neural Networks SPIE Proc. Vol. 1709*, Orlando, Fl., April 1992.

[BL88] D.S. Broomhead and D. Lowe. Multivariable functional interpolation and adaptive networks. *Complex Systems*, 2:321–355, 1988.

[CDS90] Yann Le Cun, J.S. Denker, and S. A. Solla. Optimal brain damage. In *Advances in Neural Information Processing Systems -II*, pages 598–605, 1990.

[CGDB91] S. Chakravarthy, J. Ghosh, L. Deuser, and S. Beck. Efficient training procedures for adaptive kernel classifiers. In *Neural Networks for Signal Processing*, pages 21–29. IEEE Press, 1991.

[CGE89] J.M. Combes, A. Grossman, and Ph. Tchamitchian (Eds.). *Wavelets: Time-Frequency Methods and Phase Space*. Springer-Verlag, 1989.

[DH73] R. O. Duda and P. E. Hart. *Pattern Classification and Scene Analysis*. Wiley, New York, 1973.

[GBC92] J. Ghosh, S. Beck, and C.C. Chu. Evidence combination techniques for robust classification of short-duration oceanic signals. In *SPIE Conf. on Adaptive and Learning Systems, SPIE Proc. Vol. 1706*, pages 266–76, Orlando, Fl., April 1992.

[GDB92] J. Ghosh, L. Deuser, and S. Beck. A neural network based hybrid system for detection, characterization and classification of short-duration oceanic signals. *IEEE Jl. of Ocean Engineering*, 17(4):351–363, october 1992.

[GH89] J. Ghosh and K. Hwang. Mapping neural networks onto message-passing multicomputers. *J. of Parallel and Distributed Computing*, 6:291–330, April, 1989.

[Gho91] J. Ghosh et al. Adaptive kernel classifiers for short-duration oceanic signals. In *IEEE Conf. on Neural Networks for Ocean Engineering*, pages 41–48, August 1991.

[Gis90] H. Gish. A probablistic approach to the understanding and training of neural network classifiers. In *Proc. Int'l Conf. on ASSP, Albuquerque,NM*, pages 1361–1364, April 1990.

[GLF81] T.D. Garvey, J.D. Lowrance, and M.A. Fischler. An inference technique for integrating knowledge from disparate sources. In *Proc. of IJCAI*, pages 319–325, 1981.

[Hec86] D. Heckerman. Probabilistic interpretation for MYCIN's uncertainty factors. In L.N Kanal and J.F. Lemmer, editors, *Uncertainty in Artificial Intelligence*, pages 167–196. North-Holland, 1986.

[HK91] E. Hartman and J.D. Keeler. Predicting the future: Advantages of semi-local units. *Neural Computation*, 3(4):566–78, 1991.

[KL91] A. Khotanzad and J. Liou. Recognition and pose estimation of 3-d objects from a single 2-d perspective view by banks of neural networks. In *Proc. ANNIE 91*, pages 479–484, Nov 1991.

[LHDN91] J. Lee, J.-N. Hwang, D.T. Davis, and A.C. Nelson. Integration of neural networks and decision tree classifiers for automated cytology screening. In *Proc. IJCNN*, pages I:257–262, Seattle, July 1991.

[Lip89] R. P. Lippmann. Pattern classification using neural networks. *IEEE Communications Magazine*, pages 47–64, Nov 1989.

[LK88] Sukhan Lee and Rhee M. Kil. Multilayer feedforward potential function network. In *Proceedings of the Second International Conference on Neural Networks*, pages 161–171, 1988.

[Mak91] J. Makhoul. Pattern recognition properties of neural networks. In *Proc. 1st IEEE Workshop on Neural Networks for Signal Processing*, pages 173–187, Sept 1991.

[MD89] J. Moody and C. J. Darken. Fast learning in networks of locally-tuned processing units. *Neural Computation*, 1(2):281–294, 1989.

[NL91] K. Ng and R.P. Lippmann. Practical characteristics of neural network and conventional pattern classifiers. In *Advances in Neural Information Processing Systems -III*, pages 970–976, 1991.

[RL91] M.D. Richard and R.P. Lippmann. Neural network classifiers estimate bayesian a posteriori probabilities. *Neural Computation*, 3(4):461–483, 1991.

[SB75] E.H. Shortcliffe and B.G. Buchanan. A model of inexact reasoning in medicine. *Mathematical Biosciences*, 23:351–379, 1975.

[SCSP91] P.A. Shoemaker, M.J. Carlin, R.L. Shimabukuro, and C.E. Priebe. Least squares learning and approximation of posterior probabilities on classification problems by neural network models. In *Proc. 2nd Workshop on Neural Networks, WNN-AIND91,Auburn*, pages 187–196, February 1991.

[Shl90] S. Shlien. Multiple binary decision tree classifiers. *Pattern Recognition*, 23(7):757–63, 1990.

[WK91] S. M. Weiss and C.A. Kulikowski. *Computer Systems That Learn.* Morgan Kaufmann, 1991.

[Wol92] D. H. Wolpert. Stacked generalization. *Neural Networks*, 5:241–259, 1992.

[ZMW92] X. Zhang, J.P. Mesirov, and D.L. Waltz. Hybrid system for protein secondary structure prediction. *J. Molecular Biology*, 225:1049–63, 1992.

Techniques in the Application of Chaos Theory in Signal and Image Processing

Woon S.Gan

Acoustical Services Pte.Ltd.
29 Telok Ayer Street
Singapore 048429
Republic of Singapore

I. HISTORY OF CHAOS

Chaos occurs only duing nonlinear phenomena. It is deterministic in nature and originates from nonlinear dynamical systems. Hence to trace the history of chaos one has to start with nonlinear dynamical systems. The history of nonlinear dynamical systems begins with Poincare' [1]. He introduced the mathematical techniques of topology and geometry to discuss the global properties of solutions of these systems. In the early 1900's Birkhoff adopted Poincare's point of view and realized the importance of the study of mappings. He emphasized discrete dynamics as a means of understanding the more difficult continuous dynamics arising from differential equations.

The geometric and topological techniques gradually paved the way for mathematicians to fields of study such as differential topology and algebraic topology which became objects of study in their own right. Meanwhile, the study of the dynamical systems themselves subsided, except in the Soviet Union where mathematicians such as Lyapunov, Pontryagin, Andronov and others continued their research in dynamics.

Around 1960, the study of nonlinear dynamical systems revived, mainly due to Moser and Smale in the United States, Peixoto in Brazil and

Kolmogorov, Arnold and Sinai in the Soviet Union. More recently, dynamical systems have been boosted by the techniques arising from a variety of fields. Computer graphics has helped demonstrate the fascinating and beautiful dynamics of simple systems. Feigenbaum, a physicist, has shown the universality of the dynamics of low-dimensional discrete systems. The Lorenz [2] system from meteorology proved the existence of stably chaotic systems.

The discovery of chaos changes our understanding of certain random phenomena which are actually deterministic in nature. Random phenomena are unpredictable but deterministic phenomena are predictable. Many complex deterministic phenomena which are chaotic in nature can be represented only by a few simple dynamical equations.

II. EXAMPLES OF CHAOS

During the last decade, it has been understood that turbulence is chaotic in nature. Other examples of chaos are

A. Thermal convection in fluids, giving rise to the Lorentz model. It was in this system that chaotic-dynamics was first appreciated theoretically with the work of Lorentz [2]. The Lorentz model is of such importance historically and there has been so much work done on it, that the Lorentz equations have become one of the important examples for chaotic dynamics.

B. Supersonic panel flutter, important for supersonic aircraft and rockets, studied by Kobayashi [3].

C. Some chemical reactions, and in particular the Belousof-Zhabotinsky reaction, exhibit chaotic dynamics as discussed by Epstein [4] and Roux [5].

D. Chaos in laser power output has been studied by Atmanspacher and Scheingraber [6] for the case of a continuous-wave dye laser.

E. Cardiac dysrhythmias or abnormal cardiac rhythms have been discussed by Glass et al [7]. It has chaotic behaviour. In addition to the dynamics of the heart, its very structure has several manifestations of self-similar geometrical structures called fractals. Fractal structures are commonly the result of non-linear dynamics, and, although the dynamics governing growth and development of the heart are unknown, fractal structures are detailed in the vascular network for the heart. Furthermore the cardiac impulse itself is transmitted to the ventricles via an irregular fractal network. Many such fractal structures in physiology are reviewed by West and Goldberger [8].

F. There are instances that nonlinear electrical circuits also exhibit chaotic dynamics. One example is the oscillator described by Van der Pol and Van der Mark [9]. For decades this has been a model for nonlinear vibrations. Nonlinear circuits have provided analog devices for modeling many types of nonlinear equations.

G. Several types of standard chaotic behaviour have been observed in simple plasma systems as reported by Cheung and Wong [10] and Cheung et al [11].

H. Ecology and biological population dynamics provide a simple and instructive example of a dynamical system exhibiting chaotic dynamics. This can be described by the "logistic equation". This equation may describe the variations in nonoverlapping biological populations from one year to the next. This equation and its importance were pointed out in early review by May [12].

I Vibrations of buckled elastic systems have provided experimental examples of double-well potential systems and chaotic behaviour. These systems have been studied theoretically and experimentally by Moon and Holmes [13] and Holmes and Whitley [14] as realizations of Duffing's equation which is one of the classical systems studied in chaotic nonlinear oscillations.

J. Chaotic dynamo models have been proposed for representing the geomagnetic field reversals and have been studied by Cook and Roberts [15]. A review has been given by Bullard [16].

K. Harth [17], Nicolis [18] and Skarda and Freeman [19] have found that EEG data suggests that chaotic neural activity plays a role in the processing of information by the brain.

L. By constructing a special computer for the purpose of studying the stability of planetary orbits over long time scales, Sussman and Wisdom (reported by Lewin [20]) have found the orbit of pluto to be chaotic on a time scale of about 20 million years.

III. PROPERTIES AND REPRESENTATION OF CHAOS

A chaotic dynamical system is a deterministic system that exhibits random behaviour through its sensitive dependence on initial conditions. Chaos also has fractal characteristics, that is, it has self-similarity.

There are several nonlinear functions that exhibit chaos and can be used as the representation of chaos:

A. LOGISTIC MAP

$$x_{n+1} = \gamma x_n(1 - x_n) \tag{1}$$

Depending on the value of γ, the dynamics of this system can change dramatically, exhibiting periodicity and chaos.

B. TENT (TRINAGULAR) FUNCTION

$$x_{n+1} = \gamma(1 - 2|0.5 - x_n|)$$

C. SAWTOOTH FUNCTION

$$x_{n+1} = \begin{cases} \gamma x_n, x_n \le \frac{1}{\gamma} \\ \frac{\gamma}{\gamma-1}[x_n - \frac{1}{\gamma}], x_n > \frac{1}{\gamma} \end{cases}$$

D. SINE FUNCTION $x_{n+1} = \gamma \sin(2\pi x_n)$

where γ = bifurcation parameter

Chaos originates from dynamical systems, ie, $f(x,t) = \dot{x}$, $x(t_0) = x_0$

or discrete-time dynamical systems,ie $x_{n+1} = f(x_n)$

where x and x_n are vectors, in general. A one dimensional discrete dynamical system may be represented by the map

$$x_{n+1} = f(x_{n)} = f^{n+1}(x_0), \quad n = 0, 1, 2, \cdots$$

where f: $[a,b] \rightarrow [a,b]$

and $f^{n+1}(x_0) = f(f^n(x_0)) = f(x_n) = x_{n+1}$ (2)

E. CORRELATION PROPERTIES OF CHAOTIC SEQUENCE

The analysis of chaotic sequences is very difficult due to their nonlinear nature. Whenever appropriate, we will take advantage of the similarities of these deterministic sequences to random sequences having smilar properties although chaotic sequences are completely deterministic they can easily be assumed to be samples of a random process with certain probability distribution. This assumption, by no means implies that the consecutive members of the sequnce are statistically independent. In fact, if we start with the assumption that the nth member of a chaotic sequence x_n , is a random variable, then all the consecutive members of the sequence, $x_{n+1}, x_{n+2}, x_{n+3}, \cdots$, will be random variables that are functions of x_n. Moreover, if the chaotic map is invertable, then all the preceding members of the sequence, $x_{n-1}, x_{n-2}, x_{n-3}, \cdots$, , are also functions of the random variable x_n. In the analysis that follows, we will suppose that one of the members of a chaotic sequence is a random variable with a pdf identical to the invariant measure of the map. The rest of the members of the sequence are then taken to be functions of a random variable.

F. AUTOCORRELATION FUNTION OF A LOGISTIC MAP

Recall that the logistic map is of the form

$$x_{n+1} = \gamma x_n(1 - x_n)$$

where $0 \leq x_n \leq 1$ and γ is the bifurcation parameter. Consider a change of variable in the logistic map by scaling and translating x_n:

$$y_n = 2x_n - 1 \tag{3}$$

The new sequence $\{y_n\}$ will then take on values in the range [-1,1]. Thus using (3),(1) transforms to

$$\frac{y_{n+1}+1}{2} = \gamma \frac{y_n+1}{2}\left(1 - \frac{y_n+1}{2}\right)$$

$$y_{n+1} = \tfrac{1}{2}\gamma(1 - y_n^2) - 1, \quad -1 \leq y_n \leq 1 \tag{4}$$

(4) is identical in chaotic characteristics to (1). When γ=4, the original logistic map is chaotic. Substituting this value of γ in (4) yields

$$y_{n+1} = 1 - 2y_n^2, \quad -1 \leq y_n \leq 1 \tag{5}$$

Henceforth, we will refer to (4) as the shifted logistic map.

Another mathematical tool which helps in understanding chaotic dynamical systems is the invariant measure ρ_x. It detemines the density of the iterates of a unimodular map and is defined as

$$\rho(x) = \lim_{N\to\infty} \frac{1}{N} \sum_{i=0}^{N-1} \delta(x - f^i(x_0))$$

$$= \lim_{N\to\infty} \frac{1}{N} \sum_{i=0}^{N-1} \delta(x - x_i) \tag{6}$$

The invariant measure can be thought of as the probability density of the iterates. It has the same form as the empirical distribution of $\{x_n\}$ as if the iterates were random samples from a certain population. It relates the concept of deterministic chaos to the familiar probability theory.

As an example, to find the invariant measure for the logistic map with γ=4,

$$f_4(x) = 4x(1 - x) \tag{7}$$

we can simplify the calculations by introducing the variable

$$x = \tfrac{1}{2}[1 - \cos(2\pi y)] \equiv h(y), \; 0 \leq y \leq 1 \tag{8}$$

Then since $x_{n+1} = f_4(x_n) = 4x_n(1 - x_n)$

we have

$$\tfrac{1}{2}[1 - \cos(2\pi y_{n+1})] = [1 - \cos(2\pi y_n)]$$

$$[1 + \cos(2\pi y_n)] = \tfrac{1}{2}[1 - \cos(4\pi y_n)]$$

(8) has the solution

$$y_{n+1} = 2y_n mod 1 \equiv g(y_n)$$

That is, g is a sawtooth funtion. Hence the invariant measure of g is given by

$$\rho(x) = 1 \tag{9}$$

Let us call this invariant measure $\rho_g(x)$, to distinguish it from that of the logistic map, $\rho_l(x)$. Using (6) and (8) we obtain

$$\rho_l(x) = \lim_{N\to\infty} \frac{1}{N} \sum_{n=0}^{N} \delta(x - x_n)$$

$$= \lim_{n\to\infty} \frac{1}{N} \sum_{n=0}^{N} \delta[x - h(y_n)] \tag{10}$$

Assuming, ergodicity

$$\rho_l(x) = \int_0^1 \delta[x - h(y)]\rho_g(y)dy = \frac{2}{|h'[y(x)]|} = \frac{1}{\pi\sqrt{x(1-x)}} \tag{11}$$

Hence, the invariant measure of the logistic map with γ=4 is the Beta probability density with parameters $\alpha = \frac{1}{2}$ and $\beta = \frac{1}{2}$.
Then the invariant measure of the shifted logistic map is also of the same form but shifted and scaled appropriately:

$$p(y) = \frac{1}{2}\frac{1}{\pi\sqrt{\frac{y+1}{2}(1-\frac{y+1}{2}}} = \frac{1}{\pi\sqrt{1-y^2}}, \quad -1 \le y \le 1 \tag{12}$$

A random variable Y having this pdf has a mean and a variance given by

$$E[Y] = 0 \tag{13}$$

$$Var(Y) = E[y^2] = \int_{-1}^{1} y^2 p(y) dy$$
$$= \int_{-1}^{1} y^2 \frac{1}{\pi\sqrt{1-y^2}} dy = \frac{1}{2} \tag{14}$$

respectively

G. THE AUTOCORRELATION FUNCTION

The autocorrelation funtion of a discrete random process y_n is defined by $R_y[n, m] = E[y_n, y_m]$

If y_n is a zero mean random process, then the autocorrelation function of y_n is the same as its auto-covariance function. We will assume that y_n is a stationary random process which follows the pdf given by (12). We will not assume independence between the consecutive members of the sequence.

Now we will obtain the autocorrelation function of the shifted logistic map. This can be done by the use of the following change of variable:

$$y_n = \cos(\theta),\ 0 \le \theta \le \pi, -1 \le y_n \le 1 \tag{15}$$

θ is uniformly distributed over $[0, \pi]$. Using

$$E[\cos m\theta] = \frac{1}{\pi}\int_0^{\pi} \cos(m\theta) d\theta = \delta(m)$$
$$= \begin{cases} 1, m = 0 \\ 0, m = \pm 1, \pm 2, \pm 3, \cdots, \end{cases} \tag{16}$$

and the fact that, for $m = 1, 2, 3, \cdots,$ we can write $\cos^{2m}(\theta)$ and $\cos^{2m-1}(\theta)$ as

$$\cos^{2m}(\theta) = \frac{1}{2^{2m}}[\sum_{k=0}^{m-1} 2[\begin{matrix} 2m \\ k \end{matrix}]\cos(2(m-k)\theta) + [\begin{matrix} 2m \\ m \end{matrix}]] \quad (17)$$

$$\cos^{2m-1}(\theta) = \frac{1}{2^{2m-1}}\sum_{k=0}^{m-1}[\begin{matrix} 2m-1 \\ k \end{matrix}]\cos((2m-2k-1)\theta) \quad (18)$$

we conclude that

$$E[y_n^{2m}] = \frac{1}{2^{2m}}2\sum_{k=0}^{m-1}[\begin{matrix} 2m \\ k \end{matrix}]E[\cos(2(m-k)\theta) + [\begin{matrix} 2m \\ m \end{matrix}]]$$

$$= \frac{1}{2^{2m}}[\begin{matrix} 2m \\ m \end{matrix}]$$

$$E[y_n^{2m-1}] = \frac{1}{2^{2m-2}}\sum_{k=0}^{m-1}[\begin{matrix} 2m-1 \\ k \end{matrix}]E[\cos(2m-2k-1)\theta)] = 0$$

That is,

$$E[y_n^m] = \{ \begin{matrix} \frac{1}{2^m}[\begin{matrix} 2m-1 \\ k \end{matrix}], m - even \\ 0, m - odd \end{matrix} \quad m \geq 0 \quad (19)$$

we will assume $m \geq 0$ and with

$$y_{n+m} = -\cos(2^m \theta) \tag{20}$$

Then

$$E[y_m, y_{n+m}] \equiv E\{\cos(\theta)\cos(2^m\theta)\} = \tfrac{1}{2}E\{\cos[(2^m + 1)\theta] + \cos[(2^m - 1)\theta]\}$$

Invoking (16) yields

$$R_y(m) = E\{y_n, y_{n+m}\} = \tfrac{1}{2}\delta(m) \tag{21}$$

Thus, the autocorrelation function of the shifted logistic map is a delta function.

H. THE CROSS-CORRELATION FUNCTION

The cross-correlation of two discrete random processes x_n and y_n is defined by

$$R_{xy}(n, m) = E\{x_n, y_m\}$$

If x_n and y_n are zero mean random processes, then the cross-correlation function of x_n and y_n are the same as cross-covariance funtion. Let x_n and y_n be two independent sequences generated by the shifted logistic map. Hence, we will assume that they are stationary random processes which follow the pdf given by (12). We will not assume independence between the consecutive members of any one sequence, but the members of one sequence is assumed to be independent of those of the other sequence.

Since x_n and y_n are independent,

$$R_{xy}(n, m) = E\{x_n\}E\{y_m\}$$

As shown in (19), $E\{x_n\} = E\{y_n\} = 0$. Hence

$$R_{xy}(m) \equiv R_{xy}(x_n, y_{n+m}) = 0, \quad m = 0, \pm 1, \pm 2, \pm 3, \cdots \tag{22}$$

(22) proves that the two sequences are thereotically orthogonal to each other. The orthogonality is a very desirable feature in the application of chaotic signals for digital communicaiton.

An estimate of the autocorrelation function of y_n is given by

$$r_y(m) = \frac{1}{N} \sum_{n=0}^{N-1} y_n y_{n+m} \tag{23}$$

The estimate is unbiased since the mean value of $r_y(m)$ can be shown, using (21) to be

$$E[r_y(m)] = \frac{1}{N} \sum_{n=0}^{N-1} E[y_n y_{n+m}] = R_y(m) = \frac{1}{2}\delta(m) \tag{24}$$

The estimate of the cross-correlation function of x_n and y_n is given by

$$r_{xy}(m) = \frac{1}{N} \sum_{n=0}^{N-1} x_n y_{n+m} \tag{25}$$

This estimate is unbiased since the mean value of $r_{xy}(m)$ can be shown to be

$$E[r_{xy}(m)] = \frac{1}{N} \sum_{n=0}^{N-1} E[x_n] E[y_{n+m}] = 0 = R_{xy}(m) \tag{26}$$

IV. APPLICATION OF CHAOTIC THEORY TO SIGNAL PROCESSING [21]

Chaotic theory can be applied to signal processing such as noise reduction, signal detection and signal separation. Chaotic theory utilizes nonlinearities underlying the signal generating process to model and characterize the system. Thus it has potential for those regimes where the assumptions inherent in conventional linear signal processing techniques do not hold. The technique is based in the determinations of dynamical equations of motions that approximate the underlying generating process. The purpose is to find a single set of differential equations which describe the global motion of the system over its attractor. This set of differential equations of motion for the system has good predictive power and can be used in signal processing applications.

Here one assumes that motion in the original phase space is governed by a set of coupled ordinary differential equations (ODES). We attempt to find an approximation to this set of ODES in the reconstructed phase space, capturing the underlying dynamics in a simple set of global equations, which are valid globally on the data attractor and which have the form: $dy[t]/dt = F(y[t])$.

To determine an approximate ODE for the ith vector component of F, we assume a basis set of the components of vector argument, the simplest case being polynomials, with

$$F_i(y[t]) = a_{i1}p_1(y[t]) + a_{i2}p_2(y[t]) \\ +\cdots + a_{iM}p_M(y[t]) \tag{27}$$

where the p_M are the terms of an Oth order, d-dimensional polynomial constructed from the elements of $y[t]$ and the a_{iM} are the coefficients of these terms. One parameter that must be chosen a priori is the order O of the approximation, which determines the number of coefficients required.

For basis sets such as simple polynomials that are linear in the coefficients, the system of equations can be written as a set of matrix equations, $yA_i = D_i$. In the representation, the Kth row of Y consists of the p_M generated by the vector y[k] as described above, A_n is the matrix of the unknown coeffcients for the ith element of F and D_i is a column matrix of approximations to the local derivatives in the ith direction of the phase space projectory at each data point y[k]. These local derivates are estimated using a three-point quadratic formula, a good compromise between accuracy consideration and robustress to noise.

To obtain A_i (and therefore F), we invect this matrix equation using a least-square minimization method such as the singular value decomposition (SVD). For basis sets that are nonlinear in the coefficient or that have nonlinear contraints, a nonlinear iterative optimization procedure such as NPSOL must be used.

The algorithm is useful in extracting dynamical information from a data set. This algorithm has potential applicaitons to real-world signal processing probelms. Once we have obtained a set of ODE's which provide a sufficient degree of predictive ability they can be used to perform signal separation or noise reduction by utilizing a predict-and-subtract algorithm. In the simplest version, we use the approximate dynamical equations F to integrate forward one or more sampling time steps from the kth data point y[k], generating an estimated phase space trajectory and obtaining $F^p(y|k|)$, a p step estimate of $y[k+p]$. We then form the difference vectors

$$Z[k+p] = y[k+p] - F^p(y[k]) \tag{28}$$

This general scheme performs the essential signal separation or noise reduction of the method, although in practice we use more elaborate techniques to obtain optimal results.

A key issue is that, rather than modeling the signal as a chaotic process and attempting to remove a noise component we can similarly attempt to model a noise process as a chaotic system and extract from it an underlying non-chaotic signal. This latter application has resulted in examples of (often broad-band) signal extraction for large, negative SNR levels. The general method described above has been used to analyze several real-world data sets over the past three years as follows.

In passive acoustics, noise modeling and separation of signals of interest has been performed in three cases. In the first, human voice masked by broad-band mechanical noise(from an air conditioner) has been recovered, resulting in legible speech when the import SNR is as low as -30dB. In the second case, the acoustic signature of helicopters has been separated from wind noise,enhancing the detection range and capability of acoustic system. Another acoustic application is to sonar processing and analysis. Band-filtered passive SONAR noise data has been successfully modeled using a 5-dimensional, 3rd order set of differential equations and spectially sharp signals have been extracted when the input SNR was as low as -25dB (narrowband).

In addition to acoustic processing these techniques have been exploited on HF data, using a dynamic model of Ionospheric fluctuations, derived from received data, to mitigate the effects of multipath fading. Finally, dynamical models of sunspot counts have been generated that, when tested on historical data, provide more accurate prediction than existing traditional models.

Hence, it is possible to estimate a set of global differential equations describing the time evolution of a system in reconstructed phase space. This method provides a good approximations to the original dynamics with surprisingly few numbers of data points for noise-free data. Equations for data with increased noise levels can be extracted with comparable accuracy by increasing the number of data points used. Hence an optimal method of data sampling for estimation of dynamics may be to sample at very high sampling rates for short bursts, widely spaced in time.

V. APPLICATION OF CHAOTIC THEORY TO NONLINEAR NOISE AND VIBRATION MEASUREMENT AND ANALYSIS [22]

A. IDENTIFICATION OF CHAOS

It is necessary for a noise and vibration signal to be tested first for the existence of chaos. Chaos means exponential sensitivity to initial conditions and therefore, occurs by definition, if there is a positive Lyapunov characteristic exponent (LCE). The LCE associated with a trajectory gives the average rates at which nearby trajectories diverge. Another tool in testing for chaos is to compute power spectra. If the motion is quasiperiodic the spectrum of any coordinate is discrete, whereas chaotic motion will exhibit broadband power spectra. The temporal behaviour of a function y(t) is quasiperiodic if its Fourier transform consists of sharp spikes, ie, if

$$y(t) = \sum_{j=1}^{n} C_j e^{j\omega_j t} \tag{29}$$

Quaisperiodic motion is regular. That is, quasi periodicity, is associated with a negative or zero Lyapunov exponent. Quasiperiodic motion can certainly look very complicated and seemingly irregular, but it cannot be truly chaotic in the sense of exponential sensitivity to initial conditions. In particular, the difference between two quasiperiodic trajectories is itself quasi periodic and so we cannot have the exponential separation of initially close trajectories that is the hallmark of chaos since quasi periodicity implies order, it follows that chaos implies non-quasiperiodic motion. Thus chaotic motion does not have a purely discrete Fourier spectrum as in (29) but must have a broadband, continuous component in its spectrum as in Fig 1

Fourier analysis is therefore a very useful tool in distinguishing regular from chaotic motion and furthermore it is generally much cheaper computationally than Lyapunov exponents.

Another test for chaos is to plot the probability density function of the signal to find the existence of multi-maxima which is a characteristic for chaotic behaviour. The examples of amplitude probability density functions and waveforms connected with them are shown in Fig 2 below:

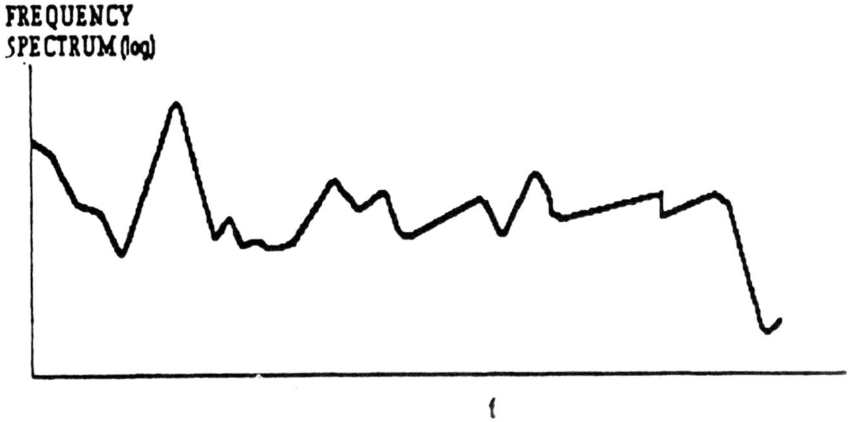

Fig. 1 Typical frequency spectrum of some coordinate of a chaotic system

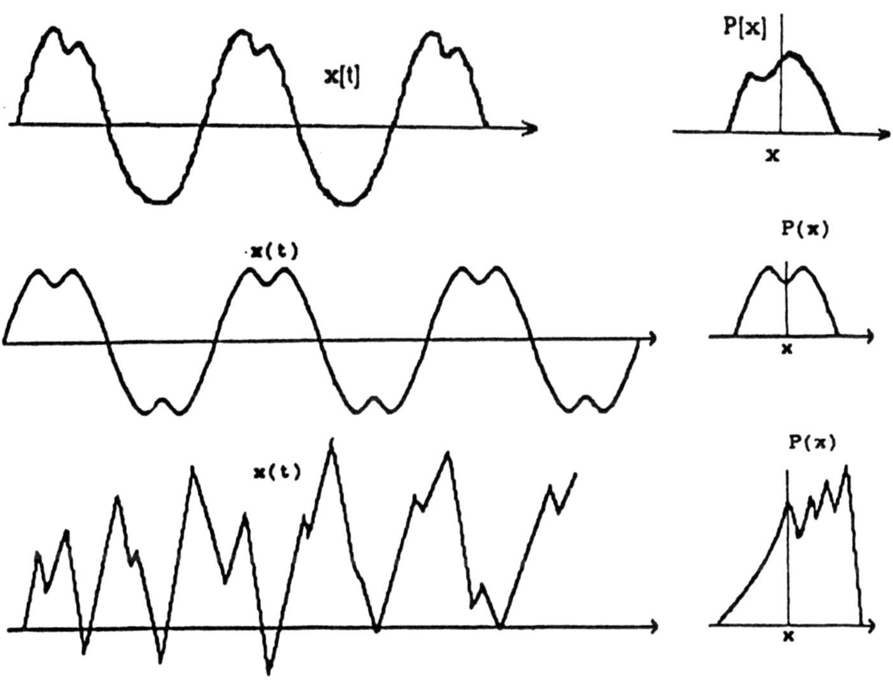

Fig.2 Amplitude probability density functions and connected waveforms

But it has to be emphasized that the sure way of identify chaotic behaviour is to compute the LCE.

B.FILTERING OF CHAOTIC SIGNAL EMBEDDED IN A RANDOM NOISE

In the analysis of nonlinear industrial noise using chaotic theory, one has to separate the noise into two portions: the chaotic signal and the random noise. Various methods can be used as shown below:

1. Maximum Likelihood Processing. In the signal separation problem, we observe the state of a chaotic system through an observation function, h, and in the presence of another signal,i.e..,

$$\begin{aligned} y_n &= h(s_n) \\ o_n &= y_n + w_n \end{aligned} \tag{30}$$

where $y_n \in R^k$ is the output from the chaotic system and w_n is the other signal. The signal separation problem is to estimate both y_n and w_n given the observations. $O_{0:N-1} = \{O_0, O_1, \cdots, O_{N-1}\}$. Three categories of signal separation problems:

a. When both the state update function, f, and the observation function,h, are known.

b. When neither the state update nor the observation function is known but we have available a "reference observation". This observation is the result of observing the nonlinear dynamical system without the presence of another signal but in a case in which the initial conditions of the nonlinear dynamical system are different from those for the case in which we observe the signal plus interference.

c. When neither the state update nor the observation function is known. When both the state update and the observation functions are known and the other signal is white Gaussian noise, with a noise correlation matrix $\sigma_w^2 I$, then it is possible to determine bounds on the performance of signal separation algorithms. In this case the maximum likelihood solution is given by a trajectory $\{\hat{S}_0, \hat{S}_1, \cdots, \hat{S}_{N-1}\}$ that obeys the constraints of the known dynamics, ie $\hat{S}_{n+1} = f(\hat{S}_n)$ and minimizes the difference between the observed signal and the predicted observations i.e. $\sum \left| h(\hat{S}_n) - O_n \right|^2$. This is equivalent to estimation of the initial condition $\hat{S}_0$ which maximizes

$$\log(P_r(O_{o:N-1} \mid \hat{S}_0)) = \sum_{n=0}^{N-1} \left| h(f^n(\hat{S}_0)) - O_n \right|^2 + C2\sigma_w^2 \tag{31}$$

Analysis of the likelihood function (31), for a chaotic system shows some interesting properties, for the logistic map. The likelihood function contains multiple narrow maxima. The narrowness of the maxima is related to the positive Lyapunov exponents of the system. The presence of multiple maxima occurs because the nonlinear dynamics folds state space trajectories together.

2. Signal Separation Using Markov Model. The signal processing method used is hidden Markov models. We assume that a sequence of observations are given: $O_{0:N} = \{O_0, O_1, \cdots, O_N\}$, where each observation is the sum of the output of the chaotic system and another signal, i.e. $O_n = y_n + w_n$ and we wish to generate the best estimates for y_n and w_n given the observations. For our initial work, we assume that the other signal, w_n can be modeled as white Gaussian noise with variance σ_w^2. We define two signal estimation algorithms - one based on a maximum likelihood state sequence estimation approach and one based on a maximum aposteriori approach.

a. The maximum likelihood signal estimation approach first estimates the most likely state sequence given the observation, i.e.

$$\hat{q}_{1:N} = \arg \max_{q_{1:N}} P_r(q_{1:N} | O_{1:N} \tag{32}$$

This is computed using the Viterbi algorithm. The signal y_n is then estimated as the expected value of y_n given the observations and the most likely state sequence, i.e.,

$$\begin{aligned} \hat{y}_n &= E[y_n | O_n \hat{q}_n] \\ &= m(\hat{q}_n) + \frac{\sigma^2(\hat{q}_n)}{\sigma^2(\hat{q}_n)+\sigma_w^2}(O_n - m(\hat{q}_m)) \end{aligned} \tag{33}$$

where $m(\hat{q}_n)$ and $\sigma^2(\hat{q}_n)$ are the mean and variance of the most likely state at time index n.

b. In the maximum aposteriori approach we attempt to estimate the signal y_n as the expected value of y_n given the observations, i.e.,

$$\hat{y}_n = E[y_n | O_{1:N}] = \sum_{q_{1:N}} E[y_n | O_n q_n] P_r(q_{1:N} | O_{1:N}) \tag{34}$$

where the summation is performed over all possible state sequnce, $q_{1:N}$, and where $E[y_n | O_n, q_n]$ is computed according to Eq. (32)

We note that in the case of a linear dynamical system, driven by white Gaussian noise, both the maximum likelihood and the maximum aposteriori approaches courage to Kalman smoothing as the number of states goes to infinity. An application of above method is to helicopter noise buried in chaotic noise.

3. Application of Convolutions to Signal Separation. Convolution can be applied to signal separation. Convolution is filtering and we will consider for

the effects of convolution on the two parameters commonly used in the description of chaotic signals - the Lyapunov exponents and the fractal dimension of the attractor. In order to determine the effects of filtering on Lyapunov exponents, we represent the time series Z[n] as the scalar observation of a composite system of the original nonlinear dynamics and the filtering dynamics:

$$\begin{aligned} x[n+1] &= F(x[n]) \\ w[n+1] &= Aw[n] + bG(x[n]) \\ Z[n] &= c^T w(n) \end{aligned} \tag{35}$$

The matrices A,b and c are chosen to represent a minimal realization of the system and w[n] the state of the filter at time n. We also assume that the overall composite system is minimal in the sense that there is no pole zero cancellations between any linear component of the original nonlinear system and the cascaded linear system. The invariance of the Lyapunov exponents under smoothing invertible coordinate changes allows us examine certain properties of the filtered signal in this augmented state space with the assurance that the results carry over to the embedded state space.

Convolution also effects the capacity dimension. The filtering of noisy chaotic data to reduce noise will cause errors in fractal dimension estimates. The effect of convolution on the capacity dimension can be examined using the time delay construction. The time delay construction defines transformation of IR^N, the state space of the original nonlinear system, to IR^N, the space consisting of the reconstructed vectors. The effect of filtering on the capacity dimension of the observed signal Z[n] depends upon the nature of this transformation.

C. COMPUTATION OF FRACTIONAL HARMONICS

It has been noted by Wei Rongjue etal [23] that chaos is caused by the presence of fractional subharmonics during their experiment on nonpropagating solitons and their transition to chaos. Hence to perform spectrum analysis of chaotic signal, it would be necessary to compute the fractional subharmonics. This can be done by using the theory of multiple scale expansion valid for the solitons. This requires a fractal analysis by applying a multiscale second-order statistical method. For a fractal surface, there are relationships among fractal dimension, scaling, power spectrum, area size and intensity difference. Fractional subharmonics has fractal nature and one needs to estimate the fractal dimension and transform the scale of frequency spectrum to scale of fractal dimension. The fractal dimension can be determined following the work of Pentland [24] who computed the Fourier transforms of an image, determined the power spectrum and used a linear regression technique on the log of the power spectrum as a function of frequency to estimate the fractal dimension.

VI. NONLINEAR NOISE AND VIBRATION SIGNAL PROCESSING - THE FRACTUM [25]

During the last decade, there has been an increased activity in nonlinear signal processing because most of the real world signals and systems are nonlinear in nature and linearized signals and systems ar only linearized, classroom cases. Here we concentrate on the application of nonlinear signal processing techniques to noise and vibration signals. In nonlinear noise and vibration systems, a frequently encountered signal is the chaotic signal. A wellknown chaotic noise signal is aerodynamic noise which is due to turbulence. Hence we will limit the nonlinear noise and vibration signal only to chaotic signal.

A. TECHNIQUES IN NONLINEAR SIGNAL PROCESSING

In this paper we will limit the scope of signal processing only to spectrum analysis. For linear signal, Fourier transform is used to compute the autocorrelation function and hence the power spectrum. For nonlinear signal,

Fourier series are not applicable and other time series representations are necessary. For chaotic signals, which are fractal in nature, we look for time series with fractal properties. We will test on two functions: (a) Weierstrass non-differenticable function, (b) radial basis function.

1.Weierstrass Function. Fourier series involves a linear progression of frequencies. The Weierstrass function on the other hand, involves a geometric progression:

$$V_{MW}(t) = \sum_{n=-\infty}^{\infty} A_n R^{nH} \sin(2\pi r^{-n} t + \Phi_n \qquad (36)$$

where A_n is a Gaussian random variable with the same variance for all n, n is a random phase uniformly distributed on $(0, 2\pi), R_n = 1/f_n, f_{n=}$ discrete frequencies and H=2-D where D=fractal dimension for the case of fractional Brownian Motion (fBm). The fractal nature of the Weierstrass function means it is self-similar and nowhere differentiable.

The usual procedure in spectrum analysis is to calculate the power spectral density (PSD) using Fourier transform and correlation function. This is correct only for linear noise and vibration signals. For chaotic signals which are also fractal and nonstationary in nature, we will first represent the time series by a fractal function instead of the usual Fourier series and compute its PSD usng the Wigner-Ville theorem which is applicable to nonstationary signals. We call the resulting power spectrum "fractum" to differentiate it from the power spectrum of stationary, linear signal.

The first step is to test whether the signal is chaotic in nature. This can be done by computing Lyapunov exponent. Lyapunov exponent can be defined as:

$$\lambda(x_0) = \lim_{n\to\infty} \frac{1}{n} \ln \left| \frac{df^n(x)}{dx} \Big|_{x_0} \right| \qquad (37)$$

where $f(x_n) = x_{n+1}$ =one-dimensional map. A positive Lyapunov exponent will confirm that the signal is chaotic. A chaotic signal will have fractal characteristics. Hence the next step is to compute its fractal dimension. We choose the Hausdorff definition of fractal dimension which gives

$$D = \frac{\ln N}{\ln 1/r''} \tag{38}$$

where N= number of self-similar parts and r"=size of ruler.

The next step is to compute the power spectral density (PSD) using the Wigner-Ville theorem. Before that we have to calculate the covariance function which is the 2 point autocorrelation function defined by

$$R'_w(\tau) = \int_{-\infty}^{\infty} V(t)V(t+\tau)dt \tag{39}$$

where $V = V_{MW}(t)$ for our case.

The Wigner-Ville spectrum of a nonstationary process $f(t)$ with covariance function $R'_f(t,s)$ is given by

$$W_f(t,\omega) = \int_{-\infty}^{\infty} R'_f(t+\tfrac{\tau}{2}, t-\tfrac{\tau}{2})e^{-j\omega\tau}d\tau \tag{40}$$

The required power spectral density (PSD), the Wigner-Ville spectrum will be obtained by substituting (36) and (39) into (40). We call this resulting spectrum, the "fractum" because it represents the characteristics of a fractal signal without using the linear representation of Fourier series.

2. Radial Basis Function. Radial basis functions (RBF) can be used for extrapolation as well as interpolation and are attractive for nonlinear modeling such as chaotic modeling. We require that the interpolation be exact at the known data points. Then

$$y_i = \sum_{j=1}^{m} \lambda_j \phi(\left|\underline{x}_i - \underline{x}_j\right|) \text{ for} \tag{41'}$$
$$i = 1, \cdots, m$$

where ϕ is the radial basis function. Writing the matrix Φ with elements

$$\Phi_{ij} = \phi(\left|\underline{x}_i - \underline{x}_j\right|) \tag{42}$$

we can rewrite (41) as in linear equations in m unknowns:

$$\underline{\Phi}\underline{\lambda} = y \tag{43}$$

where y and λ are the vectors with elements y_i and λ_i, $i = 1, \cdots, m$. Everything is known except $\lambda_1 \cdots \lambda_m$. Solving (43) therefore, determines f completely where f is the radial basis approximations to g and is defined by

$$f(\underline{x}) = \sum_{i=1}^{m} \lambda_i \phi(\left|\underline{x} - \underline{x}_i\right|) \tag{44}$$

and suppose that from experiment, values $y_1 \cdots y_m$ of y have been found at $x_1 \cdots x_m$. Then we have $y_i = g(x_i)$ for $i = 1, \cdots, m$.

Computationally, the significant part of the problem is that of solving the linear equations (43) for λ. The size of the matrix Φ is the number m of data points and so the computational effort, which is of order $\cdots$, may be large. Fortunately, this calculation is only performed once for a particular set of data points and Φ . The work involved to interpolate for any given point is then considerably less, of order m.

As well as the difficulty of long computation time, there is a risk that as m grows, Φ will become ill-conditioned. In fact, well conditioned Φ results from choices of ϕ including

$$r^{// \, 2(k+1)} \log(r^{//}) \qquad , k \geq 0$$
$$|r^{//}|^{2k+1} \qquad , k \geq 1$$
$$(r^{// \, 2} + d^2)^{\pm 1/2} \qquad , 0 < d << 1$$

In the latter case, d is often chosen to be of the order of the separation of the data points. Notice that any desired degree of smoothness can be obtained. For example $r''^{\,k}\log(r)$ is a C^{k-1} function of $r'' \geq 0$. The choice $r''^{\,2}\log(r'')$ can give singular systems of equations in some (fairly contrived) situations, even in one dimension. This is because $r''^{\,2}(\log r'') = 0$ at r''=1 as well as at $r'' = 0$. The problem is unlikely to arise in practice, and in this respect at least, r'' is probably completely safe.

For our purposes, we shall assume that the number of data points required is sufficiently small (up to a few hundred with current workstations) that numerical and computational difficulties do not nominate the problem.

As a simple example, the values $y_1, \cdots, y_m$ of y_i have been found experimentally at position $x_1, \cdots, x_m$. r''_{ij} =distance between points i and j can be measured. Let $\phi(ij) = r''^{\,3}_{ij}$ known. So both y_i and ϕ_{ij} are knowns.

The next step now is to introduce fractal and chaotic characteristics into radial basis function. To start with the chaotic signal usually takes the form of a time series. To construct a dynamical model from a time series, we apply the phase space reconstruction technique to the chaotic data sequnces. The general technique of this approach is to generate several different scalar signals $v_k(t)$ from the original $v(t)$ in such a way as to reconstruct an m-dimensional space, where, under some conditions, we can obtain a good representations of the attractor of the dynamical system.

The easiest and most popular way to do that is to use time delays. We write

$$V_k(t) = V(t + (k-1)\tau), k = 1, \cdots, m \tag{45}$$

where τ is the time delay. In this manner, an m-dimensional signal is generated, which can be represented by the vector.

$$\underline{x}(t) = (x_1(t), x_2(t), \cdots, x_m(t)) \tag{46}$$

Note that, on varying the set of variables whcih can be constructed from x(t), we get in principle the same geometric information. If d is the dimension of manifold containing the attractor, Takens showed that m=2d is sufficient to embed the attractor by the Whitney embedding theorem.

The choice of the delay time needs extra care. If τ is too small, then x(t) and $x(t+\tau)$ are too close, which means they are not "independent" enough to seve as independent coordinates. On the other hand, if τ is too big, chaos makes x(t) and $x(t+\tau)$ disconnected. Autocorrelation function and mutual informations are usually used to determine such a delay.

Next we try to determine the possibility of a nonlinear model for chaotic noise and vibration signals by uisng a more rigorous analysis, i.e. fractal dimension. Measuring dynamical invariants such as fractal dimension and Lyapunov exponent has been a widely used approach in detecting chaos. The correlation dimension definition of fractal dimension is used:

$$d_c = \lim_{r'' \to 0} \frac{l_n(c(r''))}{\ln r''} \tag{47}$$

where c(r")=cumulative correlation. The determination of the correlation dimension is usually found by plotting c(r") versus r on a ln-ln graph for different values of the embedding dimension,m.

Having measured the correlation dimension we next assume that the dynamics can be written as a map in the form

$$\underline{v}(t+\tau) = F(\underline{v}(t)) \tag{48}$$

where the current state is $\underline{v}(t)$ and $\underline{v}(t+\tau)$ is a future state. Since the only new component in vector $\underline{v}(t+\tau)$ is the point $v(t+\tau)$, the dynamical system (48) is equivalent to the probelm of prediction of $v(t+\tau)$ from the vector $\underline{v}(\underline{t})$. That is

$$v(t+\tau) = \phi(\underline{v}(t)) \tag{49}$$

To reconstruct the dynamical system (48), we need to produce $\hat{v}\,(t+\tau)$ of $v(t+\tau)$. In other words, we need to approximate the mapping ϕ by an approximation $\hat{\phi}$ and we use the radial basis function as an approximation for ϕ. Hence the radial basis function is determined within the framework of a phase space reconstruction for chaotic time series.

To use the Wigner-Ville theorem, we have to calculate the 2 point autocorrelation function for the radial basis approximation in (44), bearing in mind that the x's are functions of time. The Wigner-Ville theorem will give us the fractum.

VII APPLICATION OF CHAOS TO INDUSTRIAL NOISE ANALYSIS [26]

A. RELATIONSHIP BETWEEN CHAOS AND NOISE

Here noise means classical noise and we shall use industrial noise for the purpose of analysis. We use the model that noise consists of two parts: the deterministic part or chaotic part and the random part. Deterministic part can be described in terms of a complex system of nonlinear differential equations. Conventional techniques of industrial noise analysis, using Fourier spectra and correlation functions etc. which are linear system methods, are using linear approximation for describing industrial noise. In fact the deterministic part of industrial noise which is chaotic in nature should be analyzed by using properties for characterizing chaos such as Lyapunov exponents, Kolmogorov entropy and dimensions of the attractors. To start with our study, we shall distinguish regular and chaotic signal in terms of power spectra. A chaotic signal does not have a purely discrete Fourier spectrum as in (29) where

C=Fourier coefficient and ω=angular frequency but must have a broadband, continuous component in its spectrum as in Fig 1. Fourier analysis is therefore a very useful tool in distinguishing regular from chaotic signal and furthermore, it is generally much cheaper computationally than Lyapunov exponents. Such power spectra are especially useful in identifying the period doubling route to chaos. Decaying correlations, like broadband power spectra, are characteristic of chaotic evolution.

B. ANALYSIS OF INDUSTRIAL NOISE BY TECHNIQUE IN CHAOS

Lyapunov exponent is an essential tool in studying chaotic signal. Lyapunov exponent is the rate of the exponential separation with time of initially close trajectories. To illustrae this with the example of an industrial noise, we start with the plotting of the power spectra of the industrialnoise. The industrial noise can be represented by the time series $\{x_n\}$ generated by the logistic map given by

$$x_{n+1} = 4r' x_n(1 - x_n), 0 \le x_0, r' \le 1$$

where r'=map parameter which will dramatically affect the behaviour of the map.

An important property of chaos is its very sensitive dependence on initial condition. We illustrate this by considering the case r'=1 of the logistic map. In this case the transformation $x_n = \sin^2 \pi\theta_n$ plus the identity $\sin 2\pi\theta_n = 2\sin\pi\theta_n \cos\pi\theta_n$ reduces the mapping to the form $\theta_{n+1} = 2\theta_n$ which has the explicit solution $\theta_n = 2^n\theta_0$. Since we can add any integer to θ without changing the value of x, we can write

$$\theta_n = 2^n\theta_0 \pmod 1 \tag{50}$$

as the solution of the logistic map for r'=1.

It is easy to see that the mapping equation (50) has the property of very sensitive dependence on initial conditions: if we change the initial speed θ_0 to $\theta_0 + \varepsilon$ then θ_n changes by $2^n\varepsilon = \varepsilon e^{n \log 2}$. In other words, there is an exponential separation with time n of initially close trajectories. The rate of exponential separation, namely log2, is called the Lyapunov exponent and the fact that it is positive in this example means that we have very sensitive dependence on initial conditions, ie chaos.

This example also illustrates that we can determine the Lyapunov exponent of the chaotic time series or chaotic industrial noise from its logistic map.

Another essential property of chaos is Kolmogorov entropy. The Kolmogorov entropy of a chaotic signal is defiend in such a way that, in most instances, it is equal to the sum of the positive Lyapunov exponents. It provides an estimate of the average time over which accurate predictions can be made about a chaotic signal, before long-term predictability is lost due to the sensitivity to initial conditions. Entropy is a measure of the amount of informations necessary to determine the state of the system.
We define the entropy as

$$s = -\sum_{i=1}^{N} P_i \log_2 P_i \tag{51}$$

where N=no of states in system and P=probability.

Consider a trajectory $\vec{y}(t) = (y_1(t), y_2(t), \cdots, y_N(t))$ and partition the phase space into n hypercubes of side ε. Let $P(i_0, i_1, \cdots, i_n)$ be the joint probability that the point $\vec{y}(0)$ lies in the ith cell, $\vec{y}(\tau)$ lies in the ith cell, $\cdots$, and $\vec{y}(n\tau)$ lies in the ith cell. Then,

$$K_n = -\sum_{i_0 \cdots i_n} P(i_0, i_1, \cdots, i_n) \log_2 P(i_0, i_1, \cdots i_n) \tag{52}$$

is a measure of the amount of information necessary to specify the trajectory to within a precision ε, assuming only the probabilities $P(i_0, i_1, \cdots, i_n)$ are known a priori.

For non-chaotic systems, K=0, i.e. there is no loss of information because initially close points on a trajectory remain close together as time evolves. For chaotic systems, however, initially close points separate exponentially on average and therefore joint probabilities for cell occupations decrease exponentially with time. Thus K>0 for chaotic systems. For truly non-deterministic random systems, initially close points take on a statistical distribution over all the allowed new cells. Thus if $p(i_0) \approx \varepsilon$, then $P(i_0, i) \approx \varepsilon^2$ and so $K \rightarrow \infty$ as $\varepsilon \rightarrow 0$ for pure randomness. The K-entropy is therefore useful not only for distinguishing regular from chaotic behaviour, but also for distinguishing chaos from noise.

C. ANALYSIS OF INDUSTRIAL NOISE USING PROBABILITY DENSITY FUNCTION

Investigations of the probability density function of industrial noise show that this function can be characterized not only by one maxima curves: Let $\bar{x}(t, \omega) = [x,(t, \omega), \cdots, x_n(t, \omega)]^T$ be a random process with a probability density function $P(x_1, \cdots, x_n, t/x_{10}, \cdots, x_{n0})$. Random process $x_i(t, \omega)$, $i = 1, 2, \cdots, n$ is called the process with bifurcation if the probability density function $P(x_i, t/x_{i0})$ given by equation:

$$P(x_i, t/x_{i0}) = \int_{-\infty}^{\infty} \cdots \int P(x_1, \cdots, x_n, t/x_{10}, \cdots x_{n0}) \\ dx_1 \cdots dx_{i+1} \cdots dx_n \tag{53}$$

has two maxima for any $t > t_0$ where t_0 is constant. The probability density function with multi-maxima is a characteristic for chaotic behaviour of the industrial noise.

D. PREDICTION OF INDUSTRIAL NOISE PATTERN

A random industrial noise is unpredictable. However, the chaotic portion of an industrial noise is deterministic and hence is predictable. The Kolmogorov entropy of the chaotic portion of an industrial noise provides an estimate of the average time over which accurate predictions can be made about a chaotic signal, before long-term predictability is lost due to the sensitivity to initial conditions. Since Kolmogorov entropy is the sum of the positive Lyapunov exponents, Lyapunov exponent is a property useful for predicting chaotic industrial noise patterns.

VIII. A NEW TYPE OF ACOUSTICAL IMAGING-THE ACOUSTICAL CHAOTIC FRACTAL IMAGES FOR MEDICAL IMAGING [27]

In recent years, chaos theory has found application in many fields such as the earth science [28,29], turbulence [30], laser science [31] etc. However, not much work has been done which relate the chaos theory with inverse problems [32,33,34] maybe because the inversion is purely mathematical and physically non-realistic. Here we apply chaos theory to the inhomogeneous medium, taking account of wave nature and diffraction. A theory of diffraction tomography is then formulated which yields chaotic fractal velocity images.

A.FORWARD PROBLEM

The purpose here is to determine the scattered field. The nonlinear wave equation is used here. Various approaches to the solution of the equation have been attempted. One dimensional solutions, such as those of Blackstock [35] can provide some useful information for medical ultrasound systems but are unable to reproduce the fine detail and phase variations seen in "real" pressure fields. Highly diffracted and focused pressure fields require more rigorous treatment. Smith and Beyer [36] commented on the "lack of appropriate theoretical analysis" when they published nonlinear measurements on a focussed acoustic source operated at 2.3MHz. One of the most significant theoretical advances came in 1969 when Zabolotskaya and Khokhlov [37] published a solution of the nonlinear wave equation for a confined sound beam in which it was assumed that the shape of the wave varies, slowly both along the beam and transversely to it. In 1971 Kuznetsov [38] extended their treatment to include absorption and the resulting equation is now widely known as the KZK equation. He also obtained solution which is known as the parabolic approximation to the nonlinear wave equation and is equivalent to the paraxial approximation used in optics. The KZK equation accounts for diffraction, absorption and nonlinearity and is valid for circular apertures that are many wavelengths in diameter and will accept orbitrary source conditions.

1. General Wave Equation. The KZK equation is a nonlinear wave equation for a scalar potential Φ, in consideration of the dynamics of a viscous heat conducting fluid. It is correct to the second order with terms for diffraction, absorption and nonlinearity.

$$\frac{\partial^2\Phi}{\partial t^2} - c^2\nabla^2\Phi = \frac{\partial}{\partial t}[2\alpha c^2 k^2 \nabla^2\Phi + (\nabla\Phi)^2 + \frac{B}{2A}\frac{1}{c^2}(\frac{\partial\Phi}{\partial t})^2] \quad (54)$$

where c= speed of sound, α=absorption coefficient, k=wave number. The left-hand side of (54) is the three dimensional linear Helmholtz wave equation. Of the three terms on the right-hand side, the first term is a linear term and accounts for absorption, the second term is due to convective nonlinearity in the equation of state.

2. Parabolic Approximation. Kuznetsov [38] also showed that (54) could be simplified by approximation, in the case of a quasi-plane wave field and the Laplacian (∇^2) can be replaced by the transverse Laplacian (∇_1^2). A circular aperture that is many wavelengths in diameter (ie. ka is large) falls in this category since most of the energy is confined to a beam in the axial direction. This is known as the parabolic (or paraxial) approximation and is equivalent to the Fresnel approximation that is sometimes used in the diffraction integral for nearfield calculations. Kuznetsov [38]'s parabolic approximation can be expressed in a normalised form:

$$[4\frac{\partial^2}{\partial\tau\partial\sigma} - \nabla_1^2 - 4\alpha R_0\frac{\partial^3}{\partial\tau^3}]\overline{P} = 2\frac{R_0}{l_D}\frac{\partial^2}{\partial\tau^2}\overline{P^2} \quad (55)$$

where $\overline{P} = (\underline{P}/\underline{P}_0)$ is the acoustic pressure normalised by the source pressure and $\tau = (\omega t - kz)$ is retarded time, ie. includes a phase term for a plane wave travelling in the z direction, R_0=Rayleigh distance=$ka^2/2$, l_D =shock distance and a=aperture radius. In this equation σ is the Rayleigh distance and ζ is the radial coordinate normalised by the aperture radius, i.e. $\sigma = 2\tau/ka^2$ and $\zeta = r/a$.

A trial solution is then assumed in the form of a Fourier series (for the time wave form) with amplitude and phase that are functions of the spatial coordinate, i.e.

$$p(\sigma,\zeta,\tau) = \sum_{n=1}^{\infty} q_n(\sigma,\zeta,\tau)\sin[n\tau + \Psi_n(\sigma,\zeta,\tau)]$$

or

$$p(\sigma,\zeta,\tau) = \sum_{n=1}^{\infty} g_n(\sigma,\zeta,\tau)\sin(n\tau) + h_n(\sigma,\zeta,\tau)\cos(n\tau) \quad (56)$$

where $g_n = q_n \cos\varphi_n, h_n = q_n \sin\varphi_n$ and n is the harmonic number with n=1 representing the fundamental frequency, q=Fourier solution amplitude and ψ=Fourier solution phase. Substituting the trial solutions (56) in (55) and collecting terms in $\sin(n\tau)$ and $\cos(n\tau)$ gives a set of coupled differential equations for g_n and h_n:

$$\frac{\partial g_n}{\partial\sigma} = -n^2\alpha R_0 g_n + \frac{1}{4n}\nabla_1^2 h_n + \frac{nR_0}{2l_0}\Big[\frac{1}{2}\sum_{k=1}^{n-1}(g_k g_{n-k} - h_k h_{n-k}) - \sum_{p=n+1}^{\infty}(g_{p-n}g_p + h_{p-n}h_p)\Big] \quad (57)$$

$$\frac{\partial h_n}{\partial\sigma} = -n^2\alpha R_0 h_n + \frac{1}{4n}\nabla_1^2 g_n + \frac{nR_0}{2l_0}\Big[\frac{1}{2}\sum_{k=1}^{n-1}(h_k g_{n-k} - g_k h_{n-k}) - \sum_{p=n+1}^{\infty}(h_{p-n}g_p + g_{p-n}h_p)\Big] \quad (58)$$

Eqs (57) and (58) form the basis of the numerical solution which can be implemented in a FORTRAN program.

B. FRACTAL STRUCTURE AS A DIFFRACTION MEDIUM

Sound propagation is related to the elastic properties of the medium. Most solids have tensorial elasticity and can support transverse as well as longitudinal sound waves. Hence, it is necessary to investigate the nature of their elasticity first. To attack this problem, fractals is formulated as a growth problem and we use the Diffusion Limited Aggregation (DLA) model [39], a growth model. For DLA, one needs to calculate the growth probabilities. Here one focuses on multi-fractality's relation to transport properties of the fractal medium.

Let P(r,t)=probability of finding the random walker on sites at a fixed distance r from the starting point. The probability P(r,t) to find the walker at l at time t is a Gaussian,

$$P(l,t) = P(o,t)\exp(-l^2/4Dt) \tag{59}$$

where D=fractal dimension. The moments of the probability density $< P^q(r,t) >$ can be written as a convolution integral:

$$< P^q(r,t) >= \int_{-\infty}^{\infty} Q(r/l)P^q(l,t)dr \tag{60}$$

where $Q(r/l)$=probability of finding the sites separated by a chemical distance l and Euclidean distance r. The chemical distance is the shortest path between two sites on the cluster. In the general case, the qth moment $< P^q(r,t) >$ can be written as

$$< P^q(r,t) >= \frac{1}{N_r} \sum_{i=1}^{N_r} P_i^q(r,t) \tag{61}$$

where the sum is over all N_r sites located a distance r from the origin (N_r) may include many configurations or a single configuration with a very large number of cluster sites). The sum equation (61) can be separated into sums over different l values (N_m values of l_m):

$$< P^q(r,t) >= \frac{1}{N_r}\{\sum_{i=1}^{N_1} P_i^q(l_1,t)+\sum_{i=1}^{N_2} P_i^q(l_2,t)+\cdots\}$$

$$= \frac{1}{N_r}\sum_{m} \{N_m \times \frac{1}{N_m}\sum_{i=1}^{N_m} P_i^q(l_m,t)\}$$
$$= \frac{1}{N_r}\sum N_m < P^q(l_m,t) > \tag{62}$$

This covers all the scattering points within the fractal medium. In this problem it is assumed that the random walker starts at the origin D and after t time steps can be found at r[x] with very different probabilities at different sites.

For the scattering of sound by a fractal medium one needs to treat all sites of the fractal as starting points and the various parameters like sound velocity, attenuation coefficients etc. have to be modified for fractals. Fractal media are characterized by not having a very characteristic length scale and they have a very inhomogeneous density distribution. One can therefore expect to find very different physical properties in materials with fractal structure compared to the ordinary solids. Furthermore, real fractals are disordered and highly irregular. In some sense they can be regarded as ideally disordered materials. In conventional diffraction tomography theory, one considers only scattering by one point by ignoring the object size. This is known as Born approximation. Here the object size is taken into account as consisting of several scattering points and all sites of the fractal are considered as scattering points. We call this type of diffraction "fractal diffraction".

1. Wave Scattering Modified by the Fractal Medium The expression for the scattered acoustic pressure wavefield is modified by the correlation coeffient

which contains the fractal dimension of the medium. We have obtained scattered wavefield amplitude fluctuation as

$$p(\sigma,\zeta,\omega t-kz)=\sum_{n=0}^{\infty} q_n(\sigma,\zeta,\omega t-kz)$$
$$\sin[n(\omega t-kz)+\Psi(\sigma,\zeta,\omega t-kz)] \tag{63}$$

For diffraction of sound wave by a fractal medium, one needs to consider all sites of the fractal as scattering points. For this reason, the correlation coefficient is chosen as (65). By modifying (63) by (62), then the autocorrelation function for the amplitude fluctuation is given by the following formula:

$$\overline{P_1(t)P_2(t)}=\int_0^{R_1}\int_0^{R_2}\iiiint_{-\infty}^{+\infty} P_1(\sigma_1,\zeta_1,\omega_1 t-kz_1)$$
$$P_2(\sigma_2,\zeta,\omega_2 t-kz_2)\cdot\langle P^q(r,t)\rangle\, d\sigma_1 d\sigma_2 dz_1 dz_2 d\omega_1 d\omega_2 \tag{64}$$

where the coordinates of the receivers are $(R_1,0,0)$ and $(R_2,0,0)$.

The power spectral density (PSD) of the scattered field = Fourier transform of autocorrelation function= $\int_{-\infty}^{\infty}\overline{P_1(t)P_2(t)}e^{-j2\pi ft}dt$ (65)

where f=frequency. The overall amplitude of the acoustic pressure of the scattered field is proportional to the square not of the PSD.

C. INVERSE PROBLEM

The purpose here is to obtain sound velocity field in the medium from the scattered sound pressure field. The method of nonlinear iteration will be used. The aim is to obtain velocity images under diffraction tomography format. Our purpose is to apply to medical imaging. The nonlinearity at tomographic inversion here is related to heterogeneity of the human tissue. For instance, the problem of inverse scattering in a homogeneous background is linear because straight rays are involved. The inverse scattering with small disturbances belongs to quasi-linear as raypaths are smooth curves of small curvature. The vital difficulties in the inverse scattering problems are the typical nonlinearity caused by the strong disturbances which cannot be solved by direct employment of Born or Rytov approximations. In order to study the characteristics of nonlinear inversion, one needs instructions from the theory of nonlinear systems. In nonlinear dynamics, chaos means a state of disorder in a nonlinear system. Usually chaotic solutions of nonlinear system are considered only in forward problem such as nonlinear oscillation etc. In this work, one is dealing with chaos in the inverse problem of nonlinear iteration instead of occuring in the solution of differential equation. First of all, the scattered wavefield (acoustic pressure) during the forward problem will be needed in the inverse problem. This will be the square root of the P.S.D given by (65) and (64).

1. Reconstruction Algorithm As a start, the following inhomogeneous planar wave equation is used:

$$[\nabla - \frac{1}{c^2(x)}\frac{\partial^2}{\partial t^2}]u(x,t) = \alpha' \tag{66}$$

The following iteration formulae are introduced:

$$c_k^{-2}(x) = c_{k-1}^{-2}(x) + \gamma'\ _k(x \tag{67}$$

and $$u_k(X,T) = u_{k-1}(X,T) + \upsilon_k(X,T) \tag{68}$$

with $$\lim_{k\to\infty} c_k(x) = c(x) \tag{69}$$

where u=scattered wavefield, v and γ' are disturbances and γ' is proportional to α. Putting (67) and (68) into (66) yields

$$[\nabla - \frac{1}{c_{k-1}^2(x)}\frac{\partial^2}{\partial t^2}]\, \tilde{u}\,(x,t) = \gamma'\ _k(x)\frac{\partial^2 u(\tilde{x},t)}{\partial t^2} \tag{70}$$

The solution of (70) becomes

$$u_k(x,t) = u_{k-1}(x,t) + \int\int G_{k-1}(x,x',t,t')u(x',t')\gamma'\ _k(x')dx'dt' \tag{71}$$

where the Green's function satisfies

$$[\nabla - \frac{1}{c_{k-1}^2(x)}\frac{\partial^2}{\partial t^2}]G_{k-1}(x,x',t,t') = -\frac{\partial^2}{\partial t^2}\delta(t-t')\delta(x-x') \tag{72}$$

Now, one puts $\nu_k = \mu\nu_{k-1}$ for slow iterations, then

$$u_k(x,t) - u_{k-1}(x,t) = \int G_{k-1}(x,x',t,t')[u_{k-1}(x,t')+ \mu\nu_{k-1}(x,t')]\gamma'\ _k(x)dxdt' \tag{73}$$

where μ is a small number, $0 < \mu < 1$. (73) and (67) can be used for successive iterations as follows. The initial scattered wavefield (acoustic pressure) can be obtained from the square root of the P.S.D given by (67) and (66). Then γ_1, can be found from (73) by setting $\nu_k = 0$. Following iteration is to calculate u_k, G_k and ν_k, then to solve (73) for γ_k. The iteration produces a sequence of velocity estimates $c_k(x), k = 1, 2, \cdots$.

2.<u>Chaotic Solutions</u> The iteration formulae (67) and (68) are the so-called Poincare' maps. In fact they are a type of standard map. The characteristics of the nonlinear iteration depend upon the Poincare' maps together with the iteration parameters. Complicated Poincare' maps or nonlinear variation of the iteration parameters can cause chaos iteration and disorder output sequences. The inner entropy for a system given by (67) corresponding to inversion errors increases with k. In other words, the output sequence $c_k(x)$ would become disorder when k as well as the inner entropy become larger. When k>5 the output suddenly goes to disorder and irregular, giving rise to chaos. The irregularity is caused by the nonlinearity of the Poincare' map due to small errors existing in the data. To plot the Poincare' map given by $c_k^{-2}(x)$ versus $c_{k-1}^{-2}(x)$, one needs to find $\gamma_k(x)$ and $\gamma_1(x)$ can be found from (73) by setting $v_k = 0$. For numerical computation of the Poincare' map, the following parameters have to be known and in this paper for human tissue: x, l, D, R_0, a, l_0.

Presently works are being, carried out (a) on the computation of the Poincare' map and to prove numerically the existence of chaos for certain limit of the values of parameters, (b) computer simulation of the reconstructed velocity images and this will be the acoustical chaotic fractal images.

D. CONCLUSIONS

Chaotic fractal images do exist in acoustical imaging especially when the medium is highly inhomogeneous and fractal. The most likely candidates of human tissue for the observation of chaotic fractal images are the human heart and the human brain which have fractal structure [40,41]. The advantage of chaotic fractal images are their high sensitivity to the change in initial parameters and this makes it useful for the detection of early stage cancerous tissue. It would be more sensitive than the B/A nonlinear parameter diffraction tomography [42] as this is limited only to the quadratic term.

IX. APPLICATION OF CHAOTIC THEORY TO VIBRATION-THE FRACTON

There are a number of mathematical and physical models which exhibit chaotic vibrations [43]. But in this section, we will concentrate only on chaotic vibration in plates and beams.

Fractons have been discovered in quantum physics [44] in percolation and the vibrational excitations in fractals are called fractons by Alexander and Orbach [44]. In contrast to regular phonons, fractons are strongly localized in space. Chaotic vibration has fractal characteristics and we call the fractal mode in calssical vibration the fracton.

We start with coupled vibrations. Consider N be the number of mass points located at the sites of a fractal embedded in a d-dimensional hypercubic structure, where neighbour particles are coupled by springs. Denoting the matrix of spring constants between nearest neighbour mass points i and j by $k_{ij}^{\alpha\beta}$, the equation of motion reads

$$\frac{d^2u_i^{\alpha}(t)}{dt^2} = \sum_j \sum_\beta k_{ij}^{\alpha\beta}(u_j^{\beta}(t) - u_i^{\beta}(t)) \tag{74}$$

where u_i^{α} is the displacement of the ith mass point along the α -coordinate.

For simplicity, we assume that the coupling matrix $k_{ij}^{\alpha\beta}$ can be considered as a scalar quantity, $k_{ij}^{\alpha\beta} = k_{ij}\delta_{\alpha\beta}$. Then different components of the displacements decouple, and we obtain the same equation

$$\frac{d^2u_i(t)}{dt^2} = \sum_j k_{ij}(u_j(t) - u_i(t)) \tag{75}$$

for all components $u_i^{\alpha} \equiv u_i, \cdots$.

The solution of (75) using standard classical mechanics yields: the ansatz $u_i(t) = A_i \exp(-j\omega t)$ leads to a homogeneous system of equations for the N unknowns A_i, from which the N real eigenvalues $\omega_\alpha^2 \geq 0, \alpha = 1, 2, \cdots, N$, and the corresponding eigenvectors $(A_1^{\alpha}, \cdots, A_N^{\alpha})$ can be determined. It is convenient to choose an orthonormal set of eigenvectors $(\varphi_1^{\alpha}, \cdots, \varphi_N^{\alpha})$. Then the general solution of (75) becomes

$$u_i(t) = Re\{\sum_{\alpha=1}^{N} c_\alpha \varphi_i^\alpha \exp(-j\omega t)\}$$

where the complex constants c_α have to be determined from the initial conditions.

If the random walker model of a fractal is used, then with the initial condition $P(i, o) = \delta_{ik_0}, P(k_0, t)$ denotes the probability of being at the origin of the walk [39]. We obtain the average probability <P(r,t)> that the walker is at time t at a site separated by a distance r from the starting point by (a) averaging over all sites $i + k_0$, which are at distance r from k_0 and (b) choosing all sites of the fractal as starting points k_0 and averaging over all of them,

$$< P(r,t) >= Re\{\sum_{\alpha=1}^{N} \psi(r,\alpha)\exp(-\varepsilon_\alpha t) \tag{76}$$

where $\psi(r,\alpha) = \frac{1}{N}\sum_{k_0=1}^{N}\frac{1}{N_r}\sum_{i=1}^{N_r}(\psi_{k_0}^\alpha)^*\psi_{i+k_0}^\alpha$ (77)

and the inner sum here is over all N_r sites i, which are at distance r from k_0 and $\varepsilon_\alpha = \omega_\alpha^2$.

It has been by [44] that

$$Z(\omega) \sim \omega^{2d_f/d_w - 1} \tag{78}$$

where $Z(\omega)$=vibrational density of states, d_f=fractal dimension, and d_w =fractal dimension of the random walk.

If $Z(\omega)$ is normalized to unity, then

$$\int_0^{2\pi/T} Z^2(\omega)d\omega = 1 \tag{79}$$

giving $A = \pm\sqrt{\frac{4d_f/d_w-1}{[(\frac{2\pi}{T})^{4d_f/dw-1}-1]}}$ (80)

From the above treatment it is easy to verify that

$$< P(r,t) >= \int_0^\infty d\omega z(\omega)\psi(r,\omega)\exp(-\omega^2 t) \quad (81)$$

The inverse Laplace transform of <P(r,t)> can be performed by the method of steepest descent, yielding

$$\psi(r,\omega) \sim \lambda(\omega)^{-d_f/2}\exp\{-[constc(d_w)r/\lambda(\omega)]^{d\phi}\} \quad (82a)$$

with $d_\phi = \frac{ud_w}{u+d_w}$ (82b)

$$c(d_w) = \cos(\pi/d_w) + j\sin(\pi/d_w) \quad (82c)$$

and $\lambda(\omega)^{-1} \sim \omega^{2/d_w}$ (82d)

For our classical case, the density of states $n(\varepsilon)$ is equivalent to the number of modes within the specified frequency range. From Stephens and Bates [45] the number of vibration modes having frequencies less than or equal to f, will be

$$n(f) = \frac{4\pi Vf^3}{3c^3} \quad (83)$$

where V= volume of enclosure and c=sound velocity.

In order to use the results of quantum case in our classical vibration, we realize that $Z(\omega)$ the vibrational density of states is analogous to n(f) in the classical case and also $Z(\omega)$ is equivalent to $n(\varepsilon)$ the energy density of states in the quantum case. We also make use of the fact that the inverse Laplace transform of <P(r,t)> is a universal result for any random walker model of a fractal and should remain the same for both quantum and classical cases. That is,

$$Z(\omega)\psi_{Quantum}(r,\omega) = n(f)\psi_{Classical}(r,\omega) \tag{84}$$

so $\psi_{Classical}(r,\omega) = \lambda(\omega)^{-d_f/2}\exp\{-[constc(d_w)r/\lambda(\omega)]^{d_\phi}\}\bullet$

$$\pm\sqrt{\frac{4d_f/d_w-1}{[(\frac{2\pi}{T})^{4d_f/d_w-1}-1]}\,\omega^{2d_f/d_w-1}/\frac{4\pi}{3c^3}f^3 V} \tag{85}$$

To simplify, we choose const=1, then the amplitude of vibration (amplitude and phase of fracton) will be

$$\psi_{Classical}(r,\omega) = \lambda(\omega)^{-d_f/2}\exp\{-[c(d_\omega)r/\lambda(\omega)]^{d_\phi}\}$$

$$\pm\sqrt{\frac{4d_f/d_\omega-1}{[(\frac{2\pi}{T})^{4d_f/d\omega-1}-1]}\,\omega^{2d_f/d_\omega-1}/\frac{4\pi}{3c^3}f^3 V\cdot\frac{4\pi}{3}r^3} \tag{86}$$

where the volume V is taken to be a sphere of radius r.

Numerical computation is performed as follows:
(i) $c = 3.1\times10^5 cm/\sec$ for velocity of longitudinal sound wave in concrete, (ii). $d_f = \ln 8/\ln 3$ for the Mandelbrot-given fractal, and (iii). $d_w = \ln 22/\ln 3$ for the Mandelbrot, given fractal. Concrete is chosen as the propagation medium and Mandelbrot-given fractal is used.

f=0.0557Hz, r=20cm

$$\psi(r,\omega) = 8.9432\times10^{12} c.g.s.units$$

f=0.0557Hz, r=30cm

$$\psi(r,\omega) = 3.2249\times10^{11} c.g.s.units$$

f=1, r=20cm

$$\psi(r,\omega) = 2.17893\times10^{-3} c.g.s.units$$

f=5Hz, r=20cm

$$\psi(r,\omega) = 6.4568 \times 10^{-36} c.g.s.units$$

f=40Hz, r=0.1cm

$$\psi(r,\omega) = 9.3611 \times 10^{12} c.g.s.units$$

f=40Hz, r=1cm

$$\psi(r,\omega) = 1.373153 \times 10 \quad c.g.s.unit$$

f=40Hz, r=5cm

$$\psi(r,\omega) = 6.016319 \times 10^{-41} c.g.s.units$$

f=40Hz, r=10cm

$$\psi(r,\omega) = 6.3 \times 10^{-91} c.g.s.units$$

f=40Hz, r=20cm

$$\psi(r,\omega) = 5.52553 \times 10^{-190} c.g.s.units$$

f=40Hz, r=30cm

$$\psi(r,\omega) = 1.149092 \times 10^{-288} c.g.s.units$$

f=40Hz, r=100cm

$$\psi(r,\omega) = 2.605138332 \times 10^{-977} c.g.s.units$$

From the above computation, we find that there is a very sensitive dependence of $\psi(\omega)$ on r which gives the size of the object especially as r becomes larger. This is due to chaotic nature's sensititve dependence on initial conditions or parameters as shown in Fig.3.

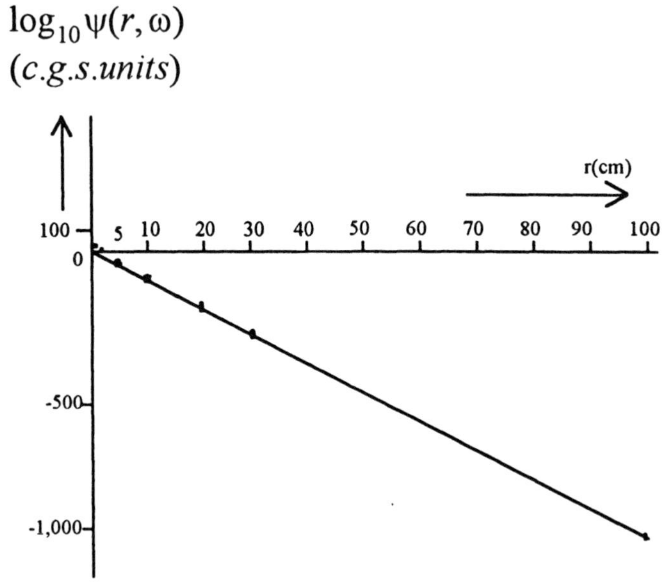

Fig.3 The dependence of the amplitude of fracton on the size of the structure

Besides sensitive dependence on initial conditions, fractons are also localized modes of vibration. This explains the mechanism that leads to the collapse of huge structure under nonlinear vibration.

X. CONCLUSIONS

Chaotic theory has many practical applications especially in the areas of signal processing and image processing. The next decade will see tremendous growth of research to enable us to have more understandings of this new field. It will penetrate many disciplines besides engineering but also in biology, medicine, geology, space research and biotechnology.

XI. REFERENCES

1. R.Devaney, An Introduction to Chaotic Dynamical Systems, Addison-Wesley, California (1989).
2. E.Lorentz, J.Atmos. Sci. **20**, pp.130 (1963).

3. S.Kobayashi, Trans.Japan Society Aeronautical Space Sciences **5**, pp.90 (1962).
4. I.Epstein, in Order in Chaos, D.Campbell and H.Rose, Eds., North-Holland, Amsterdam, pp.47 (1983).
5. J.Roux, in Order in Chaos, D.Campbell and H.Rose, Eds., North-Holland, Amsterdam, pp.57 (1983).
6. H.Atmanspacher and H.Scheingraber, Phys. Rev. **A35**, pp.253 (1986).
7. L.Glass, M.Guevara, and A.Shrier, in Order in Chaos, D.Campbell and H.Rose, Eds., North-Holland, Amsterdam, pp.89 (1983).
8. B.West and A.Goldberger, Amer.Scientist **75**, pp 354 (1987).
9. B.Van der Pol and J.Van der Mark, Nature **120**, pp.363 (1927).
10. P.Cheung and A.Wong, Phys.Rev.Lett.**59**. pp.551 (1987)
11. P.Cheung, S.Donovan, and A.Wong, Phys, Rev.Lett. **61**, pp.1360 (1988).
12. R.May, Nature **261**, pp.459 (1976).
13. F.Moon and P.Holmes, J.Sound Vib. **69**, pp.339 (1980).
14. P.Holmes and D.Whitley, in Order in Chaos, D.Campbell and H.Rose, Eds., North-Holland, Amsterdam, pp.111 (1983).
15. A.Cook and P.Roberts, Proc.Camb.Phil.Soc.**68**, pp.547 (1970)
16. E.Bullard, in AIP Conference Proceedings, S.Jorna,Eds, New York, **46**, pp.373 (1978).
17. E.Harth, IEEE Transactions **SMG-13**, pp.782 (1983)
18. J.Nicolis, J.Franklin Instit **317**. pp.289 (1984)
19. C.Skarda and W.Freeman, Behav, Brain Sci. **10**, pp 161 (1987)
20. R.Lewin, Science **240**, pp.986 (1988)
21. J. Brush and J. Kadtke. "Nonlinear Signal Processing Empirical Global Dynamical Equations", Proceedings of ICASSP, pp.V-321-V-324 (1992)
22. W.Gan, "Application of Chaotic Theory to Nonlinear Noise and Vibration Measurement and Analysis", Proceedings of Noise-Con 93, Williamsburg, Virginia, USA (1993)
23. R.Wei, B.Wang, Y.Mao, X.Zheng, and G.Miao, "Further Investigation of Nonpropagating Solitons and their Transition to Chaos", J.Acoust. Soc. Am. **88**, pp 469-472 (1990)
24. A.Pentland, "Fractal-based Description of Natural Scenes", IEEE Trans Pattern Anal. Machine Intell., **PAMI-6**, pp.666 (1984)

25. W.Gan, "Nonlinear Noise and Vibration Signal Processing - The Fractum", Proceedings of the 3rd International Congress on Air and Structure-Borne Sound and Vibration, Montreal, Canada, Vol.**2**, pp.743-746 (1994)
26. W.Gan, "Application of Chaos to Industrial Noise Analysis", Proceedings of 14th International Congress on Acoustics, Beijing, China, vol **2**, pp.E4-3 (1992)
27. W.Gan, "Acoustical Chaotic Fractal Images for Medical Imaging", in Advances in Intelligent Computing, B. Bouchon-Meunier R.Yager, and L.Zadeh, Eds., Springer Verlag (1995)
28. S.Liu, "Earth System Modelling and Chaotic Time Series", Chinese Journal of Geophysics **33**, pp.155-165 (1990)
29. Y.Chen, Fractal and Fractal Dimensions, Academic Journal Publishing Co., Beijing (1988).
30. D.Ruelle and F.Takens, "On the Nature of Turbulence", Chaos II, World Scientific, pp 120-145 (1990).
31. P.Milonni, M.Shih, and J.Ackerhalt, Chaos in Laser-Matter Interactions, World Scientific Lecture Notes in Physics, **6** (1987).
32. W.Gan, "Application of Chaos to Sound Propagation in Random Media", Acoustical Imaging, Plenum Press **19**, pp.99-102 (1992).
33. W.Gan, and C.Gan, Acoustical Fractal Images applied to Medical Imaging", Acoustical Imaging, Plenum Press **20**, pp.413-416 (1993).
34. W.Yang, and J.Du, "Approaches to solve Nonlinear Problems of the Seismic Tomography", Acoustical Imaging, Plenum Press **20**, pp.591-604 (1993)
35. D.Blackstock, "Generalised Burgers Equation for Plane Waves", J.Acoust. Soc, Am. **77**, pp.2050-2053 (1985).
36. C.Smith, and R.Beyer, "Ultrasonic Radiation Field of a Focusing Spherical Source at Finite Amplitudes", J.Acoust. Soc, Am. **46**, pp.806-813 (1969)
37. E.Zabolotskaya, R.Khokhlov, "Quasi-Plane Waves in the Nonlionear Acoustics of Confined Beams", Sov, Phys. Acoust. **15**, pp.35 (1969).
38. Y.Kuznetsov, "Equations of Nonlinear Acoustics", Sov, Phys. Acoust. **16**, pp.467 (1971).
39. H.Stanley, "Fractals and Multifractals: the Interplay of Physics and Chemistry", Fractals and Disordered Systems, A Bunde and S.Havlin, Eds., Springer-Verlag, pp 1-50 (1991).
40. B.West and A.Goldberger, Amer. Scientist **75**, pp.354 (1987).
41. C.Skarda, and W.Freeman, Behav. Brain Sci. **10**, pp.161 (1987).
42. A.Cai, Y.Nakagawa, G.Wade, and M.Yoneyama, "Imaging the Acoustic Nonlinear Parameter with Diffraction Tomography", Acoustical Imaging, Plenum Press, **17**, pp.273-283 (1989).

43. F.Moon, Chaotic and Fractal Dynamics, John Wiley & Sons, Inc, pp. 47-48 (1992).
44. S.Alexander and R.Orbach, J.Phys.Lett. **43**, pp.L625 (1982).
45. R.Stephens and A.Bate, Acoustics and Vibrational Physics, Edward Arnold (Publishers) Ltd., London, pp 641-645 (1966).

INDEX

B

C

D

E

F

G

H

Q

R

S

T

U

V

W

Z

CPSIA information can be obtained at www.ICGtesting.com
Printed in the USA
LVOW12*0710100114

368664LV00004B/47/P

9 780120 127771